Exploring Niche Tourism Business Models, Marketing, and Consumer Experience

Maria Antónia Rodrigues
Business School, Polytechnic Institute of Porto, Portugal

Maria Amélia Carvalho
Business School, Polytechnic Institute of Porto, Portugal

A volume in the Advances in Hospitality, Tourism, and the Services Industry (AHTSI) Book Series

IGI Global
PUBLISHER of TIMELY KNOWLEDGE

Published in the United States of America by
IGI Global
Business Science Reference (an imprint of IGI Global)
701 E. Chocolate Avenue
Hershey PA, USA 17033
Tel: 717-533-8845
Fax: 717-533-8661
E-mail: cust@igi-global.com
Web site: http://www.igi-global.com

Library of Congress Cataloging-in-Publication Data

Names: Rodrigues, Maria Antonia, 1977- editor. | Carvalho, Maria Amélia Machado, 1989- editor.
Title: Exploring niche tourism business models, marketing, and consumer experience / Edited by Maria Antónia Rodrigues, and Maria Amélia Machado Carvalho.
Description: Hershey, PA : Business Science Reference, [2024] | Includes bibliographical references and index. | Summary: "Exploring Niche Tourism Business Models, Marketing, and Consumer Experience provides relevant theoretical and empirical research findings, an innovative and multifaceted perspective of the niche tourist experience, and an understanding of how companies adopt business models based on sustainable paradigms and innovative technologies as a way to create value. Covering topics such as business models, rural tourism, and visitor experience, this premier reference source is an essential resource for marketing managers, product developers, niche tourism executives, marketing and tourism students, business professionals, researchers, and academicians"-- Provided by publisher.
Identifiers: LCCN 2023027750 (print) | LCCN 2023027751 (ebook) | ISBN 9781668472422 (hardcover) | ISBN 9781668472439 (paperback) | ISBN 9781668472446 (ebook)
Subjects: LCSH: Niche tourism--Management. | Niche tourism--Marketing. | Tourism--Management. | Tourism--Marketing.
Classification: LCC G156.5.N53 E97 2024 (print) | LCC G156.5.N53 (ebook) | DDC 910.68/8--dc23/eng/20230816
LC record available at https://lccn.loc.gov/2023027750
LC ebook record available at https://lccn.loc.gov/2023027751

This book is published in the IGI Global book series Advances in Hospitality, Tourism, and the Services Industry (AHTSI) (ISSN: 2475-6547; eISSN: 2475-6555)

British Cataloguing in Publication Data
A Cataloguing in Publication record for this book is available from the British Library.

Advances in Hospitality, Tourism, and the Services Industry (AHTSI) Book Series

ISSN:2475-6547
EISSN:2475-6555

Editor-in-Chief: Maximiliano Korstanje, University of Palermo, Argentina

MISSION

Globally, the hospitality, travel, tourism, and services industries generate a significant percentage of revenue and represent a large portion of the business world. Even in tough economic times, these industries thrive as individuals continue to spend on leisure and recreation activities as well as services.

The Advances in Hospitality, Tourism, and the Services Industry (AHTSI) book series offers diverse publications relating to the management, promotion, and profitability of the leisure, recreation, and services industries. Highlighting current research pertaining to various topics within the realm of hospitality, travel, tourism, and services management, the titles found within the AHTSI book series are pertinent to the research and professional needs of managers, business practitioners, researchers, and upper-level students studying in the field.

COVERAGE

- Health and Wellness Tourism
- International Tourism
- Service Management
- Hotel Management
- Sustainable Tourism
- Destination Marketing and Management
- Food and Beverage Management
- Cruise Marketing and Sales
- Service Design
- Casino Management

IGI Global is currently accepting manuscripts for publication within this series. To submit a proposal for a volume in this series, please contact our Acquisition Editors at Acquisitions@igi-global.com or visit: http://www.igi-global.com/publish/.

The Advances in Hospitality, Tourism, and the Services Industry (AHTSI) Book Series (ISSN 2475-6547) is published by IGI Global, 701 E. Chocolate Avenue, Hershey, PA 17033-1240, USA, www.igi-global.com. This series is composed of titles available for purchase individually; each title is edited to be contextually exclusive from any other title within the series. For pricing and ordering information please visit http://www.igi-global.com/book-series/advances-hospitality-tourism-services-industry/121014. Postmaster: Send all address changes to above address.

Titles in this Series

For a list of additional titles in this series, please visit:
http://www.igi-global.com/book-series/advances-hospitality-tourism-services-industry/121014

Dark Gastronomy in Times of Tribulation
Demet Genceli (Istanbul Kent University, Turkey)
Business Science Reference • © 2023 • 283pp • H/C (ISBN: 9781668465059) • US $205.00

Inclusive Community Development Through Tourism and Hospitality Practices
Vipin Nadda (University of Sunderland in London, UK) Faithfull Gonzo (University of Sunderland in London, UK) Ravinder Batta (Shimla University, India) and Amit Sharma (Shimla University, India)
Business Science Reference • © 2023 • 361pp • H/C (ISBN: 9781668467961) • US $215.00

Sustainable Growth Strategies for Entrepreneurial Venture Tourism and Regional Development
Andreas Masouras (Neapolis University, Cyprus) Christos Papademetriou (Neapolis University, Cyprus) Dimitrios Belias (University of Thessaly, Greece) and Sofia Anastasiadou (University of Western Macedonia, Greece)
Business Science Reference • © 2023 • 247pp • H/C (ISBN: 9781668460559) • US $250.00

Measuring Consumer Behavior in Hospitality for Enhanced Decision Making
Célia M.Q. Ramos (CinTurs, ESGHT, University of the Algarve, Portugal) Carlos M.R. Sousa (ESGHT, CiTUR, University of the Algarve, Portugal) Nelson M.S. Matos (CinTurs, ESGHT, University of the Algarve, Portugal) and Rashed Isam Ashqar (CinTurs, ESGHT, University of the Algarve, Portugal)
Business Science Reference • © 2023 • 295pp • H/C (ISBN: 9781668466070) • US $225.00

Global Perspectives on the Opportunities and Future Directions of Health Tourism
Oğuz Doğan (Antalya Bilim University, Turkey)
Business Science Reference • © 2023 • 319pp • H/C (ISBN: 9781668466926) • US $270.00

For an entire list of titles in this series, please visit:
http://www.igi-global.com/book-series/advances-hospitality-tourism-services-industry/121014

IGI Global
PUBLISHER of TIMELY KNOWLEDGE

701 East Chocolate Avenue, Hershey, PA 17033, USA
Tel: 717-533-8845 x100 • Fax: 717-533-8661
E-Mail: cust@igi-global.com • www.igi-global.com

Table of Contents

Detailed Table of Contents

Tourist inspiration is an emergent topic. Although studies in the psychology and marketing fields have examined inspiration theory, relevant concepts and measurement tools are lacking for tourist inspiration in the context of tourism. This research reports on two studies: Study 1 employed a literature review to conceptualize tourist inspiration and to develop a scale with satisfactory reliability and validity. In Study 2, the developed tourist inspiration scale was applied to identify the relationship between the antecedents (i.e., openness to experience, proactive personality, and existential authenticity) and consequences (i.e., transcendent tourist experience and tourist satisfaction) of tourist inspiration. The theoretical and practical implications were then presented with suggestions for future studies.

The growth of the economy and culture has made tourism a necessary part of daily life. The number of elderly and disabled people actively participating in tourism is increasing, and so is the market for accessible travel. Despite this fact, little is

Antique bookstores record the history of urban development and are crucial for urban cultural construction. Traditional business models can no longer meet antique bookstores' survival needs in the digital era. Thus, this study investigates a traditional physical bookstore, Suzhou Antique Bookstore, as the case to re-examine its functions and values as an urban cultural landmark, proposing a new sort of sustainable niche cultural tourism. Based on interviews, the authors discuss the pros and cons of the bookstore's current position using the STP model and the 7Ps marketing mix. Despite the current strategies, there is still room for improvement in digital trends. As an urban cultural benchmark, they suggest three transformation strategies for the bookstore: product selection, experience creation, and media joint. Scant studies have proposed insights into the sustainable development of antique bookstores from the perspective of urban cultural construction, especially in Asia.

Dark tourism consists of a wide range of subsections, and this chapter focuses on a less explored part of it, serial killer tourism. The demand side's fascination with death and murderers from led to diverse tourism offer types, such as museums and walking tours. This research gives an overview of these attractions and an answer to why people are visiting them. To understand the topic, three research questions were formed: What is the reason behind serial killer tourism? What type of attractions are the most attractive? and What is the attitude of the consumers. Based on the research, visitors want to be educated, to prepare to face the dark reality, but also want to be entertained. By exploring this topic, tourism professionals will get insight into the visitors' motivations, how and who to promote these attractions, and how to develop new tourism products.

Spices are always considered an integral component of the cultural heritage of the place. They provide historical, cultural, social, and geographical identities to the region. These identities are helpful in the development of tourism based on the special interest in experiences, consumption, and purchase of spices. The phenomenon can be formally named 'spice tourism'. The emerging area of spice tourism needs to be well-researched as very limited research is observed, specifically in the Indian context. Thus, identifying its dimensions, industry stakeholders, and industry framework is

Preface

The transition from mass tourism to niche tourism has been a slow process. Nowadays, it is clear that mass tourism can damage local culture, authenticity and resources (e.g., soil erosion, increased pollution, loss of natural habitat), and niche tourism is considerably important for the future of tourism companies and destination managers (Rosato et al., 2021). Thus, tourism companies and destination managers need to be proactive and adapt to market changes and challenges to hold a stronger position in the business environment in the future.

From a consumer point of view, niche tourism is linked to a specific interest, which constitutes a coherent market to be served. Niche tourism tourists are individuals who have very specific interests, needs and desires, and who seek to obtain a different experience in a certain destination (Novelli, 2005). On the other hand, from a managerial point of view, niche marketing is "a process of offering the products that are differentiated and do not have many alternatives to an individual customer or a narrowly defined group of customers with similar characteristics or needs" (Sert, 2017, p. 16). Hence, the niche is an opportunity to escape direct competition and exploit a special competence. Despite its many advantages, niche tourism does come with some risks. The first risk is that the niche market may not be profitable forever, as a profitable niche attracts competition and when competition increases, companies will struggle in differentiating from other companies (Parrish et al., 2005). Another risk is the change in customer preferences. When these preferences change niche markets can lose their relevance and easily suffer a drop in demand (Shani & Chalasani, 1992).

As a business model, the challenge is to understand how tourism companies adopt business models based on sustainable paradigms and the transition to sustainable business models as a way to create value (Rosato et al., 2021). Therefore, rather than expending marketing efforts trying to please all tourists, companies should identify the tourist segments that are particularly important to the company and target marketing efforts exclusively on the needs and wants of that specific segment. Improving business models and customer experience through innovative technologies (such as augmented reality, virtual reality, artificial intelligence, among others) are another challenge for organizations (Leone, et al., 2022). Besides, the integration between consumption and production that involves consumers and producers for the joint creation of new products and services offers new opportunities for companies (Carvalho & Alves, 2023). The motivation for engaging in these systems of co-creation is slightly different from traditional markets. Accompanying the accelerated technological advance, the co-creation processes will allow the extreme customization of niche products and services. Therefore, tourism companies and destination managers can better understand their target audience and offer them suggestions based on their budget, their family structure, and their traveling and browsing habits. Furthermore, the focus on niche tourism personalization offers greater opportunities because it attracts consumers willing to spend more money.

Thus, the basic principle of niche tourism is to produce and market tourism goods and services that few like very much, rather than goods and services that everyone likes a little (Sert, 2017).

This book, entitled *Exploring Niche Tourism Business Models, Marketing, and Consumer Experience*, provides a broad range of themes within niche tourism. As such, it is designed to offer some explanation, contextualization and discussion of each theme, supported by applied case studies and international examples, as well as, a detailed dissertation regarding some of the current contemporary debates in niche tourism and tourism management.

The book provides relevant theoretical and empirical implications for the field of tourism, providing critical insight into niche tourism practices, exploration of more lucrative segments, and new travel preferences and behaviors. This information is a valuable marketing tool for students, academics, researchers, visitor attraction managers, tour operators, travel agencies, and destination marketers.

ORGANIZATION OF THE BOOK

The book is organized into 12 chapters. The book organization starts with chapters approaching a more general view of tourism and the tourist experience followed by others focusing on niche tourism, like rural tourism, cultural tourism, dark tourism, serial killer tourism, spice tourism, and battlefield tourism.

A brief description of each of the chapters follows:

Chapter 1, titled "Tourist Inspiration and Its Empirical Study: A Mixed Methods Approach," was written by Sheng-Hshiung Tsaur and Hong-Ru Wu, both from National Chiayi University, Taiwan, and Chang-Hua Yen, from National Taichung University of Science and Technology, Taiwan. This chapter deals with an emerging topic, Tourist inspiration, and presents two studies. The first consists of a literature review to conceptualize tourist inspiration. The second is about the development of the tourist inspiration scale. The scale was applied to identify the relationship between the antecedents and consequences of tourist inspiration.

Chapter 2, titled "Assessing the Mediating Role of Destination Image on the Perceived Value and Satisfaction of People With Disabilities," was written by Marjan Kamyabi, from Cyprus International University, Turkey, and Habib Alipour and Hamed Rezapouraghdam, both from Eastern Mediterranean University, Turkey. This chapter aims to examine the effect of perceived value on customers' satisfaction, with emphasis on the mediating role of the destination image. Results show that destination image mediates the relationship between perceived value and satisfaction for people with disabilities. This study also enhances knowledge of the importance of perceived destination image by people with disabilities.

Cláudia Leitão and Bruno Maia, both from Universidade Autónoma de Lisboa, Portugal; Jorge Gomes, from Universidade de Lisboa, Portugal; and Denise Santos, from RICH, ESESFM, Portugal, wrote Chapter 3, "Tourism Industry: Leadership and Innovation." The purpose of this chapter is to analyze the impact of leadership on the relationship between innovation and performance in the Portuguese hotel sector. This chapter highlights important contributions to the hotel industry and identifies aspects that can lead to superior performance in organizations.

Chapter 4 was written by Marta Santos, Sérgio Moro, and Bráulio Alturas, all from Instituto Universitário de Lisboa, Portugal, and Paulo Rita, from NOVA Information Management School, Portugal, and is titled "Electronic Word-of-Mouth and Tourist Satisfaction in Rural Tourism in Schist Villages." This chapter explores the variation of eWOM in rural tourism and unveils the main characteristics that influence the satisfaction and the interest of consumers. The results show that eWOM has increased in rural tourism, and that the satisfaction of tourists is more influenced by the friendliness of the hosts, the variety and good breakfast or Portuguese cuisine, and the service provided.

Chapter 5, titled "Sustainable Development and Tourism," was written by Julieta Salazar-Echeagaray and Teresa Salazar-Echeagaray, both from Universidad Autónoma de Sinaloa, México, and José Vargas-Hernandez, from Tecnológico Superior de Jalisco, México. This chapter focuses on the importance of sustainable development in the tourism sector and alerts the sector to be more efficient in its activities to continue generating resources and sources of employment, protecting the environment, and implementing sustainable development.

Maximiliano Korstanje, from the University of Palermo, Argentina, wrote Chapter 6, "Methodologies in Dark Tourism Issues: Modelling the Dark Experience." The chapter focuses on the particular context of dark tourism and the dark experience, and reviews the strengths and weaknesses of dark tourism research today, as well as the future agenda for the coming years.

Chapter 7, "Dark Tourism: A Novel Trending Sector in Tourism – A Study in the Indian Subcontinent," was written by Oindrila Chakraborty, from J. D. Birla Institute, India. This chapter presents a comprehensive overview of the dark tourism industry with special reference to the Indian subcontinents.

Chapter 8 was written by Francesco Carignani and Francesco Bifulco, both from the University of Naples Federico II, Italy, and is titled "Managing Cultural Tourism: Business Models in Grave Tourism." In the context of cultural tourism and business models, this chapter explores some case studies that deal with the cultural enhancement of places still linked to ancient cults, in particular on a delimited territory, such as the UNESCO Historic Centre of Naples, in Italy.

Chapter 9, titled "Antique Bookstores Marketing Strategies as Urban Cultural Landmark: A Case Analysis for Suzhou Antique Bookstore," was written by Chenyang Xu, Apple Hiu Ching Lam, Xuechen Gao, and Dickson K. W. Chiu, all from The University of Hong Kong, Hong Kong. This chapter investigates a traditional physical bookstore, as a case of re-examining its functions and values as an urban cultural landmark, proposing a new sort of sustainable niche cultural tourism. As an urban cultural benchmark, the authors suggest three transformation strategies for the bookstore: product selection, experience creation, and media joint. Scant studies have proposed insights into the sustainable development of antique bookstores from the perspective of urban cultural construction, especially in Asia.

Titanilla Tevely and Árpád Ferenc Papp-Váry, both from the University of Sopron, Hungary, wrote Chapter 10, titled "Serial Killer Tourism: Education and Entertainment!?" This chapter focuses on a less explored part of dark tourism, serial killer tourism, gives an overview of killer tourism attractions, and explains why people are visiting these attractions. Results show that visitors want to be educated, and prepare to face the dark reality, but they also want to be entertained. By exploring this topic, tourism professionals will gain insights into the visitors' motivation, how to promote these attractions, and how to develop new tourism products.

Chapter 11, titled "Strategy for Developing Spice Tourism: A Study of the State of Maharashtra (India)," was written by Harshada Satghare, from Vishwakarma University, Pune, India. The study presented in this chapter aims to identify, develop, and promote spice tourism destinations in the state of Maharashtra. Further, the challenges of climate, spice production, global competition, and community issues were discussed for appropriate planning.

Selçuk Yücesoy, Ebru Düşmezkalender, and Yunus Özhasar, all from Eskisehir Osmangazi University, Turkey, wrote Chapter 12, titled "Battlefield Tourism." This chapter focuses on the historical development, examples, motivation, and attractions of battlefield tourism, while also exploring its popularity and ethical controversies. By providing visitors the opportunity to understand the devastating effects of war and the human suffering it causes, battlefield tourism serves as an important means for individuals to connect with the past and gain a deeper understanding of history. The chapter also offers recommendations for the future of battlefield tourism.

Maria Antónia Rodrigues
CEOS.PP, ISCAP, Polytechnic of Porto, Portugal

Maria Amélia Carvalho
ISCAP, Polytechnic of Porto, Portugal

REFERENCES

Carvalho, P., & Alves, H. (2023). Customer value co-creation in the hospitality and tourism industry: A systematic literature review. *International Journal of Contemporary Hospitality Management*, *35*(1), 250–273. doi:10.1108/IJCHM-12-2021-1528

Leone, D., Pietronudo, M. C., & Dezi, L. (2022). Improving business models through augmented reality applications: Evidence from history, theory, and practice. *International Journal of Quality and Innovation*, *6*(1), 28–42. doi:10.1504/IJQI.2022.119280

Novelli, M. (2005). *Niche tourism: Contemporary issues, trends and cases*. Elsevier Butterworth-Heinemann.

Parrish, E. D., Cassill, N. L., Oxenham, W., & Jones, M. R. (2005). The use of a niche market strategy by US textile and apparel firms. *Journal of the Textile Institute*, *96*(2), 77–85. doi:10.1533/joti.2004.0027

Rosato, P. F., Caputo, A., Valente, D., & Pizzi, S. (2021). 2030 Agenda and sustainable business models in tourism: A bibliometric analysis. *Ecological Indicators*, *121*, 106978. doi:10.1016/j.ecolind.2020.106978

Sert, A. N. (2017). Niche marketing and tourism. *Journal of Business Management and Economic Research*, *1*(1), 14–25. doi:10.29226/jobmer.2017.1

Shani, D., & Chalasani, S. (1992). Exploiting niches using relationship marketing. *Journal of Consumer Marketing*, *9*(3), 33–42. doi:10.1108/07363769210035215

Acknowledgment

The editors would like to acknowledge the support of researchers and the IGI Global editorial team involved in this project. Among editors, authors, and reviewers, the book had the participation and commitment of fifty-seven researchers.

Thirty-one authors from diverse countries or continents, namely Argentina, China, Hungary, India, Italy, México, Portugal, Taiwan, and Turkey have contributed to the book. The editors would like to thank each one of the authors for their contributions.

All chapters were reviewed by two or more reviewers, which was valuable work to enrich and improve the final result of the book. Thus, we are grateful for the support of the researchers who contributed to this book in the review process of each chapter, which included the collaboration of Apple Hiu Ching Lam (University of Hong Kong), Chang-Hua Yen (National Taichung University of Science and Technology), Chenyang Xu, Dickson K. W. Chiu (University of Hong Kong), Fábia Esteves (University of Porto), Francesco Carignani (University of Naples Federico II), Harshada Satghare (Vishwakarma University), Jonathan Simmons (University of Connecticut), Jorge Gomes (University of Lisbon), Jorge Remondes (Instituto Superior de Entre Douro e Vouga), José Santos (ISCAP/ Polytechnic of Porto), Lebogang Mathole (Tshwane University of Technology), Luis Matosas-López (Rey Juan Carlos University), Marjan Kamyabi (Cyprus International University), Maximiliano Korstanje (University of Palermo), Melanie Smith (Budapest Metropolitan University), Mustafa Çevrimkaya (Sakarya University of Applied Sciences), Nishant Sutare (Vishwakarma University), Oindrila Chakraborty (J. D. Birla Institute), Prerna Lal (International Management Institute), Sandeep Rangnath Kapse (Vishwakarma University), Sara Harvel (University of Connecticut), Selçuk Yücesoy (Eskisehir Osmangazi University), Sheng-Hshiung Tsaur (National Chiayi University).

The editors also appreciate IGI Global's support throughout the process.

Chapter 1
Tourist Inspiration and Its Empirical Study:
A Mixed Methods Approach

Sheng-Hshiung Tsaur
National Chiayi University, Taiwan

Chang-Hua Yen
National Taichung University of Science and Technology, Taiwan

Hong-Ru Wu
Ming Chuan University, Taiwan

ABSTRACT

Tourist inspiration is an emergent topic. Although studies in the psychology and marketing fields have examined inspiration theory, relevant concepts and measurement tools are lacking for tourist inspiration in the context of tourism. This research reports on two studies: Study 1 employed a literature review to conceptualize tourist inspiration and to develop a scale with satisfactory reliability and validity. In Study 2, the developed tourist inspiration scale was applied to identify the relationship between the antecedents (i.e., openness to experience, proactive personality, and existential authenticity) and consequences (i.e., transcendent tourist experience and tourist satisfaction) of tourist inspiration. The theoretical and practical implications were then presented with suggestions for future studies.

DOI: 10.4018/978-1-6684-7242-2.ch001

INTRODUCTION

The inspiration theory originated from social psychology. Thrash and Elliot (2003) conceptualized inspiration as a specific intrinsic motivational state. Inspiration is evoked by connections with external sources that are realized into new thoughts (Thrash & Elliot, 2003). Oleynick et al. (2014) defined inspiration as 'a motivational state that compels individuals to bring ideas into fruition.' Thrash and Elliot (2004) asserted that inspiration is a hybrid construct; it involves the formation of the inspired-by, or the evocation of inspiration by an object or person, and the inspired-to, or the motivation that follows evocation. Therefore, inspiration is an intrinsic motivational state in which a person transition from inspired-by to inspired-to.

Böttger et al. (2017) defined customer inspiration as a two-staged construct comprising an inspired-by and an inspired-to. The study indicated that customer inspiration is affected by environmental factors, namely the customers' inspiration source and consumer motivations. Customer inspiration affects three outcome variables: emotion, attitude, and behavior. Studies have confirmed that inspiration can lead to development of well-being (Su et al., 2021), meaning in life (Csikszentmihalyi & Seligman, 2000), and satisfaction of the need to pursue higher ideals, such as creativity, meaning, and spiritual truth (Thrash & Elliot, 2004). However, theories and literature related to inspiration in the field of tourism is still rare.

Transformative tourism has gradually developed into a trend in the industry (Zhao & Agyeiwaah, 2023). Christie and Mason (2003) defined transformative tourism as 'the practice of organized tourism that leads to a positive change in attitudes and values among those who participate in the tourist experience.' Examples of transformative tourism activities include, but are not limited to, volunteer tourism, mindful travel, pilgrimage-related trips, and study abroad (Reisinger, 2015). Studies have indicated that the meaning and responsibility tourists gained through volunteer activities affected their perceived meaning of life (Coghlan & Weiler, 2018; Robledo & Batle, 2017). Two inferences can be made about volunteer tourism experiences. The first is that external stimuli, such as living environment, evoke an inspired-by state in tourists when they seek new experiences during their travels. The second is that motivations related to inspiration can initiate a change during the tourism experience, such as motivations to encounter new people or things. Therefore, to seek new experiences, tourists' inspiration must transition from inspired-by to inspired-to. Packer and Ballantyne (2016) proposed a multifaceted model of tourist experiences, with experiences defined as belonging to 10 types, namely physical, sensory, restorative, introspective, transformative, hedonic, emotional, relational, spiritual, and cognitive. Inspiration is a crucial element of transformative experiences. Thus, transformative tourism can inspire and change tourists.

Inspiration-related themes have been examined by tourism scholars (e.g., He et al., 2021; Khoi et al., 2020; Khoi et al., 2021), who have directly applied the scales developed by Thrash and Elliot (2003) or Böttger et al. (2017). However, their scales failed to clearly demonstrate the conceptualization and scope of tourist inspiration in tourism settings (Dai et al. 2022). Inspiration plays a crucial role in the experiences of tourists during their entire journey (Dai et al., 2022). Although Tsaur et al. (2022) constructed the destination inspiration scale, the source and environment of tourist inspiration are limited to the destination, not the entire travel process. The scale developed by Tsaur et al. (2022) fails to effectively measure tourists' feelings during their journeys. Furthermore, tourist inspiration involves goals, recipients, and sources that are unlike those of other research fields. Therefore, establishing a field-specific assessment tool for tourist inspiration is necessary to effectively assess tourist inspiration and identify potential causal relationships with tourist inspiration.

To ensure that the concept of inspiration would both fit within the field of tourism and remain consistent with past studies (Thrash et al., 2014), the definition of tourist inspiration employed in this study was an intrinsic motivational state of tourists, which helps them to transition from the reception of external incentives to an internal pursuit of travel-related goals. This definition was based on the general transmission model of inspiration proposed by Thrash et al. (2010b). This study also referenced the conceptual framework of Böttger et al. (2017) and considered inspiration to comprise two stages: inspired-by and inspired-to. According to the transmission model of inspiration (Thrash et al., 2010b), tourists inspired by their external environment will transition from an inspired state to an intention and an action state, with the two states having a causal relationship.

In summary, the purpose of this study is to present the constructs and facets of tourist inspiration and develops tools suitable for assessing this inspiration. This study contributes to tourism literature by establishing a reliable and valid scale of tourist inspiration. In addition, this study proposed antecedents and consequences for tourist inspiration and developed a causal model. This study used the tourist inspiration scale to verify causal relationships within the tourist inspiration theoretical framework to bridge the research gap. The results of this study can expand current psychological understandings and provide suggestions to tourism businesses to assist them in developing tourist inspiration experiences.

LITERATURE REVIEW

Inspiration Theory

Inspiration is a unique form of intrinsic motivation and contains epistemic components. Studies have indicated that inspiration is a motivational state in which individuals are encouraged to realize their thoughts (Oleynick et al., 2014). Although inspiration is a form of intrinsic motivation, it is evoked by external sources (Thrash & Elliot, 2003). In addition, inspiration is a temporary state; it connects the deliberation phase (i.e., goal setting phase) and the implementation phase, or the goal striving phase, in the pursuit of goals (Thrash & Elliot, 2004). Thrash and Elliot (2003) proposed a tripartite conceptualization of inspiration that includes three states, namely evocation, transcendence, and motivation. Evocation refers to the external source that inspires. Transcendence refers to a positive, clear feeling of self-improvement. Motivation refers to individuals' approach motivation after they are inspired, which causes them to realize or present new thoughts. Thrash and Elliot (2004) indicated that the component process of inspiration can be categorized into two independent states, namely the activation state and the intention state.

Customer Inspiration

Böttger et al. (2017) proposed the concept of customer inspiration; they argued that customer inspiration is a temporary motivational state that can prompt customers to transition from accepting ideas presented through marketing incentives to internalising their pursuit of consumption-related targets. Customer inspiration comprises two components. In customer inspiration, being inspired-by is related to the evocation of inspiration, which facilitates the acceptance of new ideas evoked by marketing incentives, and transcendence, which denotes customers transitioning to a realization of new possibilities. This inspiration state is often described as a moment of sudden understanding or comprehension (Oleynick et al., 2014; Thrash et al., 2017). In customer inspiration, being inspired-to involves customers' internalizing the desire to pursue consumption-related targets; that is, compelling customers to form and attempt to realize new thoughts, such as buying and using products. Several scholars have since begun to apply this customer inspiration scale to investigate the relationship between customer inspiration and other constructs (e.g., Hinsch et al., 2020; Kwon & Boger, 2021). Therefore, customer inspiration is gradually becoming a crucial construct in marketing.

Tourist Inspiration

Latham et al. (2019) demonstrated that the triggers for tourist inspiration were interest, curiosity, and wonder. In addition, the results indicated that the visitors formed connections between their memories, personal identities, past experiences, and external people and objects and the sources of inspiration. Whiting and Hannam (2014) reported that artists can learn new methods or find inspiration and new ideas from tourism experiences in new or different places. Khoi et al. (2020) argued that tourist inspiration can be viewed as a cognitive appraisal of the destination. Inspiration is cultivated through the desire to be entertained, to stimulate the imagination, and to pursue creative or novel thoughts. Tourist inspiration involves tourists transitioning from accepting thoughts that initiate the desire to travel, such as images or incidents, to pursuing travel goals, such as satisfying the need for pleasure, joy, curiosity, and creativity (e.g., Khoi et al., 2020; Lindberg et al., 2014). Dai et al. (2022) indicated that inspiration is a motivator of tourists' decision-making in terms of travel behavior. Therefore, tourist inspiration is present in travel. Tourists are inspired when they encounter different environments, cultures, activities, or people during their travels.

Components of Tourist Inspiration

This study proposed that tourist inspiration can also be divided into the aforementioned two states. During travel, tourists are in the inspired-by state when they are stimulated and inspired by external environmental incentives such as atmosphere, scenery, or other tourists. They then transition into the inspired-to state when they act or realize their new thoughts. The inspired-by state occurs when tourists capture an idea evoked by an object or thing and become inspired. Inspired-by is often viewed as a sudden understanding or comprehension in our daily lives (Hart, 1998; Oleynick et al., 2014). Moufakkir and Selmi (2018) argued that the natural resources that often characterize spiritual tourism, such as forests, coasts, or deserts, can enable tourists to form spiritual connections and evoke a feeling of the Devine or inspire awe. Chen and Hsu (2021) discovered that participating in new tourism events arouses backpackers' curiosity and evokes exploratory behaviors. Mourtazina (2020) discovered that, during spiritual retreat tourism, messages conveyed by the main spiritual leader resonate with tourists and guide them through silent experiences. Therefore, this study proposed that the inspired-by state occurs when tourists are inspired by an object or thing during or after their trip. When tourists encounter people, events, or objects, they become interested or curious, which leads them to broaden their perspectives, stimulates their imagination, or incidentally causes new thoughts to form.

The inspired-to state occurs after tourists are inspired and develop the intention to pursue a specific goal; the inspired-to target may be what the tourists are inspired to do or be (Hart, 1998; Thrash & Elliot, 2004). For example, Dillette et al. (2021) discovered that the location in wellness tourism can enable tourists to contemplate their lives and change their minds and bodies. Conti and Cassel (2020) discovered that participants in nature-based tourism believe they must admire the moment and savour the beauty of the surrounding scenery. Therefore, this study proposed that the inspired-to state is an intention state in tourists that occurs after they are inspired by the external environment. When tourists are inspired during their trip, they form internal goals related to tourism, such as savouring the moment, pursuing more experiences, or making changes.

METHODOLOGY

The content of this research is divided into two stages. In Study 1, qualitative and quantitative research methods were used to conceptualize tourist inspiration, and developed tourist inspiration scales and items. In Study 2, the tourist inspiration scale constructed in Study 1 was used to examine the antecedents and consequences of tourist inspiration. The following section details the research methods and findings of both studies.

STUDY 1: SCALE DEVELOPMENT

This study followed the scale development procedures of Churchill (1979) to develop a scale of tourist inspiration. First, the components of tourist inspiration were reviewed. A scale was then developed and validated through four steps: (1) generation of items, (2) initial data collection and refinement of scales, (3) collection of secondary data and validation of scale applicability, and (4) establishment of a scale for tourist inspiration.

Item Generation

Literature Search

This study employed a literature review to obtain items related to tourist inspiration. Relevant studies have employed the inspired-by and inspired-to states to investigate the implications of inspiration (et al., Thrash & Elliot, 2004; Thrash et al., 2014). Therefore, past studies related to inspiration were identified, with the definitions and connotations of the two states being used to identify related items. Studies in the field of tourism were then selected (e.g., Khoi et al., 2020; Latham et al., 2019), with a focus on descriptions related to inspiration states or experiences during travel. In total, 23 items were selected, with 14 related to inspired-by and 9 related to inspired-to.

In-Depth Interviews

This study used in-depth interviews to obtain additional items. This study used travel-related websites and social media to recruit tourists who have participated in transformative tourism within the past 2 years and have had inspiration experiences as interviewees. Transformative tourism activities include backpacking, volunteering, educational or international learning, ecological activities, farming, learning about culture, developing well-being, religious activities, and yoga (World Tourism Organization, 2016). Before the interviews, the researchers explained the meaning of tourist inspiration; the researchers then asked questions to prompt the interviewees to recall their tourism experiences. For example, the interviewees were asked to provide information on their form of and reason for travel, the duration of the trip, and their travel partners. The interviewees were asked to provide words that described their inspiration and were further prompted with questions such as 'what inspired you?' or 'what did you feel, think, or want to do?'.

The interviews were conducted between August 2020 and October 2020, with 20 participants (10 men and 10 women) being interviewed. The interviewees were aged between 20 and 53, with an average age of 33. Due to the coronavirus pandemic and the subsequent need to avoid personal contact, all interviews were conducted through the Line or Facebook Messenger social media applications. Following the interviews, the researchers transcribed the audio recordings for subsequent data analysis.

A content analysis was performed to systematically analyze and organize the interview data (Kassarjian, 1977). Two coders, a professor with expertise in content analysis and a researcher experienced in travel experience and behavior, conducted a content analysis. Each coder assessed the content related to tourist inspiration to verify the unit of analysis. The coders simplified the meaning of each sentence before categorizing the data into states of inspiration. A total of 33 units of analysis were developed; among these, 17 units had similar features to the 23 items generated through the literature review and were, therefore, removed. Six and 10 new items were added for the inspired-by and inspired-to states, respectively, which resulted in 39 total questionnaire items (20 for inspired-by and 19 for inspired-to).

Content Validity

The content validity of the items was examined to establish expert validity. Two researchers specializing in tourism experiences and behaviors and one scholar with expertise in scale development served as expert judges to determine whether the items accurately represented the definition of tourist inspiration as well as to determine the appropriateness of the wording. The expert judges were asked to thoroughly examine the items to evaluate them for clarity, sufficient reflection of both states, and overlap. All items were required to be approved by all three judges before being included in the final questionnaire. In this step, eight items failed to achieve a consensus (four inspired-by and four inspired-to items) and were thus excluded. Accordingly, the scale of tourist inspiration was reduced to 31 items (Table 1), including 16 inspired-by and 15 inspired-to items.

Table 1. Potential indicators of tourist inspiration

Dimension	Items	Relevant Literature
Inspired-by	This trip inspired my curiosity. **	Chen & Hsu (2021); Latham et al. (2019); interviews
	This trip inspired my imagination.	Böttger et al. (2017); Khoi et al. (2020)
	I found new things on this trip.	Böttger et al. (2017); Khoi et al. (2020)
	This trip broadened my horizons.	Böttger et al. (2017); Khoi et al. (2020); interviews
	This trip inspired me spiritually. *	Moufakkir & Selmi (2018); interviews
	The things I encountered on this trip resonated with me. **	Mourtazina (2020); interviews
	This trip inspired new ideas in me.	Böttger et al. (2017); Khoi et al. (2020); Whiting & Hannam (2014); interviews
	I am interested in the things I encountered on this trip.	Figgins et al. (2016); interviews
	This trip impacted my perspectives. **	interviews
	This trip made me realize certain things.	interviews
	This trip raised my potential. **	interviews
	This trip reawakened my memories. *	Latham et al. (2019)
	This trip motivated me to work. **	Jones et al. (2014); interviews
	I am encouraged by this trip.	Jones et al. (2014); Thrash et al. (2017); interviews
	This trip inspired me.	Thrash & Elliot (2003); interviews
	This trip gave me a whole new feeling. **	interviews
	This trip impacted my thoughts. *	interviews
	Some of the things from this trip attracted me. **	interviews
	This trip gave me inspirations.	Whiting & Hannam (2014); interviews
	This trip made me develop emotional connections. *	Derrien & Stokowski (2020)

Continued on following page

Table 1. Continued

Dimension	Items	Relevant Literature
Inspired-to	I want to do something after being inspired.	Jones et al. (2014); Khoi et al. (2020); Thrash & Elliot (2003); interviews
	I am motivated to do more things after being inspired. *	Hinsch et al. (2020); Khoi et al. (2020)
	The inspiration makes me reflect.	Jones et al. (2014); Latham et al. (2019); interviews
	The inspiration motivates me to pursue certain goals.	Jones et al. (2014); Latham et al. (2019); interviews
	The inspiration makes me want to share with more people. **	Gilson (2018); interviews
	The inspiration makes me want to experience more things.	Khoi et al. (2020); interviews
	I have a desire to be inspired. *	Böttger et al. (2017)
	I have expectations after being inspired. **	interviews
	I am urged to change my perspective after getting inspired. **	interviews
	I want to make some changes after being inspired.	Dillette et al. (2021); Thrash et al. (2017); interviews
	I perceive positive feelings after being inspired. **	interviews
	I want to change my current life after being inspired. **	interviews
	I want to escape my current life after being inspired. **	interviews
	I start to enjoy life after being inspired. *	interviews
	I start to do things more cautiously after being inspired. **	interviews
	I am motivated to seize the day after being inspired.	Conti & Heldt (2020); interviews
	I want to understand more things after being inspired. **	interviews
	I have more imaginations after being inspired. *	interviews
	I want to know more things after being inspired.	interviews

Note: * The items deleted after content validity.
** The items deleted after the first confirmatory factor analysis.

Data Collection (Sample 1) and Refinement of Measures

Data Collection (Sample 1)

A 5-point Likert scale was adopted for the questionnaire items, with 1 indicating *strongly disagree* and 5 indicating *strongly agree*. The first survey was distributed to tourists who had participated in domestic transformative tourism. Forms of transformative tourism included self-guided tours, volunteer tourism, ecotourism, religious tourism, and wellness tourism. Data were collected through the recall method. Respondents were asked to select the most memorable and inspiring transformative tourism experience they had had within the past 2 years and report which form of transformative tourism they had participated in. This first survey was conducted between November 2020 and December 2020, with participants recruited through purposive sampling; a link to the questionnaire was posted on tourism-related fora and social media, such as backpacker fora or Facebook clubs.

This survey had a response rate of 93%; 365 questionnaires were returned, and after the invalid samples were excluded, 341 questionnaires were included for analysis. The reported forms of transformative tourism included self-guided tours (34.6%), religious tourism (19.3%), volunteer tourism (17.3%), ecotourism (14.7%), and wellness tourism (14.1%). Of the respondents, 51.9% were women and 48.1% were men. Most respondents were aged 30–39 (27.6%), followed by respondents aged 18–29 (22.3%). Most of respondents were unmarried (55.1%) and most had obtained a bachelor's degree (58.9%). In addition, most respondents worked in the service industry (33.1%).

Item Reduction and Confirmatory Factor Analysis

This study used iterative confirmatory factor analyses (CFAs) to assess the reliability and convergent validity of the scale. In the first CFA of the initial model, several model fit indices did not meet the acceptable threshold. To improve the scale, the researchers excluded items that fell below the acceptable range (<0.60) for the standardised factor loadings. On this basis, the researchers eliminated 7 inspired-by and 8 inspired-to items and performed a CFA for the remaining 16 items.

The CFA results after the items were eliminated indicated that the model fit was within the acceptable range (χ^2 = 290.41, df = 103, χ^2/df = 2.83, $p < .001$, GFI = 0.90, AGFI = 0.87, NFI = 0.90, CFI = 0.93, IFI = 0.93, RMSEA = 0.07, and SRMR = 0.04). As presented in Table 2, the composite reliability (CR) for each item was higher than 0.7, indicating high internal consistency (Fornell & Larcker, 1981). The factor loadings for each item were above 0.6 and the average variance extracted (AVE) for each dimension was above 0.5, which indicates that the scale had good convergent validity (Bagozzi & Yi, 1988) and is an acceptable measure of tourist inspiration.

Table 2. Confirmatory factor analysis results – sample one and sample two

Dimension and Items	Mean	Factor Loading	CR	AVE
Inspired-by			0.90 (0.90)	0.50 (0.51)
This trip inspired my imagination.	3.74 (3.83)	0.71 (0.62)		
I found new things on this trip.	4.05 (4.16)	0.68 (0.64)		
This trip broadened my horizons	4.09 (4.33)	0.70 (0.66)		
This trip inspired new ideas in me.	3.96 (4.02)	0.68 (0.71)		
I am interested in the things I encountered on this trip.	4.18 (4.32)	0.71 (0.71)		
I am encouraged by this trip.	4.01 (4.02)	0.66 (0.75)		
This trip inspired me.	4.06 (4.06)	0.75 (0.84)		
This trip made me realize certain things.	4.11 (4.03)	0.76 (0.74)		
This trip gave me inspirations.	3.88 (3.95)	0.71 (0.74)		
Inspired-to			0.88 (0.89)	0.51 (0.54)
I want to do something after being inspired.	4.07 (3.94)	0.76 (0.78)		
The inspiration makes me reflect.	4.12 (3.91)	0.72 (0.75)		
The inspiration motivates me to pursue certain goals.	4.06 (3.88)	0.70 (0.77)		
The inspiration makes me want to experience more things.	4.02 (4.24)	0.73 (0.68)		
I want to make some changes after being inspired.	3.92 (3.89)	0.68 (0.74)		
I am motivated to seize the day after being inspired.	4.07 (4.19)	0.67 (0.64)		
I want to know more things after being inspired.	4.10 (4.02)	0.76 (0.77)		

Note: CR refers to composite reliability.
AVE refers to average variance extracted.
Sample two factor analysis data are expressed in parentheses.
Sample one (n=341), sample two (n=308).

Data Collection (Sample 2) and Reanalysis of Measures

The tourist inspiration scale was developed through the aforementioned procedures and analyses. However, Churchill (1979) suggested that, after the items have been refined, different samples should be used to retest the reliability and validity of the items. Therefore, this study conducted a second questionnaire survey and analysis to ensure the reliability and validity of the scale.

Data Collection (Sample 2)

The second survey was distributed to tourists who had participated in international transformative tourism within the past 2 years. This survey was conducted between January 2021 and February 2021. For the second survey, 340 questionnaires were returned, and after the invalid samples were excluded, 308 questionnaires were included for analysis; the response rate was 91%. The reported forms of transformative tourism included self-guided tours (35.4%), study abroad (16.2%), ecotourism (13.6%), religious tourism (13.0%), wellness tourism (11.4%), and volunteer tourism (10.4%). Of the respondents, 50.3% were women and 49.7% were men. Most respondents were aged 18–29 (35.7%), followed by respondents aged 30–39 (20.9%). Most respondents were unmarried (56.8%), and most had a bachelor's degree (56.2%). Most respondents worked in the service industry (26.2%).

Confirmatory Factor Analysis

To reexamine the reliability and construct validity of the scales, this study performed a CFA. The overall model fit was within the ideal range ($\chi^2 = 257.01$, df = 103, $\chi^2/df = 2.46$, $p < .001$, GFI = 0.90, AGFI = 0.87, NFI = 0.91, CFI = 0.94, IFI = 0.94, RMSEA = 0.07, and SRMR = 0.04). Moreover, the CR and AVE for two dimensions exceeded the acceptable threshold, and the factor loadings for each item exceeded 0.6, as presented in Table 2. This indicates that the scale possessed desirable convergent validity. As listed in Table 3, the square root of the AVE for each item was greater than the correlation coefficient between two items, signifying that the scale had discriminant validity (Fornell & Larcker, 1981). After the aforementioned procedures were completed, this study finalized and developed a two-construct, 16-item tourist inspiration scale.

Table 3. Correlations of tourist inspiration scale

Tourist Inspiration	Inspired-By	Inspired-To
Inspired-by	**0.714**	
Inspired-to	0.705**	**0.734**
Notes: Diagonal values (in bold) represent the square roots of AVEs; ** $p<0.01$		

Criterion-Related Validity

Research has suggested that, although positive emotion is conceptually and empirically distinct from inspiration (Oleynick et al., 2014), inspiration states can induce positive emotion (Thrash et al., 2010a). In addition, Böttger et al. (2017) demonstrated that customer inspiration triggers an emotional response of delight. Accordingly, the present study hypothesises that tourist inspiration states induce positive emotions. Thus, positive emotion was employed as the predicted criterion variable and was measured with reference to the scale proposed by Chang et al. (2020). The scale items included 'I feel excited,' 'I feel happy,' 'I feel pleased,' 'I feel interested,' and 'I feel relaxed.' The Cronbach's alpha of this scale was 0.87, which indicates a favorable reliability. A Pearson correlation was conducted to examine correlations between the two constructs of the tourist inspiration scale and positive emotion. The results demonstrated that both inspired-by ($r = 0.51$) and inspired-to ($r = 0.44$) were significantly correlated with positive emotion ($p < .01$), indicating that the level of positive emotion increased when tourists' feelings of being inspired-by and inspired-to were higher. These results indicate that the tourist inspiration scale developed in this study has an acceptable criterion-related validity.

STUDY 2: EMPIRICAL STUDY

To further confirm the goodness of fit for the empirical study of the tourist inspiration scale constructed in Study 1, the researchers distributed the questionnaire survey again to explore the causal effects of tourist inspiration. Personal characteristics (openness to experience and proactive personality) and existential authenticity were included as antecedents, and transcendent tourist experience and tourist satisfaction were included as consequences.

Literature Review and Hypotheses

Antecedents of Tourist Inspiration

Personality is a crucial antecedent variable that affects the inspiration state. Openness to experiences is a positive personal trait that is widely studied in psychology and marketing (Parks-Leduc et al., 2015); it is related to the transcendence, evocation, and motivation of inspiration (Khoi et al., 2020). Seibert and Kraimer (2001) indicated that individuals with openness to experience generally seek new experiences and novel ideas because of personal curiosity. Therefore, openness to experience positively affects inspiration (Thrash & Elliot, 2004). Lengkeek (2001) indicated that tourism environments can employ diverse and inspiring attractions to cultivate tourist inspiration by means of the tourists' openness to experience. This thereby encourages individuals to use their imagination and elicits an approach motivation (Böttger et al., 2017) by, for example, leading tourists to wish to learn more about the tourism destination. Therefore, tourists' openness to experience can stimulate their curiosity and urge them to seek new experiences or novel ideas, which thereby stimulates their approach motivation. Therefore, the hypothesis was proposed:

H1: Openness to experience is positively related to tourist inspiration.

Li et al. (2010) argued that proactive personality causes individuals to actively participate in activities to improve their personal conditions, and, as indicated by Thrash and Elliot (2003), inspiration includes positive and self-improving feelings. Furthermore, Pan et al. (2018) discovered that proactive personality can improve individual creativity; inspiration includes creativity and spiritual truth, and it leads people to satisfy their needs and pursue higher ideals (Thrash & Elliot, 2004). Buil et al. (2019) argued that individuals with higher levels of proactive personality are more likely to take personal initiative and intentionally change their situations. Scholars stress the importance of novel experiences in tourist motivations (Lin & Kuo, 2016). Tourists who are proactive actively participate in tourism-related activities to improve themselves or to use the novel experiences of their trip to induce inspiration. Therefore, the hypothesis was proposed:

H2: Proactive personality is positively related to tourist inspiration.

Existential authenticity is another antecedent variable that affects the inspiration state; it is a state of being in which the individual perceives themselves as being true to their essential nature (Yi et al., 2017). Existential authenticity can be defined as a potential existential state of Being that is to be activated by tourism activities (Belhassen et al., 2008). Participation in tourism activities can enable individuals to initiate or pursue this state of being (Jiang et al., 2017). Studies have indicated that existential authenticity includes object authenticity and interpersonal authenticity (Mody & Hanks, 2020). Object authenticity refers to the connections between objects and people that add meaning to the tourist experience. Interpersonal authenticity is present when the individual's real self engages in sincere human interactions (Mody & Hanks, 2020). Heidegger (1996) indicated that connections between objects and people can enable individuals to view the objects as meaningful by, for example, considering the objects' use and how the objects relate with the individuals' sense of self. Latham et al. (2019) proposed that, when an object or thing attracts a tourist's attention, they focus on that object or thing and invest in it; this is likely the moment at which the tourists are inspired. Wang (1999) argued that the interpersonal relationships between tourists and their family members, friends, or strangers during their trip could spark a brief moment of personal realism. Furthermore, Fredrickson and Anderson (1999) indicated that connections with the tourism environment and social interactions can induce spiritual inspiration. Studies have reported that individual inspiration can originate from interactions with people, objects, behaviors, or ideas (Figgins et al., 2016; Gilson, 2018). Therefore, this study proposed that, when tourists feel that objects or people have existential authenticity during their trip, they may be inspired. The hypothesis was proposed:

H3: Existential authenticity is positively related to tourist inspiration.

Consequences of Tourist Inspiration

Transcendent customer experience refers to a feeling similar to that of self-transformation, awakening, separation from the mundane, and larger external phenomena (Schouten et al., 2007). Thrash and Elliot (2003) reported that inspiration is often related to a feeling of self-transcendence and is connected to objects or entities that are bigger than oneself. For example, Gilson (2018) reported that tourists experience personal transformations, such as in their thinking, feelings, speech, or actions, after they are inspired by the information provided on a tour. Studies have further indicated that individuals who are inspired while they are nature experience transcendent sentiments (Jacobs & McConnell, 2022) and undergo transcendent experiences (Böttger et al., 2017). Therefore, this study proposed that tourists who are inspired during their trip also undergo transcendent tourist experiences, and the hypothesis was proposed:

H4: Tourist inspiration is positively related to transcendent tourist experiences.

Böttger et al. (2017) indicated that customer inspiration is the customers' motivational state at the beginning of their customer journey, when they come to recognize their needs. By comparison, satisfaction is an evaluation that occurs after the customer has a positive experience of a product or service (Oliver, 1980). Ryan and Deci (2000) verified that intrinsic motivation causes higher levels of satisfaction. Thrash et al. (2010a) demonstrated that inspiration leads to enduring impacts on different well-being variables, such as self-actualization and life satisfaction;

inspiration is reported to be the basis for improving satisfaction and joy (Hinsch et al., 2020). Khoi et al. (2020) indicated that an increasing number of customers are actively searching for quality time that is closely related to their inspiration experience, and that quality time can be a feeling of total satisfaction during their experience (Manasseh et al., 2012). Therefore, tourist inspiration during travel can be considered a form of intrinsic motivation that could affect tourist satisfaction, and the hypothesis was proposed:

H5: Tourist inspiration is positively related to tourist satisfaction.

Figure 1 presents the research model.

Figure 1. Research model

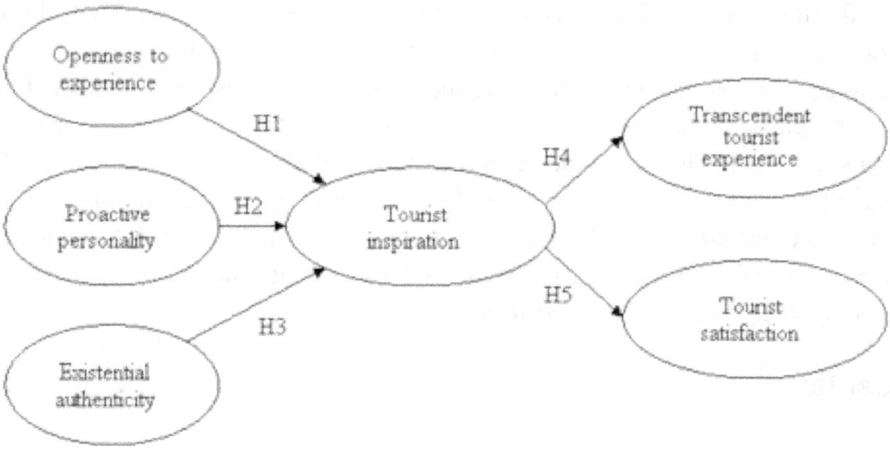

Sampling and Data Collection

The third survey distributed in this study included tourists who had participated in international transformative tourism. The sampling and survey methods were the same as those for the first and second surveys. The survey was conducted between March 2021 and April 2021. For the third survey, 446 questionnaires were returned, and after invalid samples were excluded, 408 questionnaires were included for analysis; the response rate was 91%. The reported forms of transformative tourism were self-guided tours (26.7%), study abroad (17.6%), ecotourism (17.2%), religious tourism (15.7%), wellness tourism (12.0%), and volunteer tourism (10.8%). Of the respondents, 53.0% were women and 47.0% were men. Most respondents were aged 18–29 (29.7%), followed by respondents aged 30–39 (27.9%). Most respondents were unmarried (51.2%) and most had a bachelor's degree (55.9%). Most respondents worked in the commercial and industrial sectors (24.5%).

Measures

Openness to experience was measured using five items from Yoo and Gretzel (2011), including 'I enjoy thinking deeply about things' and 'I enjoy hearing new ideas.' Proactive personality was measured using three items proposed by Bateman and Crant (1993), including 'I am always looking for better ways to do things' and 'I excel at identifying opportunities.' Existential authenticity was measured by applying seven items from relevant research in the fields of tourism (Yi et al., 2017); object authenticity included the items 'This trip gave me an understanding of local culture' and 'This trip enabled me to experience local life,' and interpersonal authenticity included items such as 'This trip has enabled me to authentically interact with the local people' and 'This trip has enabled me to authentically interact with other travellers.' Transcendent tourist experience was measured using five items modified from those proposed by Schouten et al. (2007), including 'I truly enjoyed the experience of this trip' and 'This trip experience was indescribable.' Tourist satisfaction was measured using three items proposed by Jiang et al. (2018), including 'This trip was a wise decision' and 'This trip met my travel expectations.' A 5-point Likert scale was used to measure respondents' agreement with the items on a scale of 1 (*strongly disagree*) to 5 (*strongly agree*).

Results

For the antecedent variables, openness to experience ($r = 0.48$; $p < .01$), proactive personality ($r = 0.57$; $p < .01$), and existential authenticity ($r = 0.54$; $p < .01$) were all positively correlated with tourist inspiration (Table 4). For the consequent variables, tourist inspiration was positively correlated with transcendent tourist experience ($r = 0.12$; $p < .05$) and tourist satisfaction ($r = 0.58$; $p < .01$). In addition, the assumption of multivariate normality of the observed variable, evaluated through skewness and kurtosis, were verified. The skewness (−0.85 to −0.05) and kurtosis (−0.64 to 1.68) ranges of all variables were evaluated, and all were determined to be consistent with the normal distribution (Kline, 2005).

Table 4. Means, standard deviations and correlations

Variable	Mean	SD	1	2	3	4	5	6
1. TI	4.15	0.43	0.92					
2. OE	4.08	0.56	0.48**	0.87				
3. PP	3.93	0.56	0.57**	0.45**	0.81			
4. EA	4.02	0.52	0.54**	0.30**	0.40**	0.83		
5. TTE	4.13	0.56	0.12*	0.11*	0.14**	0.11*	0.86	
6. TS	4.15	0.43	0.58**	0.25**	0.28**	0.32**	0.09	0.84

Notes: TI=Tourist inspiration; OE=Openness to experience; PP=Proactive personality; EA=Existential authenticity; TTE=Transcendent tourist experience; TS=Tourist satisfaction; The values on the diagonal are Cronbach's alphas; $*p < 0.05$; $** p < 0.01$.

Structural equation modelling (SEM) was performed using the maximum likelihood estimation to investigate these relationships. The standardized path coefficients are presented in Figure 2, with each fitness index ($\chi^2 = 309.67$, df = 162, $\chi^2/df = 1.91$, $p < .001$, GFI = 0.92, AGFI = 0.90, NFI = 0.92, CFI = 0.96, IFI = 0.96, SRMR = 0.04, and RMSEA = 0.04) indicating the fit of the model was good (Iacobucci, 2010). The path coefficients from openness to experience to tourist inspiration ($\beta = 0.21$, $t = 3.91$, $p < .01$), from proactive personality to tourist inspiration ($\beta = 0.35$, $t = 5.45$, $p < .01$), and from existential authenticity to tourist inspiration ($\beta = 0.40$, $t = 6.82$, $p < .01$) all achieved significance. Therefore, H1, H2, and H3 were supported. The results indicate that openness to experience, proactive personality, and existential authenticity all have positive effects on tourist inspiration. The path coefficients from tourist inspiration to transcendent tourist experience ($\beta = 0.15$, $t = 2.62$, $p < .01$) and from tourist inspiration to tourist satisfaction ($\beta = 0.66$, $t = 11.71$, $p < .01$) both achieved significance; hence, H4 and H5 were supported. Therefore, the results indicate that tourist inspiration positively affects transcendent tourist experiences and tourist satisfaction.

DISCUSSION

This study conducted two studies. In Study 1, past frameworks for inspiration were employed (Böttger et al., 2017; Oleynick et al., 2014) within the context of tourism to investigate the meaning and implications of tourist inspiration and to develop an assessment tool for such inspiration. This study argued that tourist inspiration is an intrinsic motivational state of tourists, which helps them to transition from the reception of external incentives to an internal pursuit of travel-related goals.

Therefore, tourist inspiration comprises two states, namely being inspired-by and being inspired-to. Being inspired-by involves tourists receiving inspiration from external incentives during their trip. The findings were consistent with those of past studies (Chen & Hsu, 2021; Moufakkir & Selmi, 2018), indicating that tourists are motivated by external incentives during trips; they become more curious, become more creative, and develop broader perspectives and ways of thinking. Being inspired-to is an intention state that occurs after tourists are inspired. Studies have demonstrated that individuals who are inspired are likely to reflect or wish to enact change (Dillette et al. 2021; Gilson, 2018). This study discovered that, when tourists are inspired during their travels, they are led to savour the moment, pursue more experiences, or enact change.

In Study 2, the causal variables of tourist inspiration were investigated. Openness to experience positively affected tourist inspiration. This result was consistent with those of past studies (Khoi et al., 2020). Tourists who are open to experience are more curious when travelling and are more accepting of new ideas and experiences. Therefore, when such tourists encounter external incentives during a trip, they are more likely to enter the inspired-by state. Studies have indicated that proactive individuals are more likely to actively participate in activities, to develop their creativity, and to pursue higher ideals (Li et al., 2010; Pan et al., 2018). This study revealed that proactive personality induced inspiration in tourists. Proactive tourists actively participate in tourism activities; because they generally possess curiosity and an interest in travel, such tourists often wish to experience more activities and pursue more goals during a trip.

In this study, the researchers discovered that existential authenticity positively affected tourist inspiration. Past studies have not directly investigated the relationship between existential authenticity and tourist inspiration; however, the results are mostly consistent with those of past studies (Fredrickson & Anderson, 1999; Latham et al., 2019). When tourists form connections with objects or people during their trips and feel a sense of existential authenticity, these objects and people lead them to participate in tourism activities and the tourists feel inspired. In addition, this study demonstrated that tourist inspiration positively related to transcendent tourist experience. Studies have indicated that an individual's inspiration experiences can initiate transcendent sentiments and transcendent experiences (Böttger et al., 2017; Lindberg et al., 2014). This study revealed that tourists who are inspired during travel change or feel awakened, indicating a transcendent tourist experience. This study also revealed that tourist inspiration positively affected tourist satisfaction. Studies have reported that inspiration can improve individual satisfaction (Hinsch et al., 2020; Manasseh et al., 2012); this study indicated that tourists who are inspired during travel have increased positive feelings toward the trip and react emotionally, which improves their level of satisfaction. Therefore, the results of this study expanded on understandings and theories of tourist inspiration and verified the causal relationship between tourist inspiration and related factors.

Theoretical Implications

Studies have indicated that tourists can obtain inspiration experiences from museums and, in doing so, develop their curiosity and interests, which then induces action (Latham et al., 2019). However, studies related to inspiration in tourism are rather sparse. Therefore, this study employed scale development procedures and qualitative and quantitative research methods to develop a tourist inspiration construct and scale and verify potential causal relationships with tourist inspiration to bridge the knowledge gap.

In Study 1, inspiration was conceptualized within a tourism perspective, and a tourist inspiration scale was developed. The tourist inspiration scale comprised 16 items related to two constructs, namely being inspired-by and inspired-to. This study tested the scale to ensure its goodness of fit, reliability, and validity, and the researchers found that it could be used to effectively assess tourist inspiration. In Study 2, the researchers verified the applicability of the tourist inspiration scale developed in Study 1 and developed a theoretical framework. Furthermore, this study proposed antecedents and consequences related to tourist inspiration and verified the relationships between the variables.

In summary, this study made crucial contributions to relevant academic theories. Past studies have formulated numerous inspiration scales to assess the types of inspiration individuals felt (e.g., Böttger et al., 2017; Jones et al., 2014; Thrash et al., 2017). Although Khoi et al. (2020) reported on inspiration in travelling customers, tourists' shopping goals and those of conventional customers differ. Furthermore, tourist inspiration and customer inspiration should be considered different concepts. Therefore, the researchers examined inspiration in tourists who are travelling and employed the elements of tourist inspiration as a basis on which a tourist inspiration construct and scale were developed (Study 1). The results of this study can expand on understandings within the field of tourism and on studies related to inspiration. Although the participants of this study are tourists who have participated in transformative tourism, the tourist inspiration scale items of this study are relatively generic. The researchers believe that it is applicable to other forms of tourism. The tourist inspiration scale developed by this study can be employed in quantitative research to verify the relationship between tourist inspiration and other constructs and can serve as a theoretical basis for future studies related to tourist inspiration.

In this study, the researchers completed a literature review and included antecedents and consequences related to tourist inspiration; this study then proposed and verified a tourist inspiration theoretical model (Study 2). The results demonstrated that openness to experience and proactive personality both positively affect tourist inspiration. In addition, past studies have not directly investigated the relationship between existential authenticity and tourist inspiration. In this study, the researchers discovered that tourists' feelings of existential authenticity during their travels positively affect tourist inspiration. Furthermore, if tourists are highly inspired during their travels, their senses of transcendent experience and satisfaction increase. This result expands on the discovery of Khoi et al. (2021) that tourist inspiration during travels can cause tourists to have transcendent experiences and can increase their tourist satisfaction. In conclusion, this study investigated causal relationships with tourist inspiration and expanded upon the tourist inspiration theory. The results provide crucial theoretical contributions to studies on tourism.

Suggestions and Recommendations

The tourist inspiration scale constructed in this study can serve as a reference for tourism businesses. The scale developed in Study 1 contained items tourists were inspired-by (such as stimulated imagination, expanded perspectives, and realization) and those the tourists were inspired-to do (such as pursuing a goal, changing, and savouring the moment). Tourism businesses can employ this scale to assess tourists' levels of inspiration and use these assessments to adjust the incentives that tourists encounter during their travels. This will enable businesses to evoke stronger inspiration. Furthermore, tourism businesses should employ different stimulants, such as products, activities, and services, that can lead to development of inspiration to initiate tourist inspiration at different levels.

The results of Study 2 demonstrated that openness to experience and proactive personality both positively affect tourist inspiration. Tourism business should provide unique tourism information that can stimulate the imaginations of tourists with these traits, and they should highlight the potential enjoyment tourists could gain from these activities and experiences. This will inspire the tourists who participate in the trip. Furthermore, this study indicates that tourists' sense of existential authenticity positively affects tourist inspiration. Tourism businesses should incorporate elements of existential authenticity when they plan travel itineraries. To achieve object authenticity, tourism businesses can use storytelling marketing to combine the history, culture, architecture, and lifestyles that tourists will encounter on a trip, and they should enable tourists to interact with the local community to lead tourists' to feel that the tour site is authentic. To achieve interpersonal authenticity, tourism businesses can hire local tour guides to provide sincere and authentic services and experiences from a local perspective. In addition, the tour guides can increase authentic interactions between tourists and locals or between tourists through different activities.

The results demonstrated that tourist inspiration positively affects transcendent experiences and satisfaction. Therefore, when tourism businesses are designing travel itineraries, they can partner with local industries to develop incentives that will evoke inspiration in tourists through, for example, digital experiences, such as augmented reality, virtual reality, or projection mapping, and through authentic services, such as creating a casual atmosphere, maintaining authenticity in interactions with tourists, or providing locally produced food and local stories during dining. Furthermore, tour guides can incorporate fun and gamified activities to inspire tourists, increase their positive feelings, and provide them with a transcendent and memorable experience.

Future Studies

This study has several limitations. First, this study included only Taiwanese tourists who have participated in transformative tourism and employed interviews and questionnaires to develop the tourist inspiration scale. However, whether the scale can be used to assess tourist inspiration in other places is dependent on the generalizability of the scale. Therefore, this study suggests that scholars conduct research in places with cross-cultural backgrounds to verify the applicability of the scale. Second, this study focused on tourists; however, the tourism system involves many key stakeholders, such as local business owners and the government. Therefore, future studies should include perspectives from stakeholders to further investigate tourist inspiration.

Finally, this study only used antecedent variables, such as personality traits and existential authenticity, and consequent variables, such as transcendent experience and satisfaction, to study tourist inspiration. Future studies can include other variables, such as transformative travel experience or tourist behavior, to complete understandings of the model of tourist inspiration.

CONCLUSION

Inspiration has been examined in various fields. However, studies on inspiration experienced by tourists during their travel have rarely investigated tourist inspiration. Therefore, tourist inspiration is an emergent topic that merits further exploration with other related constructs.

This study conceptualized tourist inspiration and defined it as an intrinsic motivational state of tourists, which helps them to transition from the reception of external incentives to an internal pursuit of travel-related goals. A tourist inspiration assessment scale was then established that includes two constructs.

The results demonstrated that the scale has goodness of fit, reliability, and validity. Additionally, this study investigated the casual model of tourist inspiration and used the established scale to identify three antecedents (openness to experience, proactive personality, and existential authenticity) and two consequences (transcendent tourist experience and tourist satisfaction). The results of the present study can extend current psychological understandings and aid tourism managers in developing tourist inspiration experiences.

REFERENCES

Bagozzi, R. P., & Yi, Y. (1988). On the evaluation of structural equation models. *Journal of the Academy of Marketing Science*, *16*(1), 74–94. doi:10.1007/BF02723327

Bateman, T. S., & Crant, J. M. (1993). The proactive component of organizational behavior: A measure and correlates. *Journal of Organizational Behavior*, *14*(2), 103–118. doi:10.1002/job.4030140202

Belhassen, Y., Caton, K., & Stewart, W. P. (2008). The search for authenticity in the pilgrim experience. *Annals of Tourism Research*, *35*(3), 668–689. doi:10.1016/j.annals.2008.03.007

Böttger, T., Rudolph, T., Evanschitzky, H., & Pfrang, T. (2017). Customer inspiration: Conceptualization, scale development, and validation. *Journal of Marketing*, *81*(6), 116–131. doi:10.1509/jm.15.0007

Buil, I., Martínez, E., & Matute, J. (2019). Transformational leadership and employee performance: The role of identification, engagement and proactive personality. *International Journal of Hospitality Management*, *77*, 64–75. doi:10.1016/j.ijhm.2018.06.014

Chang, S. Y., Tsaur, S. H., Yen, C. H., & Lai, H. R. (2020). Tour member fit and tour member–leader fit on group package tours: Influences on tourists' positive emotions, rapport, and satisfaction. *Journal of Hospitality and Tourism Management*, *42*, 235–243. doi:10.1016/j.jhtm.2020.01.016

Chen, K. Y., & Hsu, Y. L. (2021). Developing a model of backpackers' exploratory curiosity. *Tourism and Hospitality Management*, *27*(1), 1–23. doi:10.20867/thm.27.1.1

Christie, M. F., & Mason, P. A. (2003). Transformative tour guiding: Training tour guides to be critically reflective practitioners. *Journal of Ecotourism*, *2*(1), 1–16. doi:10.1080/14724040308668130

Churchill, G. A. Jr. (1979). A paradigm for developing better measures of marketing constructs. *JMR, Journal of Marketing Research*, *16*(1), 64–73. doi:10.1177/002224377901600110

Coghlan, A., & Weiler, B. (2018). Examining transformative processes in volunteer tourism. *Current Issues in Tourism*, *21*(5), 567–582. doi:10.1080/13683500.2015.1102209

Conti, E., & Cassel, H. S. (2020). Liminality in nature-based tourism experiences as mediated through social media. *Tourism Geographies*, *22*(2), 413–432. doi:10.1080/14616688.2019.1648544

Csikszentmihalyi, M., & Seligman, M. (2000). Positive psychology. *The American Psychologist*, *55*(1), 5–14. doi:10.1037/0003-066X.55.1.5 PMID:11392865

Dai, F., Wang, D., & Kirillova, K. (2022). Travel inspiration in tourist decision making. *Tourism Management*, *90*, 104484. doi:10.1016/j.tourman.2021.104484

Derrien, M. M., & Stokowski, P. A. (2020). Discursive constructions of night sky experiences: Imagination and imaginaries in national park visitor narratives. *Annals of Tourism Research*, *85*, 103038. doi:10.1016/j.annals.2020.103038

Dillette, A. K., Douglas, A. C., & Andrzejewski, C. (2021). Dimensions of holistic wellness as a result of international wellness tourism experiences. *Current Issues in Tourism*, *24*(6), 794–810. doi:10.1080/13683500.2020.1746247

Fairley, S., Gibson, H., & Lamont, M. (2018). Temporal manifestations of nostalgia: Le Tour de France. *Annals of Tourism Research*, *70*, 120–130. doi:10.1016/j.annals.2017.09.004

Figgins, S. G., Smith, M. J., Sellars, C. N., Greenlees, I. A., & Knight, C. J. (2016). "You really could be something quite special": A qualitative exploration of athletes' experiences of being inspired in sport. *Psychology of Sport and Exercise*, *24*, 82–91. doi:10.1016/j.psychsport.2016.01.011

Fornell, C., & Larcker, D. F. (1981). Evaluating structural equation models with unobservable variables and measurement error. *JMR, Journal of Marketing Research*, *18*(1), 39–50. doi:10.1177/002224378101800104

Fredrickson, L. M., & Anderson, D. H. (1999). A qualitative exploration of the wilderness experience as a source of spiritual inspiration. *Journal of Environmental Psychology*, *19*(1), 21–39. doi:10.1006/jevp.1998.0110

Gilson, J. F. (2018). Inspiring change in heritage interpretation. In S. Pulla & B. Schissel (Eds.), *Applied interdisciplinarity in scholar practitioner programs* (pp. 69–86). Palgrave Macmillan. doi:10.1007/978-3-319-64453-0_5

Hart, T. (1998). Inspiration: Exploring the experience and its meaning. *Journal of Humanistic Psychology*, *38*(3), 7–35. doi:10.1177/00221678980383002

He, M., Liu, B., & Li, Y. (2021). Tourist inspiration: How the wellness tourism experience inspires tourist engagement. *Journal of Hospitality & Tourism Research (Washington, D.C.)*. Advance online publication. doi:10.1177/10963480211026376

Heidegger, M. (1996). *Being and time: A translation of Sein und Zeit*. SUNY press.

Hinsch, C., Felix, R., & Rauschnabel, P. A. (2020). Nostalgia beats the wow-effect: Inspiration, awe and meaningful associations in augmented reality marketing. *Journal of Retailing and Consumer Services*, *53*, 101987. doi:10.1016/j.jretconser.2019.101987

Iacobucci, D. (2010). Structural equations modeling: Fit indices, sample size, and advanced topics. *Journal of Consumer Psychology*, *20*(1), 90–98. doi:10.1016/j.jcps.2009.09.003

Jacobs, T. P., & McConnell, A. R. (2022). Self-transcendent emotion dispositions: Greater connections with nature and more sustainable behavior. *Journal of Environmental Psychology*, *81*, 101797. doi:10.1016/j.jenvp.2022.101797

Jiang, J., Zhang, J., Zhang, H., & Yan, B. (2018). Natural soundscapes and tourist loyalty to nature-based tourism destinations: The mediating effect of tourist satisfaction. *Journal of Travel & Tourism Marketing*, *35*(2), 218–230. doi:10.1080/10548408.2017.1351415

Jiang, Y., Ramkissoon, H., Mavondo, F. T., & Feng, S. (2017). Authenticity: The link between destination image and place attachment. *Journal of Hospitality Marketing & Management*, *26*(2), 105–124. doi:10.1080/19368623.2016.1185988

Jones, S., Dodd, A., & Gruber, J. (2014). Development and validation of a new multidimensional measure of inspiration: Associations with risk for bipolar disorder. *PLoS One*, *9*(3), e91669. doi:10.1371/journal.pone.0091669 PMID:24670894

Kassarjian, H. H. (1977). Content analysis in consumer research. *The Journal of Consumer Research*, *4*(1), 8–18. doi:10.1086/208674

Khoi, N. H., Le, A. N. H., & Tran, M. D. (2021). Tourist inspiration and its consequences: The moderating role of neuroticism. *International Journal of Tourism Research*, *23*(5), 901–913. doi:10.1002/jtr.2452

Khoi, N. H., Phong, N. D., & Le, A. N. H. (2020). Customer inspiration in a tourism context: An investigation of driving and moderating factors. *Current Issues in Tourism, 23*(21), 2699–2715. doi:10.1080/13683500.2019.1666092

Kline, R. B. (2015). *Principles and practice of structural equation modeling*. Guilford publications.

Kwon, J., & Boger, C. A. (2021). Influence of brand experience on customer inspiration and pro-environmental intention. *Current Issues in Tourism, 24*(8), 1154–1168. doi:10.1080/13683500.2020.1769571

Latham, K. F., Narayan, B., & Gorichanaz, T. (2019). Encountering the muse: An exploration of the relationship between inspiration and information in the museum context. *Journal of Librarianship and Information Science, 51*(4), 1067–1076. doi:10.1177/0961000618769976

Lengkeek, J. (2001). Leisure experience and imagination: Rethinking Cohen's modes of tourist experience. *International Sociology, 16*(2), 173–184. doi:10.1177/0268580901016002003

Li, N., Liang, J., & Crant, J. M. (2010). The role of proactive personality in job satisfaction and organizational citizenship behavior: A relational perspective. *The Journal of Applied Psychology, 95*(2), 395–404. doi:10.1037/a0018079 PMID:20230079

Lin, C. H., & Kuo, B. Z. L. (2016). The behavioral consequences of tourist experience. *Tourism Management Perspectives, 18*, 84–91. doi:10.1016/j.tmp.2015.12.017

Lindberg, F., Hansen, A. H., & Eide, D. (2014). A multirelational approach for understanding consumer experiences within tourism. *Journal of Hospitality Marketing & Management, 23*(5), 487–512. doi:10.1080/19368623.2013.827609

Manasseh, T., Müller-Sarmiento, P., Reuter, H., von Faber-Castell, C., & Pallua, C. (2012). Customer inspiration–a key lever for growth in European retail. *Marketing Review St. Gallen, 29*(5), 16–21. doi:10.136511621-012-0159-9

Mody, M., & Hanks, L. (2020). Consumption authenticity in the accommodations industry: The keys to brand love and brand loyalty for hotels and Airbnb. *Journal of Travel Research, 59*(1), 173–189. doi:10.1177/0047287519826233

Moufakkir, O., & Selmi, N. (2018). Examining the spirituality of spiritual tourists: A Sahara desert experience. *Annals of Tourism Research, 70*, 108–119. doi:10.1016/j.annals.2017.09.003

Mourtazina, E. (2020). Beyond the horizon of words: Silent landscape experience within spiritual retreat tourism. *International Journal of Culture, Tourism and Hospitality Research, 14*(3), 349–360. doi:10.1108/IJCTHR-10-2019-0185

Oleynick, V. C., Thrash, T. M., LeFew, M. C., Moldovan, E. G., & Kieffaber, P. D. (2014). The scientific study of inspiration in the creative process: Challenges and opportunities. *Frontiers in Human Neuroscience, 8*, 436. doi:10.3389/fnhum.2014.00436 PMID:25009483

Oliver, R. L. (1980). A cognitive model of the antecedents and consequences of satisfaction decisions. *JMR, Journal of Marketing Research, 17*(4), 460–469. doi:10.1177/002224378001700405

Packer, J., & Ballantyne, R. (2016). Conceptualizing the visitor experience: A review of literature and development of a multifaceted model. *Visitor Studies, 19*(2), 128–143. doi:10.1080/10645578.2016.1144023

Pan, J., Liu, S., Ma, B., & Qu, Z. (2018). How does proactive personality promote creativity? A multilevel examination of the interplay between formal and informal leadership. *Journal of Occupational and Organizational Psychology, 91*(4), 852–874. doi:10.1111/joop.12221

Parks-Leduc, L., Feldman, G., & Bardi, A. (2015). Personality traits and personal values: A meta-analysis. *Personality and Social Psychology Review, 19*(1), 3–29. doi:10.1177/1088868314538548 PMID:24963077

Reisinger, Y. (2015). *Transformational tourism: Host perspectives*. SABI. doi:10.1079/9781780643922.0000

Robledo, M. A., & Batle, J. (2017). Transformational tourism as a hero's journey. *Current Issues in Tourism, 20*(16), 1736–1748. doi:10.1080/13683500.2015.1054270

Ryan, R. M., & Deci, E. L. (2000). Self-determination theory and the facilitation of intrinsic motivation, social development, and well-being. *The American Psychologist, 55*(1), 68–78. doi:10.1037/0003-066X.55.1.68 PMID:11392867

Schouten, J. W., McAlexander, J. H., & Koenig, H. F. (2007). Transcendent customer experience and brand community. *Journal of the Academy of Marketing Science, 35*(3), 357–368. doi:10.100711747-007-0034-4

Seibert, S. E., & Kraimer, M. L. (2001). The five-factor model of personality and career success. *Journal of Vocational Behavior, 58*(1), 1–21. doi:10.1006/jvbe.2000.1757

Su, L., Tang, B., & Nawijn, J. (2021). How tourism activity shapes travel experience sharing: Tourist well-being and social context. *Annals of Tourism Research, 91*, 103316. doi:10.1016/j.annals.2021.103316

Thrash, T. M., & Elliot, A. J. (2003). Inspiration as a psychological construct. *Journal of Personality and Social Psychology, 84*(4), 871–889. doi:10.1037/0022-3514.84.4.871 PMID:12703654

Thrash, T. M., & Elliot, A. J. (2004). Inspiration: Core characteristics, component processes, antecedents, and function. *Journal of Personality and Social Psychology, 87*(6), 957–973. doi:10.1037/0022-3514.87.6.957 PMID:15598117

Thrash, T. M., Elliot, A. J., Maruskin, L. A., & Cassidy, S. E. (2010a). Inspiration and the promotion of well-being: Tests of causality and mediation. *Journal of Personality and Social Psychology, 98*(3), 488–506. doi:10.1037/a0017906 PMID:20175626

Thrash, T. M., Maruskin, L. A., Cassidy, S. E., Fryer, J. W., & Ryan, R. M. (2010b). Mediating between the muse and the masses: Inspiration and the actualization of creative ideas. *Journal of Personality and Social Psychology, 98*(3), 469–487. doi:10.1037/a0017907 PMID:20175625

Thrash, T. M., Maruskin, L. A., Moldovan, E. G., Oleynick, V. C., & Belzak, W. C. (2017). Writer–reader contagion of inspiration and related states: Conditional process analyses within a cross-classified writer×reader framework. *Journal of Personality and Social Psychology, 113*(3), 466–491. doi:10.1037/pspp0000094 PMID:27124379

Thrash, T. M., Moldovan, E. G., Oleynick, V. C., & Maruskin, L. A. (2014). The psychology of inspiration. *Social and Personality Psychology Compass, 8*(9), 495–510. doi:10.1111pc3.12127

Tsaur, S. H., Yen, C. H., & Lin, Y. S. (2022). Destination inspiration: Scale development and validation. *Journal of Travel & Tourism Marketing, 39*(5), 484–500. doi:10.1080/10548408.2022.2148040

Wang, N. (1999). Rethinking authenticity in tourism experience. *Annals of Tourism Research, 26*(2), 349–370. doi:10.1016/S0160-7383(98)00103-0

Whiting, J., & Hannam, K. (2014). Journeys of inspiration: Working artists' reflections on tourism. *Annals of Tourism Research, 49*, 65–75. doi:10.1016/j.annals.2014.08.007

World Tourism Organization. (2016). The transformative power of tourism: A paradigm shift towards a more responsible traveller (Affiliate Members Global Reports, Volume 14). Madrid: UNWTO.

Yi, X., Lin, V. S., Jin, W., & Luo, Q. (2017). The authenticity of heritage sites, tourists' quest for existential authenticity, and destination loyalty. *Journal of Travel Research, 56*(8), 1032–1048. doi:10.1177/0047287516675061

Yoo, K. H., & Gretzel, U. (2011). Influence of personality on travel-related consumer-generated media creation. *Computers in Human Behavior, 27*(2), 609–621. doi:10.1016/j.chb.2010.05.002

Zhao, Y., & Agyeiwaah, E. (2023). Understanding tourists' transformative experience: A systematic literature review. *Journal of Hospitality and Tourism Management, 54*, 188–199. doi:10.1016/j.jhtm.2022.12.013

ADDITIONAL READING

Böttger, T., Rudolph, T., Evanschitzky, H., & Pfrang, T. (2017). Customer inspiration: Conceptualization, scale development, and validation. *Journal of Marketing, 81*(6), 116–131. doi:10.1509/jm.15.0007

Dai, F., Wang, D., & Kirillova, K. (2022). Travel inspiration in tourist decision making. *Tourism Management, 90*, 104484. doi:10.1016/j.tourman.2021.104484

He, M., Liu, B., & Li, Y. (2021). Tourist inspiration: How the wellness tourism experience inspires tourist engagement. *Journal of Hospitality & Tourism Research (Washington, D.C.)*. Advance online publication. doi:10.1177/10963480211026376

Khoi, N. H., Le, A. N. H., & Tran, M. D. (2021). Tourist inspiration and its consequences: The moderating role of neuroticism. *International Journal of Tourism Research, 23*(5), 901–913. doi:10.1002/jtr.2452

Khoi, N. H., Phong, N. D., & Le, A. N. H. (2020). Customer inspiration in a tourism context: An investigation of driving and moderating factors. *Current Issues in Tourism, 23*(21), 2699–2715. doi:10.1080/13683500.2019.1666092

Liu, B., Li, Y., Kralj, A., Moyle, B., & He, M. (2022). Inspiration and wellness tourism: The role of cognitive appraisal. *Journal of Travel & Tourism Marketing, 39*(2), 173–187. doi:10.1080/10548408.2022.2061676

Pung, J. M., Gnoth, J., & Del Chiappa, G. (2020). Tourist transformation: Towards a conceptual model. *Annals of Tourism Research, 81*, 102885. doi:10.1016/j.annals.2020.102885

Tsaur, S. H., Yen, C. H., & Lin, Y. S. (2022). Destination inspiration: Scale development and validation. *Journal of Travel & Tourism Marketing*, *39*(5), 484–500. doi:10.1080/10548408.2022.2148040

KEY TERMS AND DEFINITIONS

Existential Authenticity: A potential existential state of Being that is to be activated by tourism activities.

Openness to Experience: The extent to which a tourist's willingness to explore accepts a novel experience, divergent thinking, and ideas.

Proactive Personality: The tendency to initiate and seek constructive change in the surrounding context.

Tourist Inspiration: An intrinsic motivational state of tourists, which helps them to transition from the reception of external incentives to an internal pursuit of travel-related goal.

Tourist Satisfaction: The overall subjective judgment and evaluation that a tourist makes regarding the trip.

Transcendent Customer Experience: A feeling similar to that of self-transformation, awakening, separation from the mundane, and larger external phenomena.

Transformative Tourism: the practice of organized tourism that leads to a positive change in attitudes and values among those who participate in the tourist experience.

Chapter 2
Assessing the Mediating Role of Destination Image on the Perceived Value and Satisfaction of People With Disabilities

Marjan Kamyabi
Cyprus International University, Turkey

Habib Alipour
https://orcid.org/0000-0001-5517-3118
Eastern Mediterranean University, Turkey

Hamed Rezapouraghdam
Eastern Mediterranean University, Turkey

ABSTRACT

The growth of the economy and culture has made tourism a necessary part of daily life. The number of elderly and disabled people actively participating in tourism is increasing, and so is the market for accessible travel. Despite this fact, little is known about the determinant factors affecting people with disabilities' satisfaction in developing countries. The current study aims to examine the effect of perceived value on customers' satisfaction, with emphasis on the mediating role of the destination image. Data were collected from 250 PWDs in Northern Cyprus. A series of analyses were performed using the partial least square structural equation modeling to test the research hypothesis. The results support the hypothesis that destination image mediates the relationship between perceived value and satisfaction for people with disabilities. The study enhances our knowledge of the importance of perceived destination image by PWDs that can have implications for destination marketing to improve the image and its reconstruction for an accessible tourism market.

DOI: 10.4018/978-1-6684-7242-2.ch002

INTRODUCTION

Everybody should have the same right and opportunities to benefit from tourism. There are various reasons why not everyone can become a tourist, including economic constraints, mobility difficulties, psychological barriers, and/or lack of time. "Tourism for all" is the mission of accessible tourism, which aims to remove obstacles and limitations so that anyone can travel (Qiao et al., 2021).

PWDs make up approximately 15% of the world's population (World Health Organization, 2011), and their proportion is expected to become even greater as the world's population ages. Moreover, improvements in health care that allow for higher survival rates for PWD is another factor that enhances the growth of this number around the globe (Moura et al., 2022; WHO, 2011). Travel has become a human right issue and necessity for all people in the context of "social inclusion" (Cole, & Eriksson, 2010). However, physical disability in combination with social and cultural inequalities has resulted in discrimination and social invisibility of this population. This is more evident in developing economies with highly established tourism sectors (Akinci, 2013). The global tourism sector is seen as a driving force for destination development to attract and accommodate accessible tourism (hereafter "AT") (Rubio-Escuderos et al., 2021). Efforts to make traveling and destinations accessible to people with disabilities are gaining ground and have become a topic of international discussion recently (Popiel, 2016). AT is defined as the removal of barriers that prevent people with disabilities from fully experiencing a tourist attraction (Rubio-Escuderos et al., 2021).

Becoming a destination for AT is challenging notwithstanding its market value in terms of income generation, destination prestige, urban economic development, and job creation (Chen & Tsai, 2007). The challenges faced by destinations vary systematically and include physical design, respect for and preservation of the rights of PWDs, socio-cultural attitudes, the collaboration between different sectors, and more importantly, combining a strategic approach to AT in conjunction with the needs of the disabled population.

Therefore, destinations need to reshape their tourist landscape in order to accommodate AT demand (Oliver & Barnes, 2012). This way, the tourism sector can contribute to social inclusion and will not be accused of being able body-centric. It is argued that "the focus in disability policy needs to shift. It is not only about how to change disabled people so that they can better cope in their environment, but also vice versa, to change the environment (destination) so that it is suitable for people with impairments" (Tøssebro, 2004). Therefore, understanding the preferences, behaviors, and interests of PWDs is critical for the tourism industry, especially accessible tourism (Al-Ansi & Han, 2019), because tourists choose their destination and make their decisions based on destination perceptions rather than reality (Kani et al., 2017; Nazir et al., 2021).

Destination image has been discussed extensively in the literature (Baloglu, S., & McCleary, 1999; Chon, K. & Sparrowe, 2000); however, the topic has not attracted enough attention in the context of PWDs and AT. Destination image has also been contextualized in terms of the overall impression tourists have about a destination. This is crucial since destination image can influence customers' destination choice, decision-making process, overall behavior and intention to revisit. (Chen & Tsai, 2007).

The term "value" is widely accepted in marketing literature as a general assessment of the services customers receive from a service provider (Chen & Tsai, 2007; Hellier et al., 2003). Scholars assert that a systematic study of the relationships between a destination's image, perceived value, and tourist satisfaction can lead to an important competitive advantage (Jeong & Kim, 2020).

This study provides a theoretical framework for understanding how PWDs perceive destination image and how perceived value is related to their satisfaction, contributing to the tourism literature. Additionally, the finding of this research contributes to the advancement of AT beyond the issue of accessibility and rather paves the path for the rise of "inclusive tourism" with the side effect of accessibility for all.

Most investigations on AT have been performed in developed countries. Forthcoming research can consider developing countries to highlight the potential and challenges for AT, which may be different from those in developed countries (Kamyabi & Alipour, 2022). Accessibility for tourists with special needs can also help the region recover from the pandemic (UNWTO, 2022).

LITERATURE REVIEW

Accessible Tourism

The right of PWDs to travel should be the same as that of people without disabilities in political, material, and intellectual terms (Qiao et al., 2022). Nowadays, people with disabilities have become vocal and have undergone a tremendous evolution in terms of perception to be treated as equal to able-bodied people, especially in the context of tourism and leisure. Accessible tourism has been defined as "People with disabilities and seniors can live independently, equally, and with dignity if universal tourism products, services, and environments are made available to them: Mobility, vision, hearing and cognitive aspects of access are included in the definition" (Darcy, 2004) this research description assumes a lifelong approach to accessible tourism and its patrons as grounded in 'independence, equity and dignity'

(Darcy & Dickson, 2009). Agovino et al., (2017) stated that: "The UNWTO Global Code of Ethics for Tourism emphasizes in its article 7 that everybody should be able to freely exercise their right to enjoy tourism. As for its position at the top of the social hierarchy, it requires the possibility for everyone to have equal access to it regardless of physical, social and economic conditions". Therefore, facilitating travel and providing opportunities for PWDs is not only their human right but also such opportunity enhances the quality of life, which to some extent undermined by physical or mental shortcomings (Card et al., 2006; Qian et al., 2012).

The potential of AT in terms of its economic impact has been established in various destinations (Alén, E., Domínguez, T., & Losada, 2012; Darcy & Dickson, 2009; Özogul & Baran, 2016). However, some destinations have remained complacent to capitalize on this market (Abu Karim et al., 2021), mainly because it challenges the destination developers and tourism operators through two prongs. First, the concept of human rights of PWDs, which was ignored in the early stages of the growth and expansion of mass tourism, due to the attitude of both the supply and demand sides (Özogul & Baran, 2016). Second, the challenge of providing adequate infrastructure has frustrated destination developers because it requires harmonious cooperation between the tourism sector and local government commitment to budget and manpower. Dedicated infrastructure for people with disabilities can only be achieved if it is embedded in comprehensive plans for cities and rural areas (McKercher & Darcy, 2018; Moura et al., 2018; Pineda & Corburn, 2020). In the context of sustainable tourism, AT is the subject of research and discussion. In response to the negative impacts of mass tourism, AT has been considered a sustainable option in numerous destinations (Alén, Domínguez, & Losada, 2012; Nyanjom et al., 2018; Polat & Hermans, 2016; Sisto et al., 2021).

Perceived Value

Tourism researchers have explored the idea of perceived value, as it is probably the most important factor determining customer behavior (Al-Ansi & Han, 2019; Chen & Tsai, 2007; Ilban & Bezirgan, 2015; Javier & Bign, 2001; Phillips et al., 2013; Zhang, Morrison, & Chen, 2018). A number of reports have confirmed the relationship between service quality and perceived customer satisfaction, perceived value, and loyalty (Cockburn-Wootten & McIntosh, 2020).

Perceived value directs to consumers' general evaluation of the positive and negative characteristics of a service or product (Melian, 2016).

Perceived value has also been studied as a mediating factor in tourist behavior; however, its role in AT has not been adequately addressed. Clarifying the difference between perceived value and satisfaction is critical. Sànchez et al (2006, p. 397) claimed that: "The concept of perceived value is different from the concept of satisfaction: unlike perceived value, which is established throughout the purchase process, satisfaction is usually established after purchase or consumption. Perceived value can exist without a product having been purchased or used, whereas satisfaction depends on actual experience."

However, due to the complexity of the tourism system (Gunn, C. A., & Var, 2002), since perceived value is a dynamic variable, it cannot be evaluated directly (Stylidis et al., 2017). It is the dynamics of perceived value (i.e., before purchase, at the moment of purchase, at the time of use, and after use) (Darcy et al., 2020) that ground the perception of people with disabilities in the destination as a platform for discourse in AT.

Destination Image

Positive perception of a destination and its products has been demonstrated in previous studies (e.g., accessible attractions, adequate infrastructure and superstructure, hospitality, etc.) can be factors that lead to visits and revisit and maintain competitiveness. (Chang & Chen, 2012; Clery et al., 2017; Ritchie, J. B., & Crouch, 2003; Tao et al., 2019). Most studies have been done in various tourism contexts, but less attention has been paid to PWDs and AT. Nonetheless, the destination image and its conceptualization are the subjects of controversy and debate, without a unified definition. Josiassen et al. (2016, p. 789) point out that this is due to the ambiguity of the destination image form and construct itself, and that there is as yet no consensus on its conceptual definition and delineation. Furthermore, the complexity of the destination image illustrates the extent of its construction (Clery et al., 2017).

For this study, we used the perceived destination image by people with disabilities living at the destination. The destination image comprises two components. The first component is the "primary" image, which is the image created through a person's first-hand experience with a destination (Cockburn-Wootten & McIntosh, 2020). This study focused on the destination image perceived by people with disabilities to determine their perceived value and satisfaction with AT. Secondly, destination image is also influenced by information that comes from other sources outside one's own experience (Gillovic et al., 2018). This is because the perceived destination image by PWDs is a sound basis for making a reasonable argument for the appropriateness of Northern Cyprus to capitalize on AT. It is also important to distinguish DI from people with disabilities compared to non-disabled people, as the former have specific concerns that ordinary citizens do not. For example, infrastructure facilities, amenities, transportation, accessibility, etc., must be based on specific design and land use planning for people with disabilities, not to mention the community's sociocultural attitude toward people with disabilities – in the context of "social value" (Card et al., 2006).

Satisfaction

The context of satisfaction, either in the concept of tourism services or non-tourism products, has been discussed in business and marketing-oriented literature extensively. Satisfaction in tourism is defined as the comparison of tourists' experiences and expectations at a destination (Al-Ansi & Han, 2019). Oliver (1980) developed the expectancy-disconfirmation model to examine the construct of "satisfaction" (Milman, A., & Pizam, 1995; Ozturk et al., 2008).

Before buying a product, consumers develop expectations. They compare actual performance with expectations after they have used the product. Customers who experience positive disappointment are more likely to buy the same product again if it meets their expectations. Consumers are likely to be dissatisfied if actual performance falls short of expectations. It is likely that consumers will look for alternatives if they are disappointed with a product (Kani et al., 2017).

The purpose of this study is not to examine the concept of satisfaction in the context of consumer behavior analysis. On the other hand, the perceptions of PWDs transcend the mere consumer satisfaction discourse instead the focus is on the destination's offerings to accommodate and provide a quality of life to this population segment with implications for AT. Of course, this does not mean 'satisfaction' is an absent construct in the whole discussion. Rather by focusing on perceived value, the aim is to understand the flaws and shortcomings in the destination, which is reflected in the overall image, to overcome the structural deficiencies if any. On this note, (Jeong & Kim, 2020b) investigated the impact of the perceived value and satisfaction on loyalty and found that functional and social value (i.e., physical, infrastructural, and cultural provision) significantly increase satisfaction. Consequently, we propose the following hypothesis:

Hypothesis Development

H1. There is a positive relationship between perceived value and destination satisfaction

H2. There is a positive relationship between perceived value and destination image

H3. There is a positive relationship between destination image and satisfaction

H4. Destination image mediates the positive relationship between perceived value and destination satisfaction.

Figure 1. proposed conceptual framework

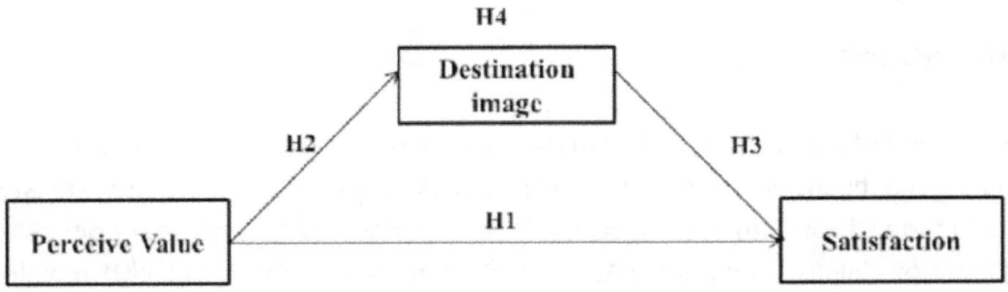

METHODOLOGY

Study Site

This study examined the relationship between perceived value and satisfaction to determine destination image for people with disabilities and data collected from local tourists with disabilities in Northern Cyprus, one of the most popular destinations in the Mediterranean. The tourism sector of Northern Cyprus contributes only $982.9 million to GDP. This is far less for an island that has the possibility to become the best tourist destination among countries that have the same level. Unfortunately, accessible tourism in North Cyprus has yet to really take off, which may be due to the lack of infrastructure, lack of coordination within the authorities, and security issues.

Sample Design and Data Collection

This study used a quantitative technique with a questionnaire survey was used to collect data from people with disabilities in North Cyprus. A cross-sectional method was employed to collect data because it is easier to get to the research objectives, Respondents are free to answer the questions without providing their personal information (Alipour et al., 2011; Alipour & Kilic, 2005; Kamyabi & Alipour, 2022). The empirical study was conducted during the period from September 2020-October 2021. The target audience was people over the age of 18 who visit or live in North Cyprus. As a result, 350 responses were collected, obtaining a response rate of 78.13%. After eliminating incomplete questionnaires, an analysis of 250 valid questionnaires was performed. Our study used several ex-ante procedural solutions to minimize common method variance (Karatepe et al., 2020; Rezapouraghdam et al., 2021).

We point out on the cover sheet of each questionnaire that there are no right or wrong answers that all data will be treated confidentially and that participation is voluntary. Then, each questionnaire contained the following information on the cover page: “By completing this questionnaire, you signify your consent” (Darban et al., 2022; Rezapouraghdam et al., 2021). In addition, the authors conducted the study involving human participants with the institution’s ethics committee (Reference number: EMUE/125).

Procedure and Pretest

A back-translation technique was used to translate the questionnaires into Turkish (Karatepe, Rezapouraghdam, & Hassannia, 2021). An English- and Turkish-speaking hospitality student translated the Turkish questionnaire back into English to compare any differences. After the translations, the committee chair evaluated the content of both versions of the questionnaire. After the committee chair approved the questionnaire, it was mailed to the pilot study participants. Based on two previous studies: hotel customer preferences, satisfaction and trustworthiness: a survey of tourists with disabilities in Taiwan, the questionnaire for the pilot study was developed (Chan, 2010). For effective tourism marketing, a model is developed that examines the influence of the tourism frame on the destination image of the destination (Haneef, 2017). The questions are divided into three parts and one general question: demographic variable, Destination image of North Cyprus from PWDs perspective, quality/adequacy of accessible transportation, accommodation, leisure and recreation facilities for PWDs, and evaluation of the perceived value of tourists with disabilities. Pilot tests were conducted to determine if the questionnaire captured the needs and satisfaction of disabled tourists. Fifteen completed questionnaires were collected for the pilot study. The results of the pilot study showed that a modified questionnaire was created and sent to people with disabilities to test and share the needs, satisfaction level, perceived value, and destination image of visitors with disabilities in relation to their experiences as guests.

Four hypotheses will be tested in this study and three latent variables exist: destination image, perceived value, and satisfaction. Two categories were used to measure the challenges faced by the disabled population. First, barriers to accessibility were evaluated using a five-point Likert scale (1 = “strongly agree”; 5 = “strongly disagree”). A five-point Likert scale was used to assess the quality of facilities for people with disabilities (1 = “very adequate”; 5 = “very inadequate”).

Data Analysis

Two stages of data analysis were performed. We first performed a confirmatory factor analysis and assessed sample adequacy. In the second stage, structural equation modeling (SEM) was used to empirically test the relationship between destination image, perceived value, and satisfaction. The developed model was analyzed with Smart PLS 3.2.8 software, while SPSS 22.0 was used to analyze the questionnaire data and perform frequency, correlation, and reliability analyzes; AMOS 22.0 was used to perform confirmatory factor analyzes and structural equation modeling (Jeong & Kim, 2020b).

Table 1. Frequency of socio-demographic variables

Variable	*f*	%	Variable	*f*	%
Gender			Occupation		
Male	147	58.8	Student	23	9.2
Female	103	41.2	Employee	94	37.6
Total	250	100.0	Self-employed	55	22.0
Age			Unemployed	27	10.8
18–25	20	8.0	Retired	51	20.4
26–30	36	14.4	Total	250	100.0
31–40	69	27.6	Years Have Been Disabled		
41–50	69	27.6	Since birth/birth defect	35	14.0
51–60	31	12.4	Less than 10 years	55	22.0
+60	25	10.0	10–20 years	71	28.4
Total	250	100.0	20–30 years	36	14.4
Type of Disability			More than 30 years	53	21.2
Cerebral Palsy	12	4.8	Total	250	100.0
Spina bifida	52	20.8	Income		
Spinal cord injury	63	25.2	None	18	7.2
Muscular Dystrophy	45	18.0	1500 TL- 3000 TL	50	20.0
Deaf	14	5.6	3000 TL- 5000 TL	69	27.6
Amputation	39	15.6	Above 5000 TL	113	45.2
Motor Neuron Disease	25	10.0	Total	250	100.0
Total	250	100.0			

Note: *f*: frequency.

FINDING

Common Method Variance

Given that self-reported surveys are used to obtain data, CMV should be checked (Podsakoff et al., 2012). The questionnaire was developed utilizing pre-existing, tried-and-true notions. To eliminate item ambiguity, seven academic and industry experts examined these constructs. The Harman single factor test was performed to address CMB issues statistically. The result of the exploratory factor analysis (EFA) indicated six constructs, with a single component accounting for 37.2% of the total variation (50%). As a result, CMV is not a severe concern in our study.

A total of 36 items were included in the round model (satisfaction 20 items, perceived value 9 items, and destination image 7 items), with first a first-order factor analysis (items as indicators) and then a structural equation for the components. The model analysis was conducted in three steps: First, the external model (measurement model), then the internal model (structural model), and finally the overall model.

Items with factor loadings less than 0.5 were removed from the first model (in this model, ten items related to satisfaction and two items from esteem were removed) and then the final model was fitted. The final model is shown in Figures 2 and 3.

Measurement Model

The measurement model was analyzed using indicator reliability (outer loadings), discriminant and convergent validity, and internal consistency. For internal consistency reliability, factor loadings, average variance extracted (AVE), Cronbach's alpha, and composite reliability (CR) were used. A Cronbach's alpha value of 0.70 and a composite reliability (CR) value of 0.60 are considered satisfactory. Values between 0.7 and 0.9 are considered satisfactory, values below 0.6 are undesirable. This model shows reasonable reliability with a composite reliability coefficient greater than 0.7 for all variables. A factor loading of 0.708 is recommended, with an AVE value greater than 0.50. Loadings less than 0.708 can be retained if the AVE value is greater than 0.50 (Chew & Jahari, 2014a; Nunnally, J. C., & Bernstein, 1994). Because the indicators in the reflexive variables are associated with a domain and are highly correlated, they can be replaced, and deleting one or more items has little effect on the content. Therefore, all constructs show satisfactory convergent validity. The values of Cronbach's alpha, CR, and AVE in Table 2 are satisfactory.

Table 2. Outer loading and validity for constructs

	OL	Cronbach Alpha	CR	rho_A	AVE
Satisfaction		**0.860**	**0.880**	**0.874**	**0.526**
S1	0.719				
S2	0.666				
S3	0.746				
S4	0.675				
S5	0.544				
S6	0.655				
S7	0.596				
S8	0.654				
S9	0.618				
S10	0.629				
Destination Image		**0.853**	**0.882**	**0.890**	**0.541**
DI1	0.831				
DI2	0.856				
DI3	0.536				
DI4	0.750				
DI5	0.576				
DI6	0.771				
DI7	0.768				
Perceive Value		**0.790**	**0.847**	**0.789**	**0.543**
PV1	0.711				
PV2	0.625				
PV3	0.635				
PV4	0.609				
PV5	0.727				
PV6	0.678				
PV7	0.665				

With a minimum acceptable value of 0.5, the convergent validity of all variables is above 0.5, and all latent variables have good convergent validity. All variables in this model have good convergent validity above 0.5.

Table 3. Descriptive statistics, HTMT ratio, and correlations

Latent Variables	Mean	SD	1	2	3
1. Destination Image	2.69	0.85	**0.736**		
2. Perceived Value	2.25	0.56	0.362	**0.665**	
3. Satisfaction	2.64	0.51	0.318	0.450	**0.653**

Discriminant validity is the third criterion for testing the validity of external models. This is the extent to which a structure is correctly distinguished from other structures based on empirical criteria. To confirm that the values of AVE were higher than the squared correlation coefficients between latent variables, we compared the values of AVE for each construct with the squared correlation coefficients between latent variables (Jeong & Kim, 2020b). It was difficult to check all variables, so the pair with the highest correlation was selected. 0.736 was the highest correlation obtained (Destination image). The results in Table 3 show that all latent variables have acceptable divergent validity. According to the results of reliability, convergent validity, and discriminant validity, external models can optimally measure the latent variables. Therefore, in the continuation of the internal (structural) model, the research is reviewed.

Structural Path Model

The next step is to test the structural model after validating the measurement model. PLS-SEM is nonparametric and does not require a normal distribution of the data (Prayag, 2009). To prevent multicollinearity, the variance inflation factor (VIF) should be lower than 5 and in some cases even lower than 3.3 (Chew & Jahari, 2014b; Kani et al., 2017b). In the mediation cases (destination image), there is no multicollinearity between the independent constructs (perceived value and satisfaction). According to Table 4, all VIF values are less than 5 or 3.3. Therefore, destination image, satisfaction, and perceived value do not cross-multiply.

Table 4. Structured model result

Construct	R^2	R^2 Adj	f^2	Q^2	VIF
Perceive Value	-	-	0.096-0.130	-	-
Destination Image	0.087	0.084	0.053	0.039	1.096
Satisfaction	0.203	0.196	-	0.065	1.096

Table 5. Direct and indirect effects

	β	t-Value	95% CI (BCa)	Testing Result
Direct effects				
Perceive Value→Satisfaction (**H1**)(+)	0.337*	4.963	{0.211,0.47}	Supported
Perceive Value→ Destination Image (**H2**)(+)	0.295*	5.769	{0.202,0.412}	Supported
Destination Image→ Satisfaction (**H3**)(+)	0.215*	2.947	{0.069,0.351}	Supported
Indirect effects				
Perceive Value→Destination Image→ Satisfaction (**H4**)	0.063*	2.544	{0.02,0.113}	Supported
Total indirect effect	0.063*	2.544	{0.02,0.113}	

*p<0.05

A significant relationship was found between the perceived value and satisfaction of people with disabilities and the destination's image. As shown in Table 5, the results can be summarized as follows: direct effect of destination image on satisfaction ($\beta = 0.215$; $p < 0.05$), perceived value on destination image ($\beta = 0.295$; $p < 0.05$), perceived value on satisfaction ($\beta = 0.337$; $p < 0.05$). Also, the indirect effect of destination image on perceived value and satisfaction ($\beta = 0.063$; $p < 0.05$) indicates that destination image fully mediates the relationship between perceived value and satisfaction. Tourists plan to revisit the destination when they perceive a positive destination image. Previous research has yielded similar results (Asgarnezhad Nouri et al., 2019; Huber et al., 2018; Javier & Bign, 2001; Kani et al., 2017b; Khan et al., 2019; Parrey et al., 2019).

DISCUSSIONS OF THE FINDINGS

Tourism is one of the most rapidly growing industries in the global economy. As a result, destinations are continuously seeking ways to attract new visitors and retain repeat customers. One of the key factors that influence a tourist's decision to visit a particular destination is their perceived value of the experience (Qiao et al., 2021). Perceived value relates to the number of benefits that a tourist believes they will receive from a particular visit, relative to the costs they need to incur (Huber et al., 2018). However, the perceived value of a destination experience may not be the same for all tourists, and this is particularly true for people with disabilities (Al-Ansi & Han, 2019). Disabilities can significantly impact a tourist's travel experience, including the accessibility and availability of services and facilities. This is where destination image can play a mediating role in shaping the perceived value of a travel experience for people with disabilities (Kamyabi & Alipour, 2022).

From the other side, destination image refers to a person's perception of a particular destination based on various factors such as marketing materials, personal experience, and word of mouth. The image that a person has of a destination can influence their perceived value of the experience, and this is especially relevant for people with disabilities (Ghorbanzadeh et al., 2021). Destination image can mediate the relationship between the accessibility of a destination and the perceived value of the travel experience. If a destination has a positive image among people with disabilities, then it is likely that they will perceive higher value in visiting that destination, even if there are still accessibility challenges (Chew & Jahari, 2014b). Moreover, a positive image of a destination can increases the satisfaction of people with disabilities. A destination with a strong image of accessibility, for example, is more likely to meet the needs of people with disabilities and provide a satisfactory travel experience. This, in turn, can increase the likelihood of repeat visitation and positive word of mouth about the destination, leading to increased tourism revenue (Wang et al., 2017). Tourism destinations specifically for people with disabilities face merciless competition, and the challenges are ever increasing (Bohdanowicz-Godfrey et al., 2019). Therefore, it is significant to understand what influences the behavior of tourists with disabilities. The main aim of this study was to examine how destination image mediates the relationship between satisfaction and perceived value for people with disabilities in the context of small islands. In addition, the authors recommend further research to confirm the relationship between these structures (Chew & Jahari, 2014; Ghorbanzadeh et al., 2021; Kim, S. H., Holland, S., & Han, 2013). In light of the results of this research, we propose the following plan in order to maximize the quality of the destination image for people with disabilities, since this strategy should increase tourist satisfaction and perceived value. When destination image is properly managed, it has a significant impact on tourist satisfaction. The present results indicate that managers should remove infrastructural barriers and improve the destination image for people with disabilities. The environmental quality of facilities should also be considered by destination managers. Sometimes, delays in repairs or non-completion of facilities jeopardize the safety of tourists with disabilities (Jeong & Kim, 2020b; Khan et al., 2019). So, when it comes to accommodating disabled tourists, the safety of the facilities should be a priority (Chew & Jahari, 2014b). When accommodating disabled tourists, facility safety should be a priority. In this study, we provide empirical evidence that destination image mediates the relationship between the perceived value and satisfaction of tourists with disabilities regarding accessible tourism, and that both large-scale and small-scale accessible tourism should be considered important for destination sustainability (Kim, S. H., Holland, S., & Han, 2013a).

Tourist satisfaction and perceived value are positively related to destination image, according to this study. Destination image has a significant impact on individual satisfaction and loyalty (Jeong & Kim, 2020b). In conclusion, the mediating role of destination image on perceived value and satisfaction for people with disabilities is an essential area of research for the tourism industry. Destinations should focus on creating a positive image of accessibility to attract more people with disabilities and improve their travel experience (Kamyabi & Alipour, 2022; Qiao et al., 2022). Further study is necessary to provide a better understanding of the complexity of the relationship between destination image, perceived value, and satisfaction for people with disabilities, which would aid destinations in meeting the needs of this growing market segment, leading to increased tourism revenue. Finally, our results may not be generalizable to other destinations due to their tourism characteristics but, similar destinations can generalize our findings.

Theoretical and Practical Implications

Eventually, the findings of current study on destination management and marketing enhance our knowledge of the significance of PWDs destination image on post-trip behavior and help destination marketing organizations to improve image reconstruction and reduce negative impacts (Jeong & Kim, 2020; Kim, Holland, & Han, 2013).

A whole series of practical implications can be drawn from the results of this study. First, to satisfy tourists, destination managers must plan, build and manage a destination's image. To reduce costs, managers should focus on the perceived value of tourists and develop new products and itineraries. A satisfied disabled tourist is more likely to recommend these destinations to friends and relatives and return there in the future. Consequently, marketers and managers need to understand tourist loyalty in order to deal with complaints (Wen et al., 2020). The second step is for tourism marketers and managers to develop promotional activities and events to attract tourists. To promote tourism, the government has organized or sponsored a number of events.

To improve the destination image, infrastructure development should be accelerated and tourist satisfaction should be prioritized. In addition, destination managers should design new products and itineraries to reduce the time and cost associated with accessible tourism. The sustainable development of the tourism industry requires a balance between tourism development and the preservation of natural and cultural heritage (Wang et al., 2017).

LIMITATION AND RECOMMENDATION FOR FUTURE STUDY

This study highlights the limitations and barriers for people with disabilities as important factors in developing the image of a destination for people with disabilities.

First of all, there are few experimental studies that examine the relationship between perceived value and tourist satisfaction. Furthermore, despite several studies on tourist satisfaction, there is no evidence that destination image influences the perceived value and satisfaction of travelers with disabilities (Akroush et al., 2016). In this study, we examined the destination image of people with disabilities in North Cyprus. It is also possible that the sample size of 250 resident tourists is not representative of the total number of disabled tourists in Northern Cyprus. Nevertheless, our sample is appropriate for islands and small countries, but we should be cautious about generalizability (Alipour et al., 2020; Alipour & Kilic, 2005). Participants may be inclined to tell the researchers what they think they want to hear instead of being honest about their perceptions, value, and satisfaction. Finally, the study might be limited in its generalizability if it only focuses on a particular destination, type of disability, or age range. In addition to the above limitations, some recommendations for future studies are given in continue. First, future studies should examine cognitive, affective, general, and subjective goal setting, as well as structural, intrapersonal, and interpersonal constraints. Second, future studies can conduct research to understand which aspects of a destination's image are most important to people with disabilities in influencing their perceived value and satisfaction with the travel experience. This will help destination managers develop targeted strategies to promote accessibility and positive perceptions of their destination. Third, later studies can be more engage with organizations representing people with disabilities and their travel concerns, to understand their unique challenges and requirements. This can be done through online forums and surveys to gather feedback on existing accessibility features and recommendations to improve overall accessibility. To sum up, Future research directions could involve comparing and contrasting the destination image of destinations with different levels of accessibility, as well as examining the relationship between destination image and perceived value/satisfaction for particular disability categories, such as physical, sensory, cognitive or invisible disabilities. Examining the role of social media and word-of-mouth in shaping destination image, along with strategies for effective marketing that promote accessibility, could also be considered. In addition, it may be valuable to explore the impact of destination image and accessibility on perceived value and satisfaction both during the pre-travel decision-making stage as well as the actual travel experience.

REFERENCES

Akinci, Z. (2013). Management of accessible tourism and its market in Turkey. *International Journal of Business and Management Studies*, *2*(2), 413–426.

Al-Ansi, A., & Han, H. (2019). Role of halal-friendly destination performances, value, satisfaction, and trust in generating destination image and loyalty. *Journal of Destination Marketing and Management, 13*(December), 51–60. doi:10.1016/j.jdmm.2019.05.007

Alén, E., Domínguez, T., & Losada, N. (2012). New opportunities for the tourism market: Senior tourism and accessible tourism. *Visions for global tourism industry: Creating and sustaining competitive strategies*, 139-166.

Alipour, H., Fatemi, H., & Malazizi, N. (2020). Is edu-tourism a sustainable option? A case study of residents' perceptions. *Sustainability (Basel), 12*(15), 5937. Advance online publication. doi:10.3390u12155937

Alipour, H., & Kilic, H. (2005). An institutional appraisal of tourism development and planning: The case of the Turkish Republic of North Cyprus (TRNC). *Tourism Management, 26*(1), 79–94. doi:10.1016/j.tourman.2003.08.017

Alipour, H., Vaziri, R. K., & Ligay, E. (2011). Governance as Catalyst to Sustainable Tourism Development: Evidence from North Cyprus. *Journal of Sustainable Development, 4*(5). Advance online publication. doi:10.5539/jsd.v4n5p32

Asgarnezhad Nouri, B., Nemati, V., & Abbasgholizadeh, N. (2019). The Effect of Perceived Value on the Destination Image, Satisfaction and Loyalty of Medical Tourists: A Case Study in Ardabil. *Journal of Health, 10*(1), 34–49. doi:10.29252/j.health.10.1.34

Chen, C. F., & Tsai, D. C. (2007). How destination image and evaluative factors affect behavioral intentions? *Tourism Management, 28*(4), 1115–1122. doi:10.1016/j.tourman.2006.07.007

Chew, E. Y. T., & Jahari, S. A. (2014). Destination image as a mediator between perceived risks and revisit intention: A case of post-disaster Japan. *Tourism Management, 40*, 382–393. doi:10.1016/j.tourman.2013.07.008

Cole, S., & Eriksson, J. (2010). Tourism and human rights. In Tourism and inequality: Problems and prospects (pp. 107–125). doi:10.1079/9781845936624.0107

Darban, G., Karatepe, O. M., & Rezapouraghdam, H. (2022). Does work engagement mediate the impact of green human resource management on absenteeism and green recovery performance? *Employee Relations, 44*(5), 1092–1108. doi:10.1108/ER-05-2021-0215

Darcy, S., & Dickson, T. J. (2009). A whole-of-life approach to tourism: The case for accessible tourism experiences. *Journal of Hospitality and Tourism Management*, *16*(1), 32–44. doi:10.1375/jhtm.16.1.32

Ghorbanzadeh, D., Shabbir, M. S., Mahmood, A., & Kazemi, E. (2021). Investigating the role of experience quality in predicting destination image, perceived value, satisfaction, and behavioural intentions: A case of war tourism. *Current Issues in Tourism*, *24*(21), 3090–3106. doi:10.1080/13683500.2020.1863924

Hellier, P. K., Geursen, G. M., Carr, R. A., & Rickard, J. A. (2003). Customer repurchase intention. *European Journal of Marketing*, *37*(11/12), 1762–1800. doi:10.1108/03090560310495456

Huber, D., Milne, S., & Hyde, K. F. (2018). Constraints and facilitators for senior tourism. *Tourism Management Perspectives*, *27*(March), 55–67. doi:10.1016/j.tmp.2018.04.003

Ilban, M. O., & Bezirgan, M. (2015). Effects of Destination Image and Total Perceived Value on Tourists' Behavioral Intentions: An Investigation of Domestic Festival Tourists. *Tourism Analysis*, *20*(November), 499–510. Advance online publication. doi:10.3727/108354215X14411980111370

Javier, S., & Bign, J. E. (2001). *Tourism image, evaluation variables and after purchase behaviour*. Academic Press.

Jeong, Y., & Kim, S. (2020). A study of event quality, destination image, perceived value, tourist satisfaction, and destination loyalty among sport tourists. *Asia Pacific Journal of Marketing and Logistics*, *32*(4), 940–960. doi:10.1108/APJML-02-2019-0101

Kamyabi, M., & Alipour, H. (2022). An Investigation of the Challenges Faced by the Disabled Population and the Implications for Accessible Tourism: Evidence from a Mediterranean Destination. *Sustainability (Basel)*, *14*(8), 4702. Advance online publication. doi:10.3390u14084702

Kani, Y., Aziz, Y. A., Sambasivan, M., & Bojei, J. (2017). Antecedents and outcomes of destination image of Malaysia. *Journal of Hospitality and Tourism Management*, *32*, 89–98. doi:10.1016/j.jhtm.2017.05.001

Karatepe, O. M., Rezapouraghdam, H., & Hassannia, R. (2020). Job insecurity, work engagement and their effects on hotel employees' non-green and nonattendance behaviors. *International Journal of Hospitality Management*, *87*, 102472. Advance online publication. doi:10.1016/j.ijhm.2020.102472

Karatepe, O. M., Rezapouraghdam, H., & Hassannia, R. (2021). Does employee engagement mediate the influence of psychological contract breach on pro-environmental behaviors and intent to remain with the organization in the hotel industry? *Journal of Hospitality Marketing & Management, 30*(3), 326–353. doi:10.1080/19368623.2020.1812142

Khan, M. J., Chelliah, S., Khan, F., & Amin, S. (2019). Perceived risks, travel constraints and visit intention of young women travelers: The moderating role of travel motivation. *Tourism Review, 74*(3), 721–738. doi:10.1108/TR-08-2018-0116

Kim, S. H., Holland, S., & Han, H. S. (2013). A structural model for examining how destination image, perceived value, and service quality affect destination loyalty: A case study of Orlando. *International Journal of Tourism Research, 15*(4), 313–328. doi:10.1002/jtr.1877

Melian, A. G. (2016). *Accessible tourism: An integrated model of the behavior of tourists with disabilities in a destination*. Academic Press.

Moura, A., Eusébio, C., & Devile, E. (2022). The 'why' and 'what for' of participation in tourism activities: Travel motivations of people with disabilities. *Current Issues in Tourism*, 1–17. doi:10.1080/13683500.2022.2044292

Nazir, M. U., Yasin, I., & Tat, H. H. (2021). Destination image's mediating role between perceived risks, perceived constraints, and behavioral intention. *Heliyon, 7*(7), e07613. doi:10.1016/j.heliyon.2021.e07613 PMID:34368481

Nunnally, J. C., & Bernstein, I. H. (1994). *Psychological theory*. MacGraw-Hill.

Nyanjom, J., Boxall, K., & Slaven, J. (2018). Towards inclusive tourism? Stakeholder collaboration in the development of accessible tourism. *Tourism Geographies, 20*(4), 675–697. doi:10.1080/14616688.2018.1477828

Oliver, M., & Barnes, C. (2012). Back to the future: The World Report on Disability. *Disability & Society, 27*(4), 575–579. doi:10.1080/09687599.2012.686781

Parrey, S. H., Hakim, I. A., & Rather, R. A. (2019). Mediating role of government initiatives and media influence between perceived risks and destination image: A study of conflict zone. *International Journal of Tourism Cities, 5*(1), 90–106. doi:10.1108/IJTC-02-2018-0019

Phillips, W. J., Wolfe, K., Hodur, N., Leistritz, F. L., Management, H., Dakota, N., Leadership, H., & Dakota, N. (2013). *Tourist Word of Mouth and Revisit. 104*(November 2011), 93–104. https://doi.org/ doi:10.1002/jtr

Podsakoff, P. M., MacKenzie, S. B., & Podsakoff, N. P. (2012). Sources of method bias in social science research and recommendations on how to control it. *Annual Review of Psychology*, *63*(1), 539–569. doi:10.1146/annurev-psych-120710-100452 PMID:21838546

Polat, N., & Hermans, E. (2016). A model proposed for sustainable accessible tourism (SAT). *Tékhne (Instituto Politécnico do Cávado e do Ave)*, *14*(2), 125–133. doi:10.1016/j.tekhne.2016.11.002

Popiel, M. (2016). Barriers in Undertaking Tourist Activity by Disabled People. *Prace Naukowe Akademii Im. Jana Długosza w Częstochowie. Kultura Fizyczna*, *15*(3), 103–110.

Prayag, G. (2009). Tourists' evaluations of destination image, satisfaction, and future behavioral intentions-the case of mauritius. *Journal of Travel & Tourism Marketing*, *26*(8), 836–853. doi:10.1080/10548400903358729

Qiao, G., Cao, Y., & Zhang, J. (2022). Accessible Tourism – understanding blind and vision-impaired tourists' behaviour towards inclusion. *Tourism Review*. Advance online publication. doi:10.1108/TR-03-2022-0129

Qiao, G., Ding, L., Zhang, L., & Yan, H. (2021). Accessible tourism: a bibliometric review (2008–2020). In *Tourism Review*. Emerald Group Holdings Ltd. doi:10.1108/TR-12-2020-0619

Rezapouraghdam, H., Akhshik, A., & Ramkissoon, H. (2021). Application of machine learning to predict visitors' green behavior in marine protected areas: Evidence from Cyprus. *Journal of Sustainable Tourism*, 1–25. Advance online publication. doi:10.1080/09669582.2021.1887878

Rubio-Escuderos, L., García-Andreu, H., & Ullán De La Rosa, J. (2021). *Accessible tourism: origins, state of the art and future lines of research.* Academic Press.

Sisto, R., Cappelletti, G. M., Bianchi, P., & Sica, E. (2022). Sustainable and accessible tourism in natural areas: A participatory approach. *Current Issues in Tourism*, *25*(8), 1307–1324. doi:10.1080/13683500.2021.1920002

Tøssebro, J. (2004). Introduction to the special issue: Understanding disability. *Scandinavian Journal of Disability Research*, *6*(1), 3–7. doi:10.1080/15017410409512635

UNWTO. (2022). *Accessible tourism identified as 'game changer' for destinations*. https://www.unwto.org/news/accessible-tourism-identified-as-game-changer-for-destinations

Wang, B., Yang, Z., Han, F., & Shi, H. (2017). Car tourism in Xinjiang: The mediation effect of perceived value and tourist satisfaction on the relationship between destination image and loyalty. *Sustainability (Basel)*, *9*(1), 22. Advance online publication. doi:10.3390u9010022

Zhang, J., Morrison, A. M., & Chen, Y. (2018). *How Country Image Affects Tourists' Destination Evaluations: A Moderated Mediation Approach.* doi:10.1177/1096348016640584

ADDITIONAL READING

Akinci, Z. (2013). Management of accessible tourism and its market in Turkey. *International Journal of Business and Management Studies*, *2*(2), 413–426.

Buhalis, D., Darcy, S., & Ambrose, I. (Eds.). (2012). *Best practice in accessible tourism: Inclusion, disability, ageing population and tourism* (Vol. 53). Channel View Publications. doi:10.21832/9781845412548

Chen, C. F., & Tsai, D. (2007). How destination image and evaluative factors affect behavioral intentions? *Tourism Management*, *28*(4), 1115–1122. doi:10.1016/j.tourman.2006.07.007

Kamyabi, M., & Alipour, H. (2022). An Investigation of the Challenges Faced by the Disabled Population and the Implications for Accessible Tourism: Evidence from a Mediterranean Destination. *Sustainability (Basel)*, *14*(8), 4702. doi:10.3390u14084702

Meira, C., Martins, I. S., & Sousa, B. B. (2021). Accessible Tourism and Digital Platforms: A Study Applied to the City of Viana do Castelo. In Handbook of Research on the Role of Tourism in Achieving Sustainable Development Goals (pp. 235-247). IGI Global. doi:10.4018/978-1-7998-5691-7.ch014

Moura, A., Eusébio, C., & Devile, E. (2022). The 'why' and 'what for' of participation in tourism activities: Travel motivations of people with disabilities. *Current Issues in Tourism*, 1–17.

Qiao, G., Cao, Y., & Zhang, J. (2023). Accessible Tourism–understanding blind and vision-impaired tourists' behaviour towards inclusion. *Tourism Review*, *78*(2), 531–560. doi:10.1108/TR-03-2022-0129

Qiao, G., Ding, L., Zhang, L., & Yan, H. (2021). Accessible tourism: A bibliometric review (2008–2020). *Tourism Review*, *77*(3), 713–730.

Rucci, A. C., Porto, N., Darcy, S., & Becka, L. (2021). Smart and accessible cities? Not always–The case for accessible tourism initiatives in Buenos Aries and Sydney. In *ICT tools and applications for accessible tourism* (pp. 115–145). IGI Global. doi:10.4018/978-1-7998-6428-8.ch006

KEY TERMS AND DEFINITIONS

Accessible Tourism: Accessible tourism refers to the provision of tourism products, services and facilities that are accessible to all people, regardless of their physical, sensory, cognitive, or other disabilities or limitations. This includes the provision of accessible transportation, accommodation, attractions, activities, and services, as well as the training of tourism industry staff to better serve people with disabilities.

Barrier-Free Tourism: Tourism that removes physical, sensory, or cognitive barriers that may prevent people with disabilities from fully participating in tourism activities or accessing tourism facilities.

Customer Experience: The overall impression or perception that a customer has of a particular product, service, or destination, based on their interactions, expectations, and emotions.

Destination Image: The overall perception or mental picture that people have of a particular destination, shaped by factors such as marketing, promotional materials, personal experiences, and word-of-mouth.

Disability: A physical or mental impairment that substantially limits one or more major life activities, such as walking, seeing, hearing, speaking, or learning.

Inclusive Tourism: Tourism that promotes the participation and enjoyment of all people, regardless of their abilities, backgrounds, or personal characteristics.

Market Segmentation: The process of dividing a larger market into smaller subgroups or segments based on shared characteristics, needs, or behaviors. In the context of accessible tourism, market segmentation may involve identifying specific groups of people with disabilities who have unique needs and preferences.

Perceived Value: The overall perceived benefit or worth that a customer or tourist receives from a particular product or service, relative to the price or effort required to obtain it.

Satisfaction: The degree of pleasure or contentment that a person experiences after consuming or experiencing a particular product or service.

Chapter 3
Tourism Industry:
Leadership and Innovation

Cláudia Ferreira Leitão
Universidade Autónoma de Lisboa, Portugal

Jorge Vareda Gomes
ADVANCE/CSG/ISEG, Universidade de Lisboa, Portugal

Denise Capela Santos
RICH, Escola Superior de Enfermagem São Francisco das Misericórdias, Portugal

Bruno Melo Maia
CICEE, Universidade Autónoma de Lisboa, Portugal

ABSTRACT

Leadership, innovation, and performance are essential factors to achieve the desired sustainable profitability of companies. The relationship between these variables is one of the keys to organizational success, although their study has proven to be complex. The purpose of this chapter is to analyse the impact of leadership on the relationship between innovation and performance in the Portuguese hotel sector. To answer this challenge, a survey was carried out to top and middle managers of four-star and five-star hotel units. The existence of a positive correlation between innovation and performance was found; however, leadership has not been shown to have a moderating effect on the relationship. The work highlights several important contributions to the hotel industry and identifies aspects that, when well implemented and developed, can lead to superior performance in organizations.

DOI: 10.4018/978-1-6684-7242-2.ch003

INTRODUCTION

Tourism companies operate in a competitive world, where innovation is an essential condition for the survival of companies (Sundbo et al., 2007). A more dynamic and rapidly changing business environment has forced the hospitality industry to resort to effective leadership processes as a way of motivating employees to obtain the desired results (Huertas-Valdivia et al., 2019). Considering the constant changes imposed by global competition, it is likely that the success and competitiveness of hotels depend on the ability of managers to promote innovation in their teams and organizations. Leadership plays a key role in promoting firm innovativeness (Khan et al., 2020). The definition of leadership has changed considerably in the last decades, the initially concept linked to the figure of "great man" fell out of use, the most recent approaches focus on the transformational dimension of the leader (Brownell, 2010). This dimension is reflected in a collaborative and relationship attitude, establishing open communication, forming and supporting the team effort and providing the necessary resources to fulfil a shared vision (Brownell, 2010; Humphreys & Einstein 2003; Stone et., 2004).

Leadership is a critical factor in organizations, as it can affect goals, visions, strategy, social environment and employee motivation (Yukl, 2013). Leadership is the ability to influence others to voluntarily make decisions that promote the short and long-term growth of companies (Nejad & Rowe, 2009).

Pioneering leadership studies originate from Ohio and Michigan Universities identified two main types of behavior among the surveyed leaders. Leadership oriented to people, as the leader is attentive to subordinates, respects their ideas and feelings and establishes mutual trust. On the other hand, task-oriented leadership, the degree to which the leader is task-oriented and directs subordinate work activities to achieve the goal (Daft, 2008).

The literature has shown that leaders influence creativity and innovation in different ways, increasing the intrinsic motivation of the follower (Jung et al., 2003), articulating an inspiring vision (Gupta & Singh, 2013; Jong & Hartog, 2007; Lee, 2008), providing support (Cheung & Wong, 2011), developing a relationship based on trust and respect (Volmer, Spurk & Niessen, 2012), enabling and sharing decision-making (Krause, 2004; Slåtten, Svensson & Sværi, 2011; Somech, 2006), delegating (Krause, Gebert & Kearney, 2007) and promoting high ethical standards (Valentine et al., 2011).

Innovation contributes to the financial performance of the hotel industry (Chang, Gong & Shum, 2011), sales growth and market value (Nicolau & Santa-Maria, 2013); increases customer loyalty and satisfaction (Enz et al., 2010; Ottenbacher & Gnoth, 2005; Victorino et al., 2005) and sustains a hotel's competitive advantage (Fraj, Matute & Melero, 2015).

Innovativeness in the hotel industry as the ability to respond faster and more flexibly to environmental changes (Fraj, Matute &Melero, 2015). Today's challenging and dynamic hotel industry requires organisations to consider innovation and differentiation in their daily practices (Nagy, 2014) in response to emergent challenges (Chen, 2011; Nagy, 2014; Ottenbacher, 2007, Sandvik, Duhan & Sandvik, 2014).

It is recognized by several authors that one of the sources of innovation for organizations is the capacity, diversity of skills and knowledge of their employees, which can generate new and useful ideas (Jong & Hartog, 2007; Slåtten, Svensson & Sværi, 2011; Subramaniam & Youndt, 2005). Also, that individual innovation significantly contributes to organizational success and effectiveness (Axtell et al., 2000, Kattara & El-Said, 2013; Tajeddini, 2010; Unsworth & Parker, 2003).

To make the right decisions, managers need to know how to balance their technical and social skills in the right combination, promoting relationships and interactions that lead to communication processes that must be effective to motivate and lead others (Page & Connell, 2009). To survive in the present business changing environment, companies realized that the ability to change and adapt was inevitable. Leading change management has become the main concern of all executives. In this context, transformational and charismatic leadership theories arise (Gill et al., 2005).

Uncertainty and constant market changes lead companies to develop innovative activities, hoping that they will contribute to improving business performance (Chen, 2017). In an empirical study of companies listed on the Fortune 1000, it was concluded that the innovation was positively related to organizational growth and profitability (Cho & Pucik, 2005). However, this relationship between innovation and performance is not deterministic, it is affected by different factors, such as, internal capital, external market and other environmental issues (Huang & Rice, 2009).

Tourism is the largest industry in the world and is one of the most dynamic and vibrant sectors of the world economy (Costa et al., 2014; Devaraja & Deepak, 2014). The World Travel & Tourism Council's (WTTC, 2019) research reveals that the Tourism sector accounted for 10.4% of global GDP and 319 million jobs, or 10% of total employment in 2018. In fact, the tourism industry is an economic driver worldwide and, in Portugal, it was responsible for 20% of the country's exports and 58% of exports in the services area (Costa et al., 2018).

This chapter analyses the relationship between innovation and performance and looks at whether different leadership styles enhance this relationship. The application of a moderation model results from the application of a survey to professionals in the four and five-star hotel sector.

THEORETICAL FRAMEWORK

Leadership

Leadership is an important organizational contextual factor that shapes the social and work environment, capable of predicting individual, group and organizational creativity and innovation in different contexts (Amabile et al., 2004; Mumford & Licuanan, 2004; Rosing et al., 2011). Leadership is the ability to influence the competence and motivation of individuals and groups to achieve specific goals (Hongdao et al., 2019; Ellemers et al., 2004). Leadership has been recognized as a social process that occurs in a group context where the leader influences the behavior of his followers so that organizational objectives are achieved (Haq & Chandio, 2017; Oke et al., 2009). The leadership style is an important management tool, its proper use can encourage close relationships with employees, improve the organizational climate and increase performance (Kozak & Uca, 2008). Managers use different leadership styles in decision making with the aim of improving the organization's performance (Bass, 2008). Despite the differences, the various definitions of leadership have four common elements (Nahavandi, 2015):

1. Leadership is a social and group phenomenon; there can be no leaders without followers. Leadership is about others.
2. Leadership involves interpersonal influence or persuasion and leaders move their followers through goals and actions;
3. Leadership is goal-oriented and action-oriented. Leaders take an active role in groups and organizations, which in turn use influence to direct their followers and achieve goals.
4. The presence of leaders in a group assumes a hierarchy. There are cases where this hierarchy is formal and well defined, keeping the leader at the top, in other cases it is informal and flexible.

The leadership style was initially conceptualized as transactional versus transformational in the 1970s and 1980s (Bennett, 2009). Burns (1978) was one of the first authors to work on the characteristics of both styles and Bass & Avolio (1990) provided the metrics for the respective assessment – the Multifactor Leadership Questionnaire (MLQ).

Transformational leaders link followers' work functions to a compelling vision of the organization's future, making followers see work as something important and meaningful, increasing their intrinsic motivating potential (Zhu, et al., 2009). Transformational leaders are recognized agents of change, visionary, trust people, value-oriented, and lifelong learning, capable of dealing with complexity, ambiguity and uncertainty (Peterson et al., 2009; Judge & Piccolo, 2004). These leaders influence and encourage their followers to be creative, innovative and motivate them to contribute more than was expected of them (Boerner et al., 2007; Hall et al., 2008). Transformational leadership focuses on the processes of transformation and change (Bass & Riggio, 2006). Rafferty and Griffin (2004) identify five dimensions for transformational leadership:

1. **Vision:** Refers to an idealized image of the future, based on the organization's values.
2. **Inspirational Communication:** Refers to positive and courageous messages about the organization, as well as statements that lead to motivation and confidence.
3. **Supportive Leadership:** Expresses concern for followers, always considering their individual needs.
4. **Intellectual Stimulation:** Stimulates the interest and awareness of employees about problems, as well as increasing their ability to see these same problems in a new way.
5. **Personal Recognition:** Always rewards for recognition and effort in achieving goals.

The conceptualisation of transformational leadership has been researched widely as a predictor of employees' creativity and innovation (Cheung & Wong, 2011; Eisenbeis, van Knippenberg & Boerner, 2008; Jung, Wu & Chow, 2008). Transformational leadership has been most strongly correlated with innovation (Rosing, Frese & Bausch, 2011). Transformational leadership was conceptualized as a style that challenges status quo and organizational norms (Eisenbeiss, van Knippenberg & Boerner, 2008), shapes followers' beliefs, values and attitudes (Castro et al., 2008), increases follower intrinsic motivation (Jung, Chow & Wu 2003), strengthens the perspective of creative self-concept (Wang & Zhu, 2011) and conveys a new vision (Lee, 2008), to stimulate the creativity of followers, groups and organizations and innovation (Eisenbeis, van Knippenberg & Boerner, 2008; Jung, Wu & Chow, 2008; Michaelis, Stegmaier & Sonntag 2009). Several authors address the topic of transformational leadership in the hospitality industry (Gill et al., 2006; Chiang & Jang, 2008; Erkutlu, 2008; Hinkin & Schriesheim, 2008; Scott-Halsell et al., 2008; Patiar & Mia, 2009; Khalili, 2016; Liang et al., 2017).

On the opposite side, transactional leaders have a traditional view of the organization and use power to ensure the execution of tasks. Transactional leadership presupposes two dimensions (Lai, 2011):

1. **Contingent Reward:** Is a motivation-based system that is used to reward those employees that meet their goals. It can provide a positive reinforcement for a job well done.
2. **Management-by-Exception:** This can be active or passive. Active leaders are always watching to evaluate performances of employees. Passive management only assess after the task has been done and will only let you know about problems after they occurred.

Research in hospitality has shown that the application of transactional leadership can result, but with less job satisfaction, less commitment to organization, low quality of service and low performance (Boerner et al., 2007). Transactional leadership is based on the concept of exchange between the leader and the follower - the leader provides followers with the necessary resources and rewards in exchange for motivation, productivity and effective task execution (Bass, 2008; Wang et al., 2011).

The hotel industry is highly customer-oriented and faces times of intense competition (McManus, 2013). The complex and changing environment of the hospitality industry presents a tremendous set of incentives, pressures and demands that have proven to be stressful, especially for frontline personnel (Kara et al., 2013). A key element of success for a hospitality company is the employee's motivation to reach their maximum potential, be engaged, embrace, change and make good technical decisions (Bennett, 2009). As noted earlier, transformational leadership distinguishes itself from the transactional one by its approaches in focus and behavior.

Effectiveness of transformational leaders is often enhanced by their charisma and the strong relationships they establish (Humphreys & Einstein. 2003). The hotel sector has registered a growing interest in transformational leadership as a key factor for the effectiveness of its activity (Brownell, 2010). Effectively, the most relevant leadership style in the hospitality industry in recent decades has been transformational (Bass, 2000; Bennis & Thomas, 2002; Avolio & Gardner, 2005). As early as the 1990s, Tracey and Hinkin (1996) had studied the results of transformational leadership in a hotel management company by exploring the measurement qualities and practical utility of two leadership assessment tools, including transforming the questionnaire's leadership scales of Multifactor Leadership (Bass & Avolio, 1990). Gill et al. (2006) questioned hotel and restaurant employees and found that with managers who exhibit behaviors of transformational leaders, employees had less work stress. Other survey that examined hotel staff revealed that the values shared and inspired by leaders were among the most important factors for their motivation and satisfaction (Clark et al 2009).

Innovation

Innovation is a change in the status quo, involving discovering new things and new ways to sell them (Oke et al., 2009). Innovation is also seen as an essential component for competitiveness, it is incorporated into organizational structures, processes, products and services and is essential for the survival of organizations (Gunday et al., 2011). Innovation has become a critical competency for leaders operating in a world surrounded by challenges that require new thinking and solutions. Innovation is increasingly being acknowledged as a strategic imperative for sustainability and differentiation (Skarzynski & Gibson, 2008; Morris et al., 2011; Lowe & Marriott, 2007; Snyder & Duarte, 2003).

The recent literature considers innovation as a process with dynamic, social, complex and other characteristics, in which combinations or connections between variables are created so that new ideas emerge and manifest themselves as new technologies, applications, markets and organizational practices aimed at creation of value (Ungerer, 2011). Innovation is recognised as a means to convert opportunities to new business ideas and increase an organisation's profitability and competitiveness by offering differentiated products and services (Chen, 2011; Ottenbacher, 2007; Slåtten, Svensson & Sværi, 2011; Sundbo, Orfila-Sintes & Sørensen, 2007). The key to innovation comes from the need for firms to achieve better business performance and an increase in competitive advantage (Gunday et al., 2011). Innovation has a considerable impact on corporate performance generating a better market position that transforms into competitive advantage and superior performance (Walker, 2008). Research has demonstrated that employees' innovative behaviour depends highly on their interaction with others and on the environmental contextual factors in the organisation (Axtell et al., 2000; West & Sacramento, 2012; Zhou & Shalley 2003). Many studies focus on the relationship between innovation and performance, the results of these studies indicate that the greater the degree of innovation, the greater the corporate performance (Garg et al., 2003; Wu et al., 2003).

In certain environments and considering different contextual variables such as leadership, organizational support and climate, these have been identified as motivators of individual creativity and innovation (Herrmann & Felfe, 2013; Jong & Hartog, 2007; Sokol et al., 2015; West & Sacramento, 2012). Creativity has been described as a complex outcome of person and situation interaction in an organisation (Amabile et al., 2004). Creativity and innovation are considered the most relevant capabilities for all organizations that wish to seek competitive advantage (Gisbert-López et al., 2014). Previous research has focused on the antecedents of creativity and innovation, namely on personal (ie, leadership) and contextual factors (favorable climate to innovation) (Černe et al., 2013; Wang et al., 2014). In a context of change, organizations need to be more innovative, namely at individual, group and organizational level, in order to improve its global degree of competences (Mumford et al., 2002). Innovative behavior is that which intentionally generates and promotes the realization of new ideas in a function, work group or organization and represents the key to competitiveness (Janssen & Van Yperen, 2004). The direct or indirect relationship between transformational leadership and innovative employee behavior has been previously identified in the academic literature (Kahai et al., 2003; Shin and Zhou, 2007). Research findings suggest that the success behaviors of innovation leaders are indeed different from the leadership behaviors that are deemed to be sufficient in conventional leadership development initiatives (Elkins & Keller, 2003; Govindarajan & Trimble, 2005; Hamel, & Labarre, 2011).

Innovation leadership seems to be a new trend that deals with new complexities in the realization of value and the role of innovation in dealing with them. Innovative leaders are those who consistently create and drive the organization to make changes. Innovative leaders are recognized with a set of distinctive characteristics, namely (Metcalf & Morelli, 2015):

1. Clarify and align vision with strategic initiatives.
2. Create effective teams and help colleagues.
3. alliances and partnerships.
4. Anticipate and respond to challenges and opportunities.
5. Develop robust and resilient solutions.
6. Develop and test hypotheses, evaluate, learn, and continually improve.

A study by Carmeli and Waldman (2010), examined the importance of leadership in innovation in the strategic alignment of the organization with its environment and in the improvement of various economic, relationship and product performance results. The results suggest that leadership in innovation has significantly improved the company's performance.

Performance

In today's world, where change becomes the main determinant, the survival of organizations and their ability to developing high performance depends on your ability to understand environmental changes and create innovations that respond to those changes (Kalmuk & Acar, 2015).

Performance is one of the most debated concepts and for which there has never been an agreement between the researchers (Jenatabadi, 2015). Recently, the definition of organizational performance has focused on an organization's ability to efficiently use the available resources to achieve the objectives set by the company, considering the relevance of all its users (Peterson, 2003). This definition highlights the three main elements, "efficiency", "effectiveness" and "relevance". The organization must be able to align performance with organizational objectives (effectiveness), organizational resources (efficiency), and, with the stakeholders' expectations (relevance). For Osaze and Anao (2000) organizational performance means the degree of fulfilment that the organization's goals are being achieved. The control of this performance can be categorized in two dimensions (Tseng, 2010): Internal performance - related to product quality, costs and profit levels and benchmarked performance - Comparing the company's performance, product quality, costs, operations and customer satisfaction in reference to the sector. Maltz et al. (2003) proposed five performance indexes, namely, financial performance, market/customer, process, people development, and future. Robinson et al., (2005) suggest that cultural approaches are important for improving performance and sustaining innovation in terms of technologies, processes, and products, namely, through knowledge management and organizational learning.

Leadership, Innovation, and Performance Interrelationship

Leadership is an important organizational function and considered critical in the creative achievements of employees in the hospitality industry as well (Chen, 2011; Enz & Siguaw, 2003; Slåtten, Svensson & Sværi, 2011; Wong & Pang, 2003). Leadership is essential for the success and innovativeness of a firm (Hongdao et al., 2019). Leaders need to be creative not only for survival, but also to compete quickly in an ever-changing world; therefore, leadership plays an active role, influencing, adapting, moving, to lead and innovate (Buekens, 2013; Vargas, 2015). According to Jung et al. (2003), leaders can influence the followers' innovation process in both direct and indirect ways through motivation and higher-level needs. It is extremely important to have the right type of leadership to drive innovation efficiently and effectively. Unique leadership resources are the hallmark of companies capable of successfully managing different types of innovative activities (Oke et al., 2009). Failing to innovate can increase the threat to sustainability; therefore, organizations and their leadership consider it extremely important to promote a climate in which innovation can be created among employees (Shanker et al., 2017).

Ottenbacher, Gnoth and Jones (2006) claimed that employees are at the heart of change and differentiation in the hotel industry because of their critical role as the organisation's ambassadors. Macey and Schneider (2008) argued that high levels of employee involvement in innovation contribute to better organizational performance. In their study, Harter et al. (2002) concluded that building an environment that enhances and supports employee innovation can significantly increase the possibility of business success. Employees, when applying their thoughts and actions, play an important role in the continuous innovation process, which is crucial for improving the achievement of better organizational profitability, growth and market value (De Jong & Den Hartog, 2010). Consistently, several exploratory studies suggest that an innovation environment promotes innovative work behaviors and organizational performance (e.g., Crespell & Hansen, 2009; King, et al., 2007; Nybakk & Jenssen, 2012). Other authors have pointed out how innovative behavior at work can help organizations gain competitive advantage and improve organizational performance (e.g., Janssen & Van Yperen, 2004; Yuan & Woodman, 2010; Shih and Susanto, 2011). Morales et al, (2008) argue that innovation is essential to improve organizational performance and show that organizations that focus on innovation are more successful in ensuring greater market share which can lead to high income and profitability.

Slåtten, Svensson and Sværi (2011) in a survey of frontline service employees in the hospitality industry in Norway, concluded that empowering leadership has a positive impact on employee creativity and innovation. Still, Slåtten and Mehmetoglu (2011), using a survey of 279 hotel and restaurant employees, found that autonomy at work is related to employee engagement, which, in turn, is closely related to innovative behavior. Ispas (2012) observed that improvements in employee performance result from positive interactions with managers and the quality of customer service. Fraj, Matute and Melero (2015) found in a survey of 232 directors, environmental managers and owners with a minimum of two stars, that orientation towards hotel learning and innovation are determinants of a proactive environmental strategy and competitiveness.

The theory of resources and capabilities also states that organizations need the capabilities, resources and technologies to implement an innovation strategy that will be a challenge for competitors to emulate, and that allows organizations to have sustainable competitive advantages and achieve greater organizational performance (Bommer & Jalajas, 2004; Calantone et al., 2002). Competitiveness in the hospitality sector is particularly dependent on innovation to achieve lower costs and higher quality results (Ottenbacher & Gnoth, 2005). Innovation in the hospitality sector is essentially of an intangible nature. Therefore, they are difficult to monitor and evaluate in terms of frequency and time of execution (efficiency) and also their contribution to customer satisfaction and profitability (effectiveness) (Ottenbacher & Gnoth, 2005). The incorporation of technologies in hotels has promoted productive efficiency and a greater capacity for differentiation, factors that improve the service provided and translate into competitive advantage (Orfila-Sintes et al., 2005). Although, interviews conducted with hospitality managers indicate that the most critical factors in the development of new services in the hospitality sector are the employee's motivation (Ottenbacher & Show, 2002).

In the tourism and hospitality sector, competitiveness depends on the level of innovation in terms of high-quality and low-cost production of its services, which meet or exceed the customer's need with a certain level of novelty and sophistication

(Hjalager, 2002). Current research trends reveal that transformational leadership plays a vital role in a company's innovation capacity (Amankwaa et al., 2019). Vaccaro et al. (2012) found that transformational leadership is the antecedent of the innovation company.

The intellectual stimulation component of transformational leadership explicitly focuses on employee creativity and innovation. Transformational leaders motivate followers to experiment, take risks, and think outside the box continuously to perform tasks and innovations. Ford (2002) suggests that creativity and innovation depend on leadership and argued that leaders who are concerned with the effectiveness of the current system and promote actions to instigate change, creativity and dynamic capabilities. Transformational leadership attributes, such as coaching, training, group cohesion, knowledge sharing, psychological training, supportive behavior, and emphasis on extra-role performance, all contribute to the company's innovation.

The literature refers to a relationship between leadership and innovation and it is also agreed that the transformational leadership style is significantly related to creativity and organizational innovation (Al-Husseini & Elbeltagi, 2012; Hu, et al., 2012; Tipu et al., 2012). On the other hand, several studies show that innovation is positively related to superior performance (Oke et al., 2012; Nybakk & Jenssen, 2012; Durán-Vázquez et al., 2012). Also, several research studies using the Multifactor Leadership Questionnaire (MLQ), to measure the behaviors involved in transformational and transactional leadership, positively relate transactional and transformational leadership to indicators of leadership effectiveness, such as subordinate satisfaction, motivation and performance (Oke el al., 2009).

METHODS

Therefore, the objective of this chapter is to research the relationship of the variable's leadership, innovation and organizational performance. A conceptual model is proposed to analyze leadership as a moderator variable in the relationship between innovation (independent variable) and performance (dependent variable) (Figure 1).

Figure 1. Conceptual model

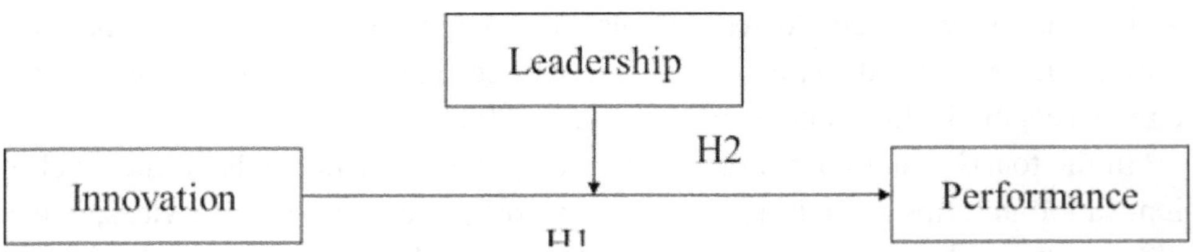

The moderator variable affects the strength of the relationship between the dependent and independent variable in correlation (Judd, 2015). In this way it is intended to verify the following hypotheses:

H1: Innovation positively influences the hotel's performance.

H2: Leadership positively influences the relationship between innovation and hotel performance.

Data Collection Instrument

To test the proposed hypotheses, a quantitative study was carried out and primary data were collected from the answers to a questionnaire. The questionnaire consists of two parts. The first part aims to collect socio-demographic information and the second part consists of 37 questions, divided by the three variables.

The first 12 question are related to leadership and supported in the item-scale of Yukl (2013), the following 11 ones concerned to innovation and the last 14 are related to performance. The innovation and performance questions were supported on the literature. A pre-test process was carried out to validate the consistency of the constructs.

Sample's Socio-Demographic Characterization

The target population was people who work in the Portuguese hotel industry and hold positions of leadership in 4 and 5-star hotels. Questionnaires were sent to 82 hotels and 34 responses were subsequently validated.

The dataset features of 34 answers have the following characteristics: 17 male (50%) respondents. Ages between the 26-35 years and 36-45 years, were the most representative, respectively with 38% and 35%. The graduation degree represents 47% of the sample and 65% of the people come from 4-star hotels. 73% are middle management professionals and the most relevant range of professionals has up to three years old on the organizations, 44%.

RESULTS

Calculation of Leadership, Innovation, and Performance

Table 1 presents the Cronbach's Alpha values which were calculated for each set of questions measuring each one of three latent variables from our conceptual model.

Table 1. Cronbach's alpha for sets of items measuring each latent variable

	α	N Items
Leadership	.75	12
Innovation	.85	11
Performance	.83	14

All of these values belong to the interval from .75 to .85, which means that each set of items shows an acceptable to good internal consistency (Streiner, 2003). We have also calculated Cronbach's Alpha coefficients for each removed item in each group and found out that there are no significant internal consistency improvements upon removal of any particular item. Since it was not feasible to perform a pre-test to our query (given the difficulty of collecting even a small sample), this reliability analysis confirms that the query is well built and therefore we will measure each latent variable through the simple average of all of the corresponding items. We have calculated the simple average of the corresponding items to compute the values of the three latent variables of our conceptual model: Leadership, Innovation and Performance.

Table 2 shows their descriptive and inferential statistics (for N = 34 valid observations).

Table 2. Descriptive and inferential statistics for the latent variables

						95.0% CI for μ		95.0% CI for σ
	Min	Max	Mdn	M	M SE	Lower Upper	SD	Lower Upper
Leadership	2.58	4.42	3.75	3.67	.08	3.51 3.84	.46	.37 .80
Innovation	2.36	4.82	3.55	3.45	.10	3.24 3.66	.61	.49 .92
Performance	3.14	4.86	3.89	3.92	.08	3.77 4.07	.44	.35 .78

Normality Tests

To test the three variables normality distribution hypothesis, and given that our sample size N>30, we have chosen to perform Kolmogorov-Smirnov normality tests (at 5% significance level), whose results are shown in Table 3.

Table 3. Frequency table for leadership type and corresponding class intervals

Leadership Type	Class Interval	Frequency	Percent
Transformational (1)	[4.5; 5.0]	12	35.29
Neutral (0)	[3.5 ; 4.0[	11	32.35
Non-transformational (-1)	[1.0 ; 3,5[	11	32.35

The Sig. values for the variables Innovation and Performance satisfy Sig.> .05, therefore we should not reject the null hypothesis: these variables are normally distributed. On the other hand, Leadership has Sig. = .02 <.05 which means that we should accept the alternative hypothesis: it is not normally distributed.

Innovation and Performance Linear Relationship

Given that Innovation and Performance are continuous (scale) variables, with verified normal distribution hypothesis and having no significant outliers, a simple linear regression was run to predict Innovation from Performance. Figure 2 displays the corresponding box plots, which also prove the absence of outliers.

Figure 2. Box plot of the latent variables

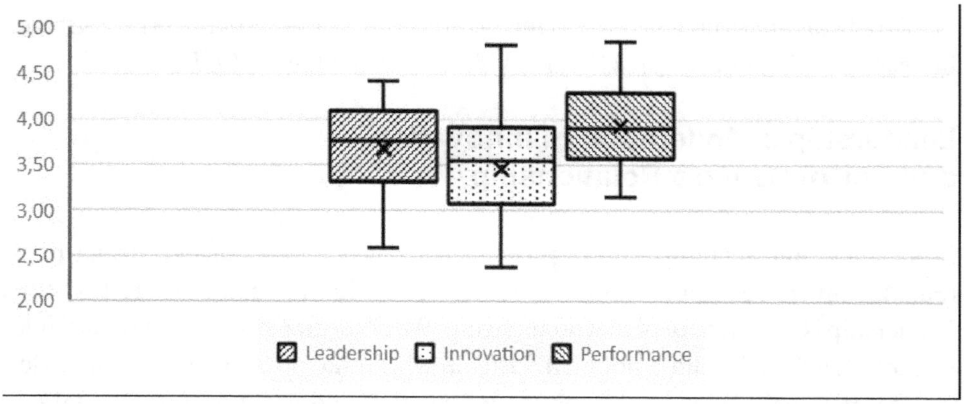

The results F (1.32) = 53.15, $p < .001$ showed that Innovation is a significant predictor of Performance, with a Pearson correlation coefficient R=.79 proving the existence of a strong positive linear correlation, and $R^2 = .62$ meaning that 62% if the total variation of Performance can be explained by the predictor Innovation. The regression constant coefficient $B_0=1.95$ is significant ($p < .001$) and its 95%

CI is [1.41, 2.50]. The regression coefficient for the predictor Innovation B_1=.57 is significant ($p < .001$) and its 95% CI is [.41; .73] (Figure 3). The regression equation is Performance = 1.95 + .57 * Innovation.

Figure 3. Scatter plot and regression line for performance and innovation

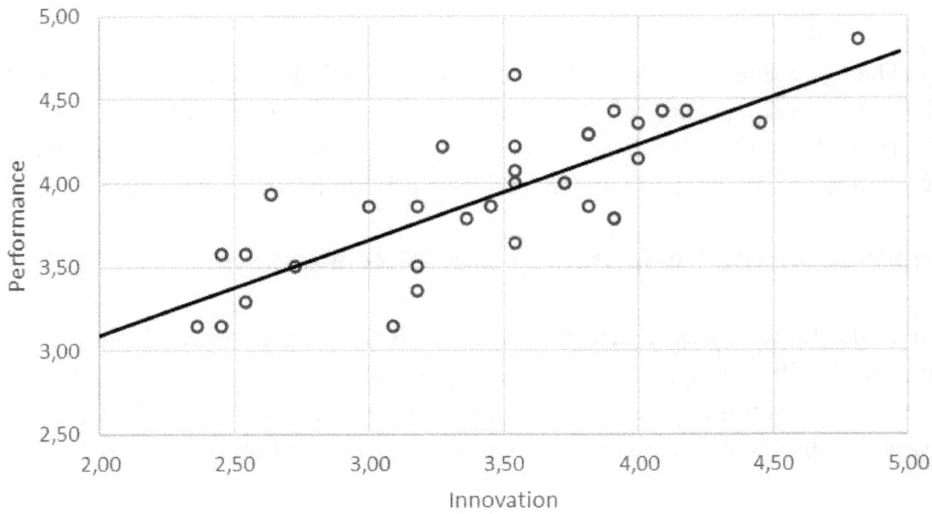

This means that a unit increase in Innovation implies an approximate Performance increase of .57 (or between .41 and .73 if we consider the 95% CI).

Leadership's Moderation of Innovation and Performance's Relationship

In the foregoing section we have proved the existence of a positive strong linear relationship between Innovation and Performance. In this section we test whether Leadership is a moderator of that relationship. We chose transforming the continuous variable Leadership into a nominal variable with three possible leadership styles: transformational (the most desirable one), neutral, and non-transformational (the least desirable one). The reason for choosing three classes instead of four (or more) is since a higher number of classes (with N = 34 observations) would generate at least one class with fewer than 10 observations (which would affect the statistical significance of the hypothesis' test), as well as adding an unnecessary level of complexity to the analysis. On the other hand, three classes seemed suitable, given the distribution of the continuous variable Leadership. The 33rd percentile of Leadership is approximately 3.5 and the 67th percentile is approximately 4.0. Therefore, we defined the following class intervals to obtain the ordinal version of Leadership.

To test the Leadership mediating hypothesis, we studied the patterns of interactions and associations between the three latent variables through a General Linear Model. In short, the outcomes of this model will tell us if the coefficients of the three linear regressions for the three Leadership groups are statistically significantly different (Table 4).

Table 4. General linear model parameter estimates for the dependent variable "Performance"

Parameter	B	Std. Error	t	Sig.	95% Confidence Interval	
					Lower Bound	Upper Bound
Intercept	1.88	.73	2.56	.02	.38	3.38
[Leadership Type =-1]	.39	.91	.43	**.67**	-1.46	2.24
[Leadership Type =0]	.10	.99	.10	**.92**	-1.92	2.12
[Leadership Type =1]	0[a]	.	.	.	.	.
Innovation	0[a]	.	.	.	.	.
[Leadership Type=-1] * Innovation	.46	.18	2.63	.01	.10	.82
[Leadership Type =0] * Innovation	.55	.19	2.90	.01	.16	.93
[Leadership Type =1] * Innovation	.60	.19	3.21	.00	.22	.99

[a]. This parameter is set to zero because it is redundant.

The reference model for the General Linear Model will be the linear regression for the Transformational Leadership Type (Leadership Type = 1). The regression line parameters for this group are B_0=1.88 (statistically significant, as Sig.=.02<0.05) and B_1=.60 (statistically significant, as Sig.=.00<0.05) with R^2=.55. The corresponding linear regression equation for the Transformational Leadership Type, which will be our reference model is: Performance = 1.88 + .60 * Innovation

Consider the Neutral Leadership Type (Leadership Type = 0). The intercept parameter is B_0=0.10 which is not statistically significant, as Sig. = .92 > .05, but the corresponding new intercept would be B_0=1.88 + 0.10=1.98. The Innovation coefficient is B_1=.55, which is statistically significant (Sig. = .01 < .05) and R^2=.42.

Therefore, the linear regression equation changed to Performance = 1.98 + .55 * Innovation

Finally, let us consider the Non-transformational Leadership Type (Leadership Type = -1). Although the intercept parameter is B_0= .39, it is not statistically significant, as Sig. = .67 > .05, but the corresponding new intercept would be B_0=1.88 + 0.39 = 2.27. But the Innovation coefficient B_1= .46 is statistically significant (Sig. = .01 < .05), R^2=.46 and the corresponding linear regression equation is Performance = 2.27 + .46 * Innovation.

As we have seen (Figure 4), the two intercepts for the neutral and non-transformational Leadership Styles are not significantly different from the reference intercept (for the transformational Leadership Style). On the other hand, although the Innovation coefficients for the three cases are different (B_1=.60, B_1=.55 and B_1=.46) we cannot claim that they are significantly different, as their 95% CI overlap: for the transformational leadership style, we have CI = [.22, .99] ; for the neutral leadership style, we have CI = [.16, .93] and for the non-transformational leadership style, we have CI = [.10, .82].

This means that the hypothesis "Leadership style is a moderator of the relationship between Innovation and Performance" could not be proved, given that there are no statistically significant differences in the regression models when we consider different Leadership style groups.

Figure 4. Scatter plot and linear regression according to leadership style

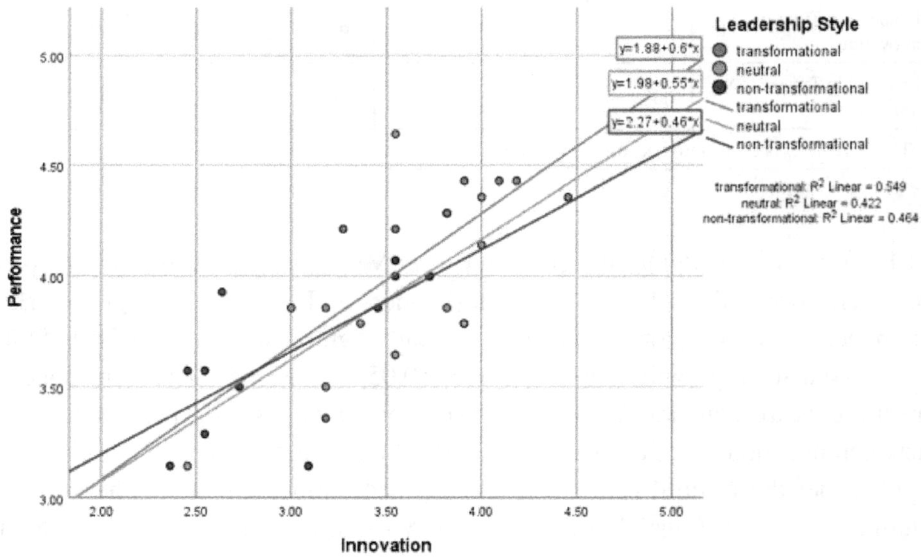

CONCLUSION

Leadership is one of the most discussed topics today, but also one of the most difficult to understand. The literature shows different definitions and views, albeit with elements in common. Transformational leadership is one style that the academy has given greater importance to recently. A good transformational leader is innovative, open to change, explores new approaches, motivates people to learn and to have a strategic and clear thinking, which would lead to good performance of teams and, consequently, to the desired business profit.

In this study, it was found that there is a strong relationship between innovation and performance, in line with what is described in the literature. The results also indicate that the improvement of innovation has positive consequences on performance. However, the generalized linear model showed that leadership is not a moderating variable in the relationship between innovation and performance, in the hotel sector. This result may be explained by the leadership style that more frequently occurs in this sector, that is, probably, still not transformational enough. The new ways of working require new approaches by administrations and managers, which should better inspire their teams to be proactive, to be aligned with the company's strategy and vision, create clear targets and encourage them to handle change as a friend.

Leaders have the authority to set specific goals and encourage employees to implement innovation. Administrators have the need to select transformational leaders or to pressure managers for a style change. In fact, performance depends on the ability to innovate. When performance increases, innovation tends to be more frequently implemented and the control of non-financial performance indicators, often not given much importance in the sector, is a must.

In practical terms, this could mean that without good leadership, and good development and implementation of innovation, it is unlikely that a company will succeed. However, it should be noted that, contrary to expectations, in the hotel sector, innovation and organizational performance showed weak values, which is why it is considered worthwhile to think about new innovative strategies, namely, technological innovations, bringing the opportunity to new environmental experiences and forms of communication with customers in this sector, to exceed clients' expectations and stimulate the increase of its performance.

REFERENCES

Al-Husseini, S., & Elbeltagi, I. (2012). The Impact of Leadership Style and Knowledge Sharing on Innovation in Iraqi Higher Education Institutions. In *Proceedings of the 4th European Conference on Intellectual Capital*. Arcada University of Applied Sciences.

Amabile, T., Schatzel, E., Moneta, G., & Kramer, S. (2004). Leader behaviors and the work environment for creativity: Perceived leader support. *The Leadership Quarterly*, *15*(1), 5–32. doi:10.1016/j.leaqua.2003.12.003

Amankwaa, A., Gyensare, M. A., & Susomrith, P. (2019). Transformational leadership with innovative behaviour. *Leadership and Organization Development Journal*, *4*(4), 402–420. doi:10.1108/LODJ-10-2018-0358

Avolio, B. J., & Gardner, W. L. (2005). Authentic Leadership Development: Getting to the Root of Positive Forms of Leadership. *The Leadership Quarterly*, *16*(3), 315–338. doi:10.1016/j.leaqua.2005.03.001

Axtell, C. M., Holman, D. J., Unsworth, K. L., Wall, T. D., Waterson, P. E., & Harrington, E. (2000). Shop-floor innovation: Facilitating the suggestion and implementation of ideas. *Journal of Occupational and Organizational Psychology*, *73*(3), 265–285. doi:10.1348/096317900167029

Bass, B. M. (2000). The Future of Leadership in Learning Organizations. *The Journal of Leadership Studies*, *7*(3), 19–40. doi:10.1177/107179190000700302

Bass, B. M. (2008). *The Bass handbook of leadership: Theory, research, & managerial applications* (4th ed.). Free Press.

Bass, B. M., & Avolio, B. J. (1990). *Transformational leadership development: Manual for the multifactor leadership questionnaire*. Consulting Psychologists Press.

Bass, B. M., & Riggio, R. E. (2006). *Transformational Leadership* (2nd ed.). Lawrence Erlbaum Associates, Inc. doi:10.4324/9781410617095

Bennett, T. (2009). A study of the management leadership style preferred by it subordinates. *Journal of Organizational Culture Communications and Conflict*, *13*, 1–15.

Bennis, W. G., & Thomas, R. J. (2002). Crucibles of leadership. *Harvard Business Review*, *80*(9), 39–45. PMID:12227145

Boerner, S., Eisenbeiss, S., & Griesser, D. (2007). Followers behaviour and organizational performance: The impact of transformational leaders. *Journal of Leadership & Organizational Studies*, *13*(3), 15–26. doi:10.1177/107179190701 3003020

Bommer, M., & Jalajas, D. S. (2004). Innovation sources of large and small technology-based firms. *IEEE Transactions on Engineering Management*, *51*(1), 13–18. doi:10.1109/TEM.2003.822462

Brownell, J. (2010). Leadership in the service of hospitality. *Cornell Hospitality Quarterly*, *51*(3), 363–378. doi:10.1177/1938965510368651

Buekens, W. (2013). Coping with the innovation paradoxes: The challenge for a new game leadership. *Procedia Economics and Finance*, *6*, 205–212. doi:10.1016/ S2212-5671(13)00133-0

Burns, J. M. (1978). *Leadership*. Harper Perennial.

Calantone, R. J., Cavusgil, T. S., & Zhao, Y. (2002). Learning orientation, firm innovation capability, and firm performance. *Industrial Marketing Management*, *31*(6), 515–524. doi:10.1016/S0019-8501(01)00203-6

Carmeli, A., & Waldman, D. (2010). Leadership, behavioral context, and the performance of work groups in a knowledge-intensive setting. *The Journal of Technology Transfer*, *35*(4), 384–400. doi:10.100710961-009-9125-3

Castro, C., Periñan, M., & Bueno, J. (2008). Transformational leadership and followers' attitudes: The mediating role of psychological empowerment. *International Journal of Human Resource Management*, *19*(10), 1842–1863. doi:10.1080/09585190802324601

Černe, M., Jaklič, M., & Škerlavaj, M. (2013). Authentic leadership, creativity, and innovation: A multilevel perspective. *Leadership*, *9*(1), 63–85. doi:10.1177/1742715012455130

Chang, S., Gong, Y., & Shum, C. (2011). Promoting innovation in hospitality companies through human resource management practices. *International Journal of Hospitality Management*, *30*(4), 812–818. doi:10.1016/j.ijhm.2011.01.001

Chen, S. (2017). The Relationship between Innovation and Firm Performance: A Literature Review. *Proceedings of the 7th International Conference on Social Network, Communication and Education, in Advances in Computer Science Research*. 10.2991nce-17.2017.132

Chen, W. J. (2011). Innovation in hotel services: Culture and personality. *International Journal of Hospitality Management*, *30*(1), 64–72. doi:10.1016/j.ijhm.2010.07.006 PMID:32287854

Cheung, M., & Wong, C. (2011). Transformational leadership, leader support, and employee creativity. *Leadership and Organization Development Journal*, *32*(7), 657–672. doi:10.1108/01437731111169988

Chiang, C. F., & Jang, S. (2008). The Antecedents and Consequences of Psychological Empowerment: The Case of Taiwan's Hotel Companies. *Journal of Hospitality & Tourism Research (Washington, D.C.)*, *32*(1), 40–61. doi:10.1177/1096348007309568

Cho, H. J., & Pucik, V. (2005). Relationship between innovativeness, quality, growth, profitability, and market value. *Strategic Management Journal*, *26*(6), 555–575. doi:10.1002mj.461

Clark, R., Hartline, M., & Jones, K. (2009). The Effects of Leadership Style on Hotel Employees' Commitment to Service Quality. *Cornell Hospitality Quarterly*, *50*(2), 209–231. doi:10.1177/1938965508315371

Costa, J., Gomes, J., & Montenegro, M. (2014). Did the context of economic crisis affect the image of Portugal as a tourist destination: Strategic question overview. *Worldwide Hospitality and Tourism Themes*, *6*(5), 392–396. doi:10.1108/WHATT-09-2014-0025

Costa, J., Montenegro, M., & Gomes, J. (2018). What challenges and opportunities will lead to success? *Worldwide Hospitality and Tourism Themes*, *10*(6), 631–634. doi:10.1108/WHATT-08-2018-0053

Crespell, P., & Hansen, E. (2009). Antecedents to innovativeness in the forest product industry. *Journal of Forest Products Business Research*, *6*(1), 1–20.

Cumming, G. (2008). Inference by eye: Reading the overlap of independent confidence intervals. *Statistics in Medicine*, 205–220. PMID:18991332

Daft, R. L. (2008). *The New Era of Management* (2nd ed.). Thomson South-Western Corporation.

De Jong, J. P. J., & Den Hartog, D. (2010). Measuring innovative work behaviour. *Creativity and Innovation Management*, *19*, 23–36. DOI:.00547.x doi:10.1111/j.1467-8691.2010

Devaraja, T. S., & Deepak, K. (2014). Role of Tour Operator in Sustainable Supply Chain Management of Tourism - A Case Study on Bharat International Travels (Bit) in Mysore City. *Global Journal for Research Analysis*, *3*(8). Advance online publication. doi:10.15373/22778160/August2014/8

Durán-Vázquez, R., Lorenzo-Valdés, A., & Moreno-Quezada, G. (2012). Innovation and CSR Impact on Financial Performance of Selected Companies in Mexico. *Journal of Entrepreneurship. Management and Innovation*, *8*(3), 5–20.

Eisenbeiss, S. A., van Knippenberg, D., & Boerner, S. (2008). Transformational leadership and team innovation: Integrating team climate principles. *The Journal of Applied Psychology*, *93*(6), 1438–1446. doi:10.1037/a0012716 PMID:19025260

Elkins, T., & Keller, R. T. (2003). Leadership in research and development organizations: A literature review and conceptual framework. *The Leadership Quarterly*, *14*(4), 587–606. doi:10.1016/S1048-9843(03)00053-5

Ellemers, N., De Gilder, D., & Haslam, S. A. (2004). Motivating individuals and groups at work: A social identity perspective on leadership and group performance. *Academy of Management Review*, *29*(3), 459–478. doi:10.2307/20159054

Enz, C., & Siguaw, J. (2003). Innovations in hotel practice. *The Cornell Hotel and Restaurant Administration Quarterly*, *44*(5–6), 115–123. doi:10.1177/001088040304400516

Enz, C., Verma, R., Walsh, K., Kimes, S. E., & Siguaw, J. (2010). Cases in innovative practices in hospitality and related services. *Cornell Hospitality Report*, *10*(10), 4–26.

Erkutlu, H. (2008). The impact of transformational leadership on organizational and leadership effectiveness: The Turkish case. *Journal of Management Development*, *27*(7), 708–726. Advance online publication. doi:10.1108/02621710810883616

Ford, C. M. (2002). The futurity of decisions as a facilitator of organizational creativity and change. *Journal of Organizational Change Management*, *15*(6), 635–646. doi:10.1108/09534810210449541

Fraj, E., Matute, J., & Melero, I. (2015). Environmental strategies and organizational competitiveness in the hotel industry: The role of learning and innovation as determinants of environmental success. *Tourism Management*, *46*, 30–42. doi:10.1016/j.tourman.2014.05.009

Garg, V. K., Walters, B. A., & Priem, R. L. (2003). Chief executive scanning emphases, environmental dynamism, and manufacturing firm performance. *Strategic Management Journal*, *24*(8), 725–744. doi:10.1002mj.335

Gill, A. S., Flaschner, A. B., & Shachar, M. (2006). Mitigating stress and burnout by implementing transformational leadership. *International Journal of Contemporary Hospitality Management, 18*(6/7), 469–481. doi:10.1108/09596110610681511

Gill, F., Rico, R., Alcover, C. M., & Barrasa, A. (2005). Change-oriented leadership satisfaction and performance in work groups: Effects of team climate and group potency. *Journal of Managerial Psychology, 20*(3/4), 312–328. doi:10.1108/02683940510589073

Gisbert-López, M. C., Verdú-Jover, A. J., & Gómez-Gras, J. M. (2014). The moderating effect of relationship conflict on the creative climate – innovation association: The case of traditional sectors in Spain. *International Journal of Human Resource Management, 25*(1), 47–67. doi:10.1080/09585192.2013.781525

Govindarajan, V., & Trimble, C. (2005). *Ten Rules for Strategic Innovators; from idea to execution*. Harvard Business School Press.

Gunday, G., Ulusoy, G., Kilic, K., & Alpkan, L. (2011). Effects of innovation types on firm performance. *International Journal of Production Economics, 133*(2), 662–676. doi:10.1016/j.ijpe.2011.05.014

Gupta, V., & Singh, S. (2013). How leaders impact employee creativity: A study of Indian R&D laboratories. *Management Research Review, 36*(1), 66–88. doi:10.1108/01409171311284594

Hall, J., Johnson, S., Wysocki, A., & Kepner, K. (2008). *Transformational Leadership: The Transformational of Managers and Associates*. University of Florida.

Hamel, G., & Labarre, P. (2011). Improving our capacity to manage. *The Wall Street Journal*.

Haq, S., & Chandio, J. (2017). Transactional Leadership and its Impact on the Organizational Performance: A Critical Analysis. *International Journal of Trend in Scientific Research and Development*, 135-139.

Harter, J. K., Schmidt, F. L., & Hayes, T. L. (2002). Business-unit-level relationship between employee satisfaction, employee engagement, and business outcomes: A meta-analysis. *The Journal of Applied Psychology, 87*(2), 268–279. doi:10.1037/0021-9010.87.2.268 PMID:12002955

Herrmann, D., & Felfe, J. (2013). Moderators of the relationship between leadership style and employee creativity: The role of task novelty and personal initiative. *Creativity Research Journal, 25*(2), 172–181. doi:10.1080/10400419.2013.783743

Hinkin, T. R., & Schriesheim, C. A. (2008). A theoretical and empirical examination of the transactional and non-leadership dimensions of the Multifactor Leadership Questionnaire (MLQ). *The Leadership Quarterly*, *19*(5), 501–513. doi:10.1016/j.leaqua.2008.07.001

Hjalager, A.-M. (2002). Repairing innovation defectiveness in tourism. *Tourism Management*, *23*(5), 465–474. doi:10.1016/S0261-5177(02)00013-4

Hongdao, Q., Bibi, S., Khan, A., Ardito, L., & Nurunnabi, M. (2019). Does what goes around really comes around? The mediating effect of CSR on the relationship between transformational leadership and employee's job performance in law firms. *Sustainability (Basel)*, *11*(12), 3366. doi:10.3390u11123366

Hu, Q., Dinev, P., Hart, T., Cooke, D. (2012). Managing Employee Compliance with Information Security Policies: The Critical Role of Top Management and Organizational Culture. *Decisions Science*, *43*(4).

Huang, F., & Rice, J. (2009). The role of absorptive capacity in facilitating open innovation outcomes: A study of Australian SMEs in the manufacturing sector. *International Journal of Innovation Management*, *13*(02), 201–220. doi:10.1142/S1363919609002261

Huertas-Valdivia, I., Llorens-Montes, F. J., & Ruiz-Moreno, A. (2018). Achieving engagement among hospitality employees: A serial mediation model. *International Journal of Contemporary Hospitality Management*, *30*(1), 217–241. doi:10.1108/IJCHM-09-2016-0538

Humphreys, J. H., & Einstein, W. O. (2003). Nothing new under the sun: Transformational leadership from a historical perspective. *Management Decision*, *41*(1/2), 85–95. doi:10.1108/00251740310452934

Ispas, A. (2012). The perceived leadership style and employee performance in the hotel industry: a dual approach. *A Review of International Comparative Management*, *13*(2), 294–304.

Janssen, O., & Van Yperen, N. W. (2004). Employees' goal orientations, the quality of leader member exchange, and the outcomes of job performance and job satisfaction. *Academy of Management Journal*, *47*(3), 368–384. doi:10.2307/20159587

Jenatabadi, H. (2015). *An Overview of Organizational Performance Index: Definitions and Measurements*. University of Malaya. doi:10.2139srn.2599439

Jong, J. D., & Hartog, D. N. (2007). How leaders influence employees' innovative behaviour'. *European Journal of European Management*, *10*(1), 41–64.

Judd, C. (2015). Moderator Variable: Methodology. International Encyclopedia of the Social & Behavioral Sciences, 672-674.

Judge, T. A., & Piccolo, R. F. (2004). Transformational and transactional leadership: A meta-analytic test of their relative validity. *The Journal of Applied Psychology*, *89*(5), 755–768. doi:10.1037/0021-9010.89.5.755 PMID:15506858

Jung, D., Chow, C., & Wu, A. (2003). The role of transformational leadership in enhancing organizational innovation: Hypotheses and some preliminary findings. *The Leadership Quarterly*, *14*(4-5), 525–544. doi:10.1016/S1048-9843(03)00050-X

Jung, D., Wu, A., & Chow, C. W. (2008). Towards understanding the direct and indirect effects of CEOs' transformational leadership on firm innovation. *The Leadership Quarterly*, *119*(5), 582–594. doi:10.1016/j.leaqua.2008.07.007

Kahai, S. S., Sosik, J. J., & Avolio, B. J. (2003). Effects of leadership style, anonymity, and rewards on creativity-relevant processes and outcomes in an electronic meeting system context. *The Leadership Quarterly*, *14*(4-5), 499–524. doi:10.1016/S1048-9843(03)00049-3

Kalmuk, G., & Acar, A. (2015). The Mediating Role of Organizational Learning Capability on The Relationship Between Innovation and Firm's Performance: A Conceptual Framework. *Procedia: Social and Behavioral Sciences*, *210*, 164–169. doi:10.1016/j.sbspro.2015.11.355

Kara, D., Uysal, M., Sirgy, M., & Leed, G. (2013). The effects of leadership style on employee well-being in hospitality. *International Journal of Hospitality Management*, *34*, 9–18. doi:10.1016/j.ijhm.2013.02.001

Kattara, H., & El-Said, O. (2013). Innovation strategies: The implementation of creativity principles in Egyptian hotels. *Tourism and Hospitality Research*, *13*(3), 140–148. doi:10.1177/1467358414522053

Khalili, A. (2016). Linking transformational leadership, creativity, innovation, and innovation-supportive climate. *Management Decision*, *54*(9), 2277–2293. doi:10.1108/MD-03-2016-0196

Khan, A., Bibi, S., Lyu, J., Garavelli, A. C., Pontrandolfo, P., & Perez Sanchez, M. A. (2020). Uncovering Innovativeness in Spanish Tourism Firms: The Role of Transformational Leadership, OCB, Firm Size, and Age. *Sustainability (Basel)*, *12*(10), 3989. doi:10.3390u12103989

King, E. B., De Chermont, K., West, M. A., Dawson, J. F., & Hebl, M. R. (2007). How innovation can alleviate negative consequences of demanding work contexts: The influence of climate for innovation on organizational outcomes. *Journal of Occupational and Organizational Psychology*, *80*(4), 631–645. doi:10.1348/096317906X171145

Kozak, M., & Uca, S. (2008). Effective factors in the constitution of leadership styles: A study of Turkish hotel managers. *Anatolia*, *19*(1), 117–130. doi:10.1080/13032917.2008.9687057

Krause, D. (2004). Influence-based leadership as a determinant of the inclination to innovate and of innovation-related behaviours: An empirical investigation. *The Leadership Quarterly*, *15*(1), 79–102. doi:10.1016/j.leaqua.2003.12.006

Krause, D., Gebert, D., & Kearney, E. (2007). Implementing process innovations the benefits of combining delegative-participative with consultative-advisory leadership. *Journal of Leadership & Organizational Studies*, *14*(1), 16–25. doi:10.1177/1071791907304224

Lai, A. (2011). *Transformational-Transactional Leadership Theory*. 2011 AHS Capstone Projects. Paper 17. http://digitalcommons.olin.edu/ahs_capstone_2011/17

Lee, J. (2008). Effects of leadership and leader–member exchange on innovativeness. *Journal of Managerial Psychology*, *23*(6), 670–687. doi:10.1108/02683940810894747

Lehman, A., O'Rourke, N., Hatcher, L., & Stepanski, E. (2013). *JMP for Basic Univariate and Multivariate Statistics: Methods for Researchers and Social Scientists* (2nd ed.). SAS Institute.

Liang, T. L., Chang, H. F., Ko, M. H., & Lin, C. W. (2017). Transformational leadership and employee voices in the hospitality industry. *International Journal of Contemporary Hospitality Management*, *29*(1), 374–392. doi:10.1108/IJCHM-07-2015-0364

Lowe, R., & Marriott, S. (2007). *Enterprise: Entrepreneurship and Innovation*. Elsevier.

Macey, W. H., & Schneider, B. (2008). The meaning of employee engagement. *Industrial and Organizational Psychology: Perspectives on Science and Practice*, *1*(1), 3–30. doi:10.1111/j.1754-9434.2007.0002.x

Maltz, A. C., Shenhar, A. J., & Reilly, R. R. (2003). Beyond the balanced scorecard: Refining the search for organizational success measures. *Long Range Planning*, *36*(2), 187–204. doi:10.1016/S0024-6301(02)00165-6

McManus, L. (2013). Customer accounting and marketing performance measures in the hotel industry: Evidence from Australia. *International Journal of Hospitality Management, 33*, 140–152. doi:10.1016/j.ijhm.2012.07.007

Metcalf, M., & Morelli, C. (2015). *The Art of Leading Change: Innovative Leaders Transformation Model.* Articles from Integral Leadership Review. http://integralleadershipreview.com/author/maureen-metcalf-and-carla-morelli/

Michaelis, B., Stegmaier, R., & Sonntag, K. (2009). Shedding light on followers' innovation implementation behaviour: The role of transformational leadership, commitment to change, and climate for initiative. *Journal of Managerial Psychology, 25*(4), 408–429. doi:10.1108/02683941011035304

Morales, V., Barrionuevo, M., & Gutiérrez, L. (2010). Transformational leadership influence on organizational performance through organizational learning and innovation. *Journal of Business Research, 65*(7), 1040–1050. doi:10.1016/j.jbusres.2011.03.005

Morris, M. H., Kuratko, D. F., & Covin, J. G. (2011). *Corporate entrepreneurship and innovation.* South-Western Cencage Learning.

Mumford, M. D., & Licuanan, B. (2004). Leading for innovation: Conclusions, issues, and directions. *The Leadership Quarterly, 15*(1), 163–171. doi:10.1016/j.leaqua.2003.12.010

Mumford, M. D., Scott, G. M., Gaddis, B., & Strange, J. M. (2002). Leading creative people: Orchestrating expertise and relationships. *The Leadership Quarterly, 13*(6), 705–750. doi:10.1016/S1048-9843(02)00158-3

Nagy, A. (2014). The orientation towards innovation of spa hotel management: The case of Romanian spa industry. *Procedia: Social and Behavioral Sciences, 124*, 425–431. doi:10.1016/j.sbspro.2014.02.504

Nahavandi, A. (2015). *The Art and Science of Leadership.* Pearson Education Limited.

Nejad, H., & Rowe, G. (2009). Strategic leadership: Short-term stability and long-term viability. *Ivey Business Journal, 73*(5), 2–6.

Nicolau, J., & Santa-María, M. (2013). The effect of innovation on hotel market value. *International Journal of Hospitality Management, 32*, 71–79. doi:10.1016/j.ijhm.2012.04.005

Nybakk, E., & Jenssen, J. I. (2012). Innovation strategy, working climate, and financial performance in traditional manufacturing firms: An empirical analysis. *International Journal of Innovation Management, 16*(2), 1–30. doi:10.1142/S1363919611003374

Oke, A., Munshi, N., & Walumbwa, F. (2009). The Influence of Leadership on Innovation Processes and Activities. *Organizational Dynamics*, *38*(1), 64–72. doi:10.1016/j.orgdyn.2008.10.005

Oke, A., Walumbwa, F. O., & Myers, A. (2012). Innovation Strategy, Human Resource Policy, and Firms' Revenue Growth: The Roles of Environmental Uncertainty and Innovation Performance. *Decision Sciences*, *43*(2), 273–302. doi:10.1111/j.1540-5915.2011.00350.x

Orfila-Sintes, F., Crespi-Cladera, R., & Martinez-Ros, E. (2005). Innovation activity in the hotel industry: Evidence from Balearic Islands. *Tourism Management*, *26*(6), 851–865. doi:10.1016/j.tourman.2004.05.005

Osaze, B. E., & Anao, A. R. (2000). *Managerial Finance*. Uniben Press.

Ottenbacher, M. (2007). Innovation management in hospitality industry: Different strategies for achieving success. *Journal of Hospitality & Tourism Research (Washington, D.C.)*, *31*(4), 431–454. doi:10.1177/1096348007302352

Ottenbacher, M., & Gnoth, J. (2005). How to Develop Successful Hospitality Innovation. *The Cornell Hotel and Restaurant Administration Quarterly*, *46*(2), 205–222. doi:10.1177/0010880404271097

Ottenbacher, M., Gnoth, J., & Jones, P. (2006). Identifying determinants of success in development of new high-contact services: Insights from hospitality industry. *International Journal of Service Industry Management*, *17*(4), 344–363. doi:10.1108/09564230610680659

Ottenbacher, M., & Shaw, V. (2002). The Role of Employee Management in NSD: Preliminary Results from a Study of the Hospitality SectorIn Proceedings of the 2002 Product Development and Management Association (PDMA) Research Conference, 109-133.

Page, S. J., & Connell, J. (2009). *Tourism: a modern synthesis* (3rd ed.). Cengage Learning.

Patiar, A., & Mia, L. (2009). Transformational leadership style, market competition and departmental performance: Evidence from luxury hotels in Australia. *International Journal of Hospitality Management*, *28*(2), 254–262. doi:10.1016/j.ijhm.2008.09.003

Peterson, S. J., Walumbwa, F. O., Byron, K., & Myrowitz, J. (2009). CEO positive psychological traits, transformational leadership, and firm performance in high-technology start-up and established firms. *Journal of Management*, *35*(2), 348–368. doi:10.1177/0149206307312512

Peterson, W., Gijsbers, G., & Wilks, M. (2003). *An organizational performance assessment system for agricultural research organizations: Concepts, methods, and procedures*. ISNAR Research Management Guidelines.

Rafferty, A., & Griffin, M. (2004). Dimensions of transformational leadership: Conceptual and empirical extensions. *The Leadership Quarterly*, *15*(3), 329–354. doi:10.1016/j.leaqua.2004.02.009

Robinson, H. S., Carrillo, P., Anumba, C., & Ghassani, A. M. A. (2005). Review and implementation of performance management models in construction engineering organizations. *Construction Innovation*, *5*(4), 203–217. doi:10.1108/14714170510815258

Rosing, K., Frese, M., & Bausch, A. (2011). Explaining the heterogeneity of the leadership–innovation relationship: Ambidextrous leadership. *The Leadership Quarterly*, *22*(5), 956–974. doi:10.1016/j.leaqua.2011.07.014

Sandvik, I. L., Duhan, D. F., & Sandvik, K. (2014). Innovativeness and profitability: An empirical investigation in the Norwegian hotel industry. *Cornell Hospitality Quarterly*, *55*(2), 165–185. doi:10.1177/1938965514520963

Scott-Halsell, S., Shumate, S. R., & Blum, S. (2008). Using a Model of Emotional Intelligence Domains to Indicate Transformational Leaders in the Hospitality Industry. *Journal of Human Resources in Hospitality & Tourism*, *7*(1), 99–113. doi:10.1300/J171v07n01_06

Shanker, R., Bhanugopan, R., Van der Heijden, B. I., & Farrell, M. (2017). Organizational climate for innovation and organizational performance: The mediating effect of innovative work behavior. *Journal of Vocational Behavior*, *100*, 67–77. doi:10.1016/j.jvb.2017.02.004

Shin, S. J., & Zhou, J. (2007). When is educational specialization heterogeneity related to creativity in research and development teams? Transformational leadership as a moderator. *The Journal of Applied Psychology*, *92*(6), 1709–1721. doi:10.1037/0021-9010.92.6.1709 PMID:18020807

Skarzynski, P., & Gibson, R. (2008). *Innovation to the core: a blueprint for transforming the way your company innovates*. Harvard Business Press.

Slåtten, T., & Mehmetoglu, M. (2011). Antecedents and effects of engaged frontline employees: A study from the hospitality industry. *Managing Service Quality*, *21*(1), 88–107. doi:10.1108/09604521111100261

Slåtten, T., Svensson, G., & Sværi, S. (2011). Empowering leadership and the influence of a humorous work climate on service employees' creativity and innovative behaviour in frontline service jobs. *International Journal of Quality and Service Sciences*, *3*(3), 267–284. doi:10.1108/17566691111182834

Snyder, N., & Duarte, D. (2003). *Strategic innovation: embedding innovation as a core competency in your organization*. John Wiley & Sons.

Sokol, A., Gozdek, A., Figurska, I., & Blaskova, M. (2015). Organizational climate of higher education institutions and its implications for the development of creativity. *Procedia: Social and Behavioral Sciences*, *182*, 279–288. doi:10.1016/j.sbspro.2015.04.767

Somech, A. (2006). The effects of leadership style and team process on performance and innovation in functionally heterogeneous teams. *Journal of Management*, *32*(1), 132–157. doi:10.1177/0149206305277799

Stone, A., Russell, R., & Patterson, K. (2004). Transformational versus servant leadership: A difference in leader focus. *Leadership and Organization Development Journal*, *25*(3/4), 349–361. doi:10.1108/01437730410538671

Streiner, D. L. (2003). Being inconsistent about consistency: When coefficient alpha does and doesn't matter. *Journal of Personality Assessment*, *80*(3), 17–22. doi:10.1207/S15327752JPA8003_01 PMID:12763696

Subramaniam, M., & Youndt, M. (2005). The influence of intellectual capital on the types of innovative capabilities. *Academy of Management Journal*, *48*(3), 450–463. doi:10.5465/amj.2005.17407911

Sundbo, J., Orfila-Sintes, F., & Sørensen, F. (2007). The innovative behaviour of tourism firms-Comparative studies of Denmark and Spain. *Research Policy*, *36*(1), 88–106. doi:10.1016/j.respol.2006.08.004

Tajeddini, K. (2010). Effect of customer orientation and entrepreneurial orientation on innovativeness: Evidence from the hotel industry in Switzerland. *Tourism Management*, *31*(2), 221–231. doi:10.1016/j.tourman.2009.02.013

Tipu, S., Ryan, J., & Fantazy, K. (2012). Transformational leadership in Pakistan: An examination of the relationship. *Journal of Management & Organization*, *18*(4), 461–480. doi:10.5172/jmo.2012.18.4.461

Tracey, J., & Hinkin, T. (1996). How transformational leaders lead in the hospitality industry. *International Journal of Hospitality Management*, *15*(2), 165–176. doi:10.1016/0278-4319(95)00059-3

Tseng, S. (2010). The Correlation between Organizational Culture and Knowledge Conversion on Corporate Performance. *Journal of Knowledge Management, 14*(2), 269–284. doi:10.1108/13673271011032409

Ungerer, M. P. (2011). *Viable business strategies; a field book for leaders*. Knowres publishing.

Unsworth, K. L., & Parker, S. (2003). Proactivity and innovation: promoting a new workforce for the new workplace. In T. D. W. Holman, C. V. Clegg, P. Sparrow, & A. Howard (Eds.), *The new workplace: a guide to the human impact of modern working practices* (pp. 175–196). John Wiley.

Vaccaro, I. G., Jansen, J. J., Van Den Bosch, F. A., & Volberda, H. W. (2012). Management innovation and leadership: The moderating role of organizational size. *Journal of Management Studies, 49*(1), 28–51. doi:10.1111/j.1467-6486.2010.00976.x

Valentine, S., Godkin, L., Fleischman, G., & Kidwell, R. (2011). Corporate ethical values, group creativity, job satisfaction and turnover intention: The impact of work context on work response. *Journal of Business Ethics, 98*(3), 353–372. doi:10.100710551-010-0554-6

Vargas, M. I. R. (2015). Determinant factors for small business to achieve innovation, high performance and competitiveness: Organizational learning and leadership style. *Procedia: Social and Behavioral Sciences, 169*, 43–52. doi:10.1016/j.sbspro.2015.01.284

Victorino, L., Verma, R., Plaschka, G., & Dev, C. (2005). Service innovation and customer choice in the hospitality industry. *Managing Service Quality, 15*(6), 555–576. doi:10.1108/09604520510634023

Volmer, J., Spurk, D., & Niessen, C. (2012). Leader–member exchange (LMX), job autonomy, and creative work involvement. *The Leadership Quarterly, 23*(3), 456–465. doi:10.1016/j.leaqua.2011.10.005

Walker, R. M. (2008). An empirical evaluation of innovation types and organizational and environmental characteristics: Towards a configuration framework. *Journal of Public Administration: Research and Theory, 18*(4), 591–615. doi:10.1093/jopart/mum026

Wang, C.-J., Tsai, H.-T., & Tsai, M.-T. (2014). Linking transformational leadership and employee creativity in the hospitality industry: The influences of creative role identity, creative self-efficacy, and job complexity. *Tourism Management, 40*, 79–89. doi:10.1016/j.tourman.2013.05.008

Wang, G., Oh, I., Courtright, S., & Colbert, A. (2011). Transformational Leadership and Performance Across Criteria and Levels: A Meta-Analytic Review of 25 Years of Research. *Group & Organization Management*, *36*(2), 223–270. doi:10.1177/1059601111401017

Wang, P., & Zhu, W. (2011). Mediating role of creative identity in the influence of transformational leadership on creativity: Is there a multilevel effect? *Journal of Leadership & Organizational Studies*, *18*(191), 25–39. doi:10.1177/1548051810368549

West, M., & Sacramento, C. (2012). Creativity and innovation: the role of team and organizational climate. In M. D. Mumford (Ed.), *Handbook of organizational creativity* (pp. 359–385). Academic Press. doi:10.1016/B978-0-12-374714-3.00015-X

Wong, S., & Pang, L. (2003). Motivators to creativity in the hotel industry: Perspectives of managers and supervisors. *Tourism Management*, *24*(5), 551–559. doi:10.1016/S0261-5177(03)00004-9

WTTC. (2019). *Travel & Tourism: Economic Impact 2019 World*. World Travel & Tourism Council.

Wu, F., Mahajan, V., & Balasujbramanian, S. (2003). An analysis of e-business adoption and its impact on business performance. *Journal of the Academy of Marketing Science*, *31*(4), 425–447. doi:10.1177/0092070303255379

Yuan, F., & Woodman, R. (2010). Innovative Behavior in the Workplace: The Role of Performance and Image Outcome Expectations. *Academy of Management Journal*, *53*(2), 323–342. doi:10.5465/amj.2010.49388995

Yukl, G. (2013). *Leadership in organizations* (8th ed.). Pearson Education.

Zhou, J., & Shalley, C. E. (2003). Research on employee creativity: a critical review and proposal for future research directions. In J. J. Martocchio & G. R. Ferris (Eds.), *Research in personnel and human resource management*. Elsevier. doi:10.1016/S0742-7301(03)22004-1

Zhu, W., Avolio, B. J., & Walumbwa, F. O. (2009). Moderating role of follower characteristics with transformational leadership and follower work engagement. *Group & Organization Management*, *34*(5), 590–619. doi:10.1177/1059601108331242

ADDITIONAL READING

Andrews, S. (2007). *An Introduction to Hospitality and Tourism.* McGraw-Hill Company.

Barrows, C., Powers, T., & Reynolds, D. (2012). *Introduction to Management in the Hospitality Industry* (10th ed.). John Wiley & Sons, Inc.

Bass, B. M. (1999). Two decades of research and development in transformational leadership. *European Journal of Work and Organizational Psychology, 8*(1), 9–32. doi:10.1080/135943299398410

Everett, R. (2003). *Diffusion of innovations* (5th ed.). Free Press.

Richard, J., Devinney, M., Yip, S., & Johnson, G. (2009). Measuring Organizational Performance: Towards Methodological Best Practice. *Journal of Management, 35*(3), 718–804. doi:10.1177/0149206308330560

Stajkovic, A. (2019). *Management and Leadership, What Can MBD Do in My Workday?* (19th ed.). Research Paradigms Applied, LLC.

KEY TERMS AND DEFINITIONS

Hospitality Industry: Is a broad category of fields within the service industry that includes lodging, food and drink service, event planning, theme parks, travel and tourism. It includes hotels, tourism agencies, restaurants and bars. Innovation is an idea, practice, or object that is perceived as new by an individual or other unit of adoption (Everett, 2003). Leadership is the art of influencing people to attain group objectives willingly. “Leadership is influence, nothing more, nothing less” (Maxwell, 2019). Organizational performance encompasses three specific areas of firm outcomes: (a) financial performance (profits, return on assets, return on investment, etc.); (b) product market performance (sales, market share, etc.); and (c) shareholder return (total shareholder return, economic value added, etc.) (Richard et al., 2009).

Transactional Leadership: Or transactional management is the part of one style of leadership that focuses on supervision, organization, and performance. Some typical tactics of this type of management include strategy, efficiency goals, economies of scale and quality differentiation. Transactional managers focus on performance related tasks and goals (Stajkovic, 2019).

Transformational Leadership: Is a theory of leadership where a leader works with teams or followers beyond their immediate self-interests to identify needed change, creating a vision to guide the change through influence, inspiration, and executing the change in tandem with committed members of a group (; This change in self-interests elevates the follower's levels of maturity and ideals, as well as their concerns for the achievement (Bass, 1999).

Chapter 4
Electronic Word-of-Mouth and Tourist Satisfaction in Rural Tourism in Schist Villages

Marta Santos
Instituto Universitário de Lisboa (ISCTE-IUL), Portugal

Paulo Rita
NOVA Information Management School, Portugal

Sérgio Moro
Instituto Universitário de Lisboa (ISCTE-IUL), ISTAR-Iscte, Portugal

Bráulio Alturas
https://orcid.org/0000-0003-0142-3737
Instituto Universitário de Lisboa (ISCTE-IUL), ISTAR-Iscte, Portugal

ABSTRACT

Consumers' decision-making processes and the way they purchase their products and services have been evolving over the years due to the influence of information technologies. Tourists are increasingly making their decisions based on online reviews made by other users, which contain descriptive comments and/or a rating system, leveraging electronic word-of-mouth (eWOM). This study aims to understand the variation of the eWOM in rural tourism as well as unveil the main characteristics that influence the satisfaction and the interest of the consumers. To that end, the content of the comments and quantitative classification of Portuguese schist villages' lodgings on the platforms of TripAdvisor and Facebook were studied using both sentiment polarity and frequency analysis. The results show that eWOM has increased in rural tourism and that the satisfaction of tourists are more influenced by the friendliness of the hosts, the variety and good breakfast or Portuguese cuisine, and the service provided.

DOI: 10.4018/978-1-6684-7242-2.ch004

INTRODUCTION

Population in urban areas has been steadily increasing worldwide compared to rural areas, emphasizing socioeconomic differences (Costa & Chalip, 2005). Many national and regional authorities promote rural tourism to counter this trend, bringing hope to rural communities. This type of tourism has been growing over the last 25 years as urban inhabitants seek wellness, quietness, and outdoor activities that are impossible in urban areas due to traffic, pollution, and lack of time (Dashper, 2014). Tourism in rural areas (TRA) in Portugal is considered a driving force for the sustainability and development of local communities. Portugal also holds substantial asymmetry in economic activities, population distribution, and cultural and social issues between urban and rural areas, justifying the investment in rural tourism (Agapito, 2012; Valente & Figueiredo, 2003).

The tourism industry has embraced technology to leverage business (Moro et al., 2017b). The increasing number of Internet users worldwide, empowered by the technological solutions offered by Web 2.0, where users are the major contributors of Internet content, has led to both social networks, from which Facebook is the most prominent example, and to online reviews' platforms, such as TripAdvisor (Moro et al., 2018). The latter is an example of a specialized platform devoted to hospitality and tourism, as it allows users to write reviews about tourist units such as hotels and restaurants, including quantitative scores on several features (e.g., service and food). The abovementioned platforms belong to social media, a new type of online media where consumers read feedback from others to make judged decisions on their next purchase. These interactions between users online are labeled eWOM, defined by Hennig-Thurau et al. (2004) as "any positive or negative statement made by potential, actual, or former customers about a product or company, which is made available to a multitude of people and institutions via the Internet" (p. 39).

Social media offers a potential source of relevant information about consumers. Therefore, scholars have devoted attention to extracting knowledge from online platforms. Some authors have analyzed the quantitative ranks granted by users (e.g., Jeong & Jeon, 2008; Vermeulen & Seegers, 2009; Moro et al., 2017a), while others analyzed the comments written (Marcheggiani et al., 2014; Calheiros et al., 2017). However, the textual comments published on those platforms hold subjectivity inherent to human language, which can induce bias during its analysis. Opinion mining has been developed to deal with knowledge extraction from written opinions. Sentiment polarity classification is a task within opinion mining devoted to classifying textual contents according to the sentiments expressed by users (Jiménez-Zafra et al., 2016). Although there are many studies on tourists' online behavior on social media, research focused on rural tourism is still scarce (Melo et al., 2017). Furthermore, the localized nature of rural context justifies specific studies for each case.

This study aims to understand the eWOM phenomenon of the successful case of Schist Villages in Portugal. Notably, the focus is the lodging offer, as online reviews considerably impact accommodation product decisions compared to other tourist products (Gretzel & Yoo, 2008). Accordingly, three main research questions are proposed:

RQ.1: How is eWOM perceived in rural tourism in Portugal?
RQ.2: What features influence most of the review scores in hospitality and consequently contribute most to rural tourists' satisfaction?
RQ.3: How can this knowledge be worthwhile for rural accommodation managers?

The following section reviews the literature on rural tourism, particularly in Portugal, and social media in hospitality and rural tourism. Section 3 presents the conceptual model and research hypotheses of this study. Section 4 develops the proposed methodology. Section 5 presents and discusses the results, while the final section summarizes the main contributions outlined in this study.

BACKGROUND

Rural Tourism in Portugal

Tourism is a leading industry within the Portuguese economy (Andraz et al., 2015). It directly and indirectly influences several economic activities such as accommodation, transport, the food and beverage industry, construction, and much more (Proença & Souziakis, 2008). The Portuguese tourist office developed an ambitious plan for 2010-2016 to turn Portugal into Europe's most dynamic and agile tourism destination (Turismo de Portugal, 2015). As a result, the country was given the World's Leading Destination 2017 award (World Travel Awards, 2017).

Tourism has been, in recent years, the main driver of the Portuguese economy, being a strategic activity for Portugal in terms of attracting investment. Therefore, the growth of activity generated by the tourism industry has grown along with business opportunities and the creation of companies to provide services to the sector, which has awakened the entrepreneurial feeling in Portugal (Duarte et al., 2023).

OECD (1994) characterizes rural areas by three main features: "(1) population density and size of settlements, (2) land use and its dominance by agriculture and forestry, (3) traditional social structures and issues of community identity and heritage" (p. 9). Fleischer and Felsenstein (2000) also indicate that these areas are regularly located in isolated regions. In Portugal, rural areas are more concentrated inland, with most tourism still depending on coastal areas. The rural population in Portugal decreased from 2004 to 2015, representing 36.5% of the Portuguese population in the latter year, a reduction of almost 10% during that period (Trading Economics, 2016).

A universal definition of rural tourism lacks consensus. The definition is influenced by economic conditions and the increasing complexity of defining location (Rosalina et al., 2021). Rural tourism is recognized as a sustainable niche, challenges such as its fragmented nature, inappropriate management, lack of knowledge, and micro interests, hinder its appreciation (Boukas, 2019). Rural areas' observed fragilities include weak infrastructures, an elderly population, and territory desertification. It is possible to prevent this migration as local economies are stimulated, with the active population facing more stability, job creation, income generation, natural landscape, and environmental protection and preservation, thus removing agricultural activity dependency (Drăgulănescu & Druţu, 2012; Heneghan et al., 2016). Thus, local governments should implement tourism-related development projects to enhance rural tourism activities, develop the local economy, and increase employment (Yang et al., 2021).

As such, rural tourism must be understood as a niche market aiming to satisfy the demand while seeking sustainability in rural communities (Marzo-Navarro et al., 2015). Rural development through tourism activity is imperative to develop the economy of deprived and neglected rural areas and improve conditions for communities (Duarte, 2010; Drăgulănescu & Druţu, 2012). Nevertheless, for rural tourism to be attractive, it also has to innovate. Innovation in rural tourism and hospitality is a complex process that involves the exchange of knowledge and resources between many actors and the interrelationships between those actors in the business environment (Madanaguli et al., 2022). A study in India revealed that infrastructure development, growing environmental conscience, support of local government and community, availability of funds with the government, and participation of the private sector are the primary factors channeling rural tourism development. Also, locals' attitudes to adapt, tourist travel motives, marketing of the destination, destination characteristics, and recommendation by others are the major dependent factors identified (Kumar et al., 2022).

Internal and external traveling factors, called push and pull factors, can influence tourists' decision-making process. The former concerns internal intentions to satisfy their inherent needs and the latter relates to destiny attributes (Devesa et al., 2010). According to Kastenholz et al. (2012), the experience that rural tourists have is valued and based on a wide range of assets, services, persons, attractions, and surroundings, assuming that particular destination features are the experience constituents.

There are diverse factors why rural areas are increasingly chosen as tourist destinations. Based on studies by Frochot (2005), Kastenholz (2004), Kastenholz et al. (1999), Molera and Albaladejo (2007), and Park and Yoon (2009), those include:

to be closer to nature and culture, to have the pleasure to seek and enjoy peace and quiet, beautiful landscapes and relaxation. According to Tung and Ritchie (2011), experience is defined as "an individual's subjective evaluation and undergoing (i.e., affective, cognitive and behavior) of events related to his/her tourist activities that begin before (i.e., planning and preparation), during (i.e., at the destination), and after the trip (i.e., recollection)" (p. 1369). Therefore, the consumer generally needs and expects an irreplaceable and unforgettable experience (Figueiredo et al., 2014) characterized by an authentic and traditional rural lifestyle and activities with services and rural lodgings (Melo et al., 2017).

Customer expectations can have a positive or negative impact on customer satisfaction. While meeting and exceeding expectations is essential, managing expectations has proven to be a more comprehensive approach to delivering a satisfying experience (Costa et al., 2023). This study focuses on the Portuguese Schist Villages. A recent study shows the existence of significant differences in the perceptions of the various stakeholders regarding motivations, adjustment to the impacts of the tourist activities, and satisfaction with the different characterizing elements of the Schist Villages destination, in Portugal, especially regarding cultural items (Moutela et al., 2020). This study confirms a previous study on rural tourism in the USA, in which the results showed that destination image directly affects visitors' perception of value and revisit intentions and indirectly affects satisfaction and recommendation intentions (Phillips et al., 2013).

Social Media and eWOM in Rural Tourism

Nowadays, people live in the social media age where the Internet has evolved to Web 2.0, which is considered the sharing interface among end users (Thevenot, 2007; Alturas & Oliveira, 2016). Consequently, many online social networks (OSN) platforms emerged, such as blogs (e.g., Travelblog), forums, social networks (e.g., Facebook), review sites (e.g., TripAdvisor), social bookmarking and wikis (Leung et al., 2013), considering electronic social media as the core asset of this Internet development (Brogan & Smith, 2009). Facebook can be a good platform for advertising, but we still need further empirical and objective observations capable of assessing the effectiveness of advertising relative to users of the online social network sites (Barreto, 2013).

In the tourism industry, denoted as Travel 2.0, this technological progress has been used as a marketing tool to manage products, services, and destinations (Xiang & Gretzel, 2010). It is considered powerful and influential because it contributes to eWOM through commenting, rating, and spreading travel experiences, resulting in collective knowledge and decision-making and purchasing (Law et al., 2015; Chen et al., 2014; Moro & Rita, 2018). Furthermore, understanding user-generated content

(UGC) is critical for managing image destinations (Timothy, 2018; Avraham & Ketter, 2017). Regarding the planning process of a trip, most people during the COVID-19 era do more profound research than usual on social media, trying to understand if the destination and tourist services are safe (Madureira & Alturas, 2022).

E-WOM, especially in social media, has become one of the most critical marketing instruments for companies in the current competitive market (Dahka et al., 2020). A feature that distinguishes eWOM from traditional WOM is the speed with which it spreads and the ease of access to it (Huete-Alcocer, 2017). Rural tourism providers ought not to undervalue the power of eWOM since it is significant, critical, and reflects customers' opinions and perceptions (Melo et al., 2017), especially since this type of tourism is a niche within the tourism market, with lower marketing investments.

There is a scarcity of studies focusing on social media in rural tourism, especially compared to other types of tourism. However, the industry has acknowledged the relevance of Internet-based business to rural tourism. Rural tourism accommodations are offered online booking platforms such as Booking.com (Gössling & Lane, 2015). Specialized online tourism platforms first emerged to respond to specific problems such as booking or providing a network for tourists to exchange opinions. However, the most significant platforms have evolved to all-in-one solutions where users can read others' opinions, book their accommodations, and finally write their reviews in an infinite loop of eWOM. Accordingly, tourists are writing and reading opinions on rural tourism, highlighting the need to understand this relatively neglected type of tourism. Furthermore, the geographically localized nature of rural tourism emphasizes the apparent lack of research on this highly relevant field. For example, Ezeuduji (2015) identified that information technologies and social media are not widespread in Sub-Saharan Africa in event-based rural tourism, affecting its competitiveness. The scenario is different in Europe, with widespread social media among tourists. However, some rural tourism unit managers are still holding back due to an apparent lack of relevant experience with social media tools. This happens in distinct countries around Europe, including the UK (Townsend et al., 2016), Austria (Kavoura & Bitsani, 2013), and Norway (Gössling & Lane, 2015). A similar scenario occurs in China (Zhou, 2014). Some rural destinations in the US, such as North Dakota, seem to have moved forward by embracing social media with known success. Independently of the geographic location, social media in rural tourism is a vibrant subject with plenty of research gaps to fill. Remarkably, there is a lack of studies analyzing the impact of social media on rural tourism. The current study aims to fill such a research gap.

CONCEPTUAL MODEL AND RESEARCH HYPOTHESES

Proposed Conceptual Model

This section presents the conceptual model developed to identify and analyze the variables in eWOM influencing rural tourism lodgings. The model is based on the previous studies by Melo et al. (2017) and Bandyopadhyay (2016). The model is tested in the Portuguese Schist Villages, a project framed in TRA.

On the one hand, this research applies to part of Melo et al. (2017)'s study to build a branch of the conceptual model. The outcome of analyzing a rural lodging in France was a category system that reflected reviewers' global service quality perception about specific elements (e.g., decoration, environment, and space). This study uses those elements more broadly, as explained in the next section. Furthermore, this study focuses specifically on the perspective of customer perception and satisfaction. On the other hand, Bandyopadhyay´s (2016) theoretical study connects many concepts that are addressed in the study (e.g., valence rating, attributes, and credibility) to explain the usefulness of online reviews and, accordingly, the factors that affect the adoption of eWOM as part of the purchasing behavior. Therefore, Bandyopadhyay (2016)'s framework serves as input in the present research, including constructs such as eWOM quality and credibility, signaling with a circle in Figure 1.

Figure 1. Framework proposed by Bandyopadhyay (2016)

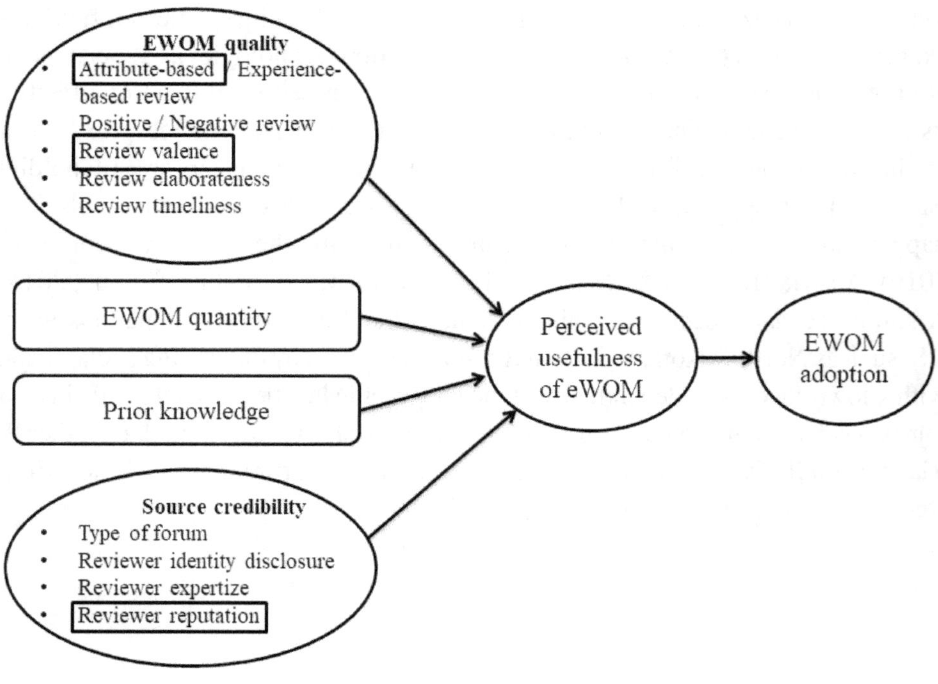

The proposed conceptual model is exhibited in Figure 3, including the highlighted constructs from Figure 1 and additional constructs identified in Table 1 to enrich the final model.

Table 1. Constructs of the conceptual model

Construct	Definition
Service Hosts Leisure activities	These constructs concern the aspects mentioned in reviewers' comments and the rating classification.
Gender	This construct is defined as male and female reviewers in TripAdvisor.
Customer satisfaction	This construct can be defined as how property managers meet, evaluate, and understand consumers' expectations and needs. Thus, the better the quality of service, the greater the satisfaction (Albacete-Sáez et al., 2006).
Reviewer reputation	Reviews' feedback measures this construct and reviews helpfulness by other people. The higher the reputation, the more credible a review is (Bandyopadhyay, 2016).
Management response	Management response is how property owners respond correctly to an online review, being an increased area of study (Levy et al., 2013).
eWOM perceived usefulness	How useful or not is electronic word-of-mouth influenced by eWOM nature (e.g., valence, timeliness, elaborateness), credibility, quantity, and previous knowledge about considered services (Bandyopadhyay, 2016)?

Research Hypotheses

EWOM quality concerns to characteristics of online reviews that impact the apparent usefulness of reviews. According to Bandyopadhyay (2016), review valence and attribute-based are two of five characteristics in his framework. On the one hand, the former is about the positive and negative ratings on the overall review, which is given by the total rank score. On the other hand, attribute-based is related to the sentiments expressed toward the service/product, which can be computed through sentiment polarity score. However, the current study takes advantage of the categories' ranks provided by TripAdvisor's reviews on a 1 to 5 scale (Figure 2).

Figure 2. TripAdvisor

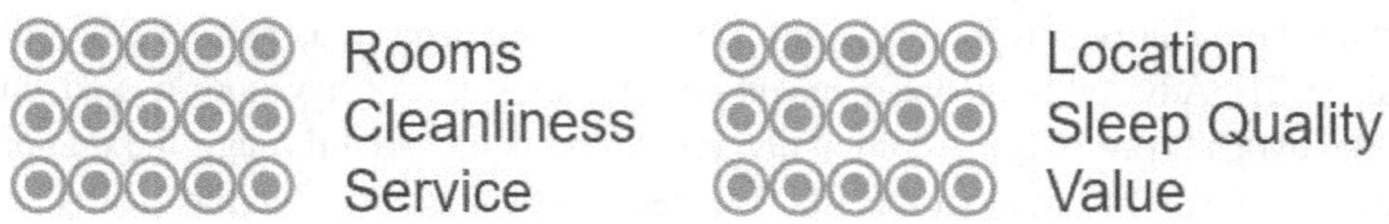

Customers' needs and satisfaction information are two crucial constructs from the previously mentioned information (Melo et al., 2017). Gender is also included. Next, the main topics the model addresses are described, based on which the subsequent hypotheses are drawn.

A choice of a destination in rural areas is influenced by many elements such as infrastructures, natural atmosphere, and service quality (Albacete-Sáez et al., 2007). The first hypothesis drawn attempts to explain if service itself as a specific feature receives higher scores when compared to other features (e.g., Figure 2):

H1: Of all TripAdvisor features, tourists value service the most.

Figueiredo et al. (2014) found that a warm welcome from hosts is important to a memorable tourist's rural experience. Moreover, Kastenholz et al. (2013), who interviewed hosts and guests from two Portuguese villages, including the Schist Villages' network, mentioned that host hospitality is considered the main traveler attraction to practice this type of tourism. In this context, the second hypothesis arose:

H2: Having friendly hosts is a relevant feature of rural tourism.

Since open-air activities encompass river beaches, sports, and outdoor animal activities, outdoor activities influence rural tourism, including family togetherness and well-being (Coyl-Shepherd & Hanlon, 2013). As a result, the following hypothesis was inferred:

H3: Most tourists who wrote about outdoor activities traveled with family.

According to Meyers-Levy and Sternthal (1991), men and women evaluate products differently with their knowledge and opinions due to the significance that both give to each product's distinct characteristics. Bandyopadhyay (2016) corroborates that gender discrepancy can affect customer satisfaction.

H4: There is a difference between genders in the score granted.

Levy et al. (2013) analyzed negative reviews (ranked only with one star) to understand how managers are coping with these reviews in the lodging market in Washington. They concluded that better-classified hotels correspond to those managed by owners who respond more frequently to negative reviews. Therefore, it is vital to understand how business owners manage their online pages (Leung et al., 2013; Litvin et al., 2008).

H5: Hosts of rural tourism units are replying to negative comments.

Based on Bandyopadhyay (2016)'s research, there are four dimensions related to source credibility: type of online platform, reviewer identity exposed, expertise, and reputation. The type of online platform dimension will not be used to formulate the subsequent two hypotheses. Nevertheless, it is perceived as more trustworthy since TripAdvisor is an independent review website, not a vendor-owned one (Bandyopadhyay 2016). Additionally, reputation is included through the number of helpful votes from other tourists. This reduces hesitations about the quality and performance of tourist products because it helps travelers decide whom to rely on and trust (Helm & Mark, 2007; Resnick et al., 2000). Furthermore, people who search for online reviews information tend to perceive lower-rated reviews as more useful (Lee et al., 2011). Therefore, reputation is more related to negative WOM rather than positive, which will be assessed by H6b.

H6 a: The TripAdvisor member's duration positively correlates with the number of helpful votes.

H6 b: Tourists who have a higher number of votes given, on average more negative reviews.

Online reviews in the tourism industry are receiving huge attention from researchers to understand consumer behavior, motivations, complaints, preferences, and profiles because eWOM issue and adoption have been increasing (Moro & Rita, 2018; Bandyopadhyay, 2016; Munar & Jacobsen, 2014; Albacete-Sáez et al., 2007; Sparks & Browning, 2010). However, this interest and evolution are not extensively reflected in other types of tourism, such as TRA. Therefore, the last research hypothesis addresses this issue:

H7: The eWOM has been increasing in rural tourism.

Once the theories and models that comprise the theoretical-conceptual framework have been identified, it was possible to define the hypotheses previously identified in Figure 3.

Figure 3. Proposed conceptual model

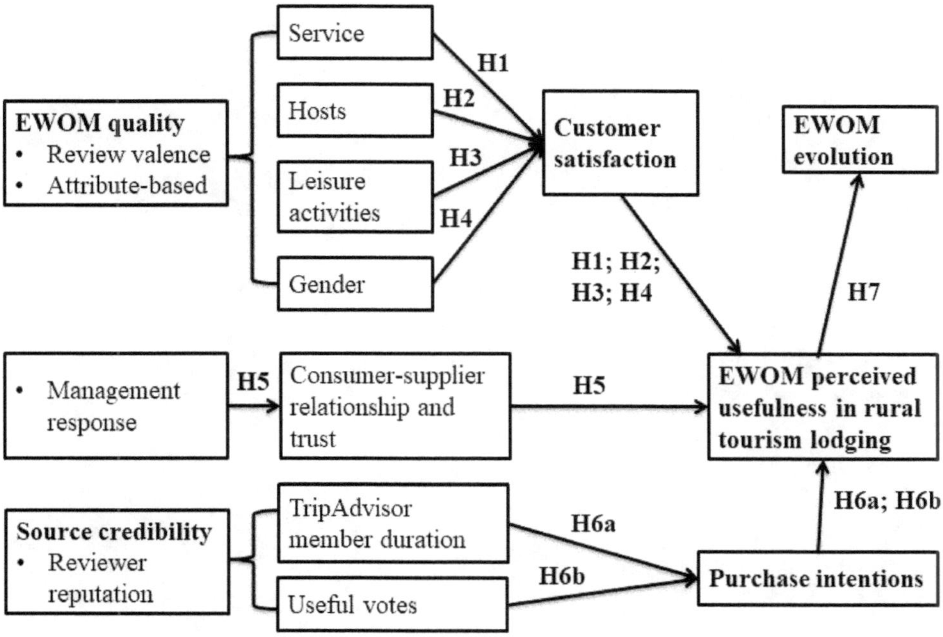

Hypotheses H1, H2, and H6 are linked to Research Question 2; hypotheses H3, H4, and H5 are linked to Research Question 3, and finally, hypothesis H7 is linked to Research Question 1.

METHODOLOGY

Case Study

Portuguese Schist Villages comprise 27 villages in 16 counties in the Central Region of Portugal, between Castelo Branco and Coimbra, covering about 5,000 km2. Schist stone is seen as a unifying element of a rich material and immaterial patrimony, resulting from the combined work between nature and man, shaping these villages and their surroundings (Aldeias do Xisto, 2023).

This is a regional sustainable development project led by ADXTUR, an Agency devoted explicitly to the tourism development of the Schist Villages, with many partnerships (public and private) operating in the territory. The brand Schist Villages represents the offer of tourist services of its associates (hotels, accommodation in rural areas, catering, tourist animation, and traditional commerce) articulated with the Calendar of Animation of the Schist Villages (Aldeias do Xisto, 2015). Table 2 highlights how tourist products and services increased from 2009 to 2014, with significant percentages.

Table 2. Evolution of the tourist resources in units

Schist Villages' Evolution of the Touristic Resources	2009	2014	%
Schist Villages	24	27	13%
Schist Villages stores	10	18	80%
Housing units	26	70	169%
Hotels	1	7	600%
Restaurants	7	13	86%
Tourist Animation Businesses	6	12	100%
Fluvial beaches	21	50	138%
Walking routes (small route)	14	45	221%
Big routes	0	1	NA
Total of the pedestrian traces (Km)	140	700	400%
BTT Centers	2	6	200%
Total cycling trails (Km)	260	900	246%
Associates (Privates)	70	179	156%
Rooms	52	492	846%
Beds	114	920	707%

Source: Aldeias do Xisto (2015)

In order to increase brand attractiveness by pursuing a communication and marketing plan, ADXTUR has implemented a consistent strategy of disclosure of Schist Villages and the promotion of its resources, equipment, animation, and services, which resulted in a constant and growing presence in the media, including online social networks. According to data from the 2015 report, the written press and online media stand out clearly, with 96.5% of published news between July 2012 and June 2015.

Schist Villages is an unavoidable mark in the Portuguese tourist panorama and inspiration for developing low economic and demographic density places. It is an excellent example of national capacities and competence when led by a dream and managed with a missionary spirit. It won many awards (e.g., Prémio 85 Anos do Diário de Coimbra; Prémio Internacionalização do Património 2014) and participated in several international fairs (e.g., International Fair for Applied Arts and Design), which was suitable for both national and international acknowledgment.

Data Collection

Several types of social media are available, from which social networks and online review platforms are among the most popular. The two most renowned are Facebook and TripAdvisor (Moro et al., 2018). Table 3 summarizes the main steps to reach the data used to analyze the hypotheses. From the first phase, a total number of 29 different accommodations were retrieved.

Table 3. Summary of the data collection process

1st phase	Went to the Schist Village website to find out how many properties were available, excluding rural hotels and camping areas.
2nd phase	Saw if those properties had a page on Facebook and TripAdvisor.
3rd phase	Collected the total number of online reviews from both online platforms.
4th phase	Discarded reviews with only a general rating classification (without descriptive content).
5th phase	Read the comments to compute the most meaningful features, according to Melo et al. (2016).
6th phase	Analyzed those features' sentiment polarity as additional features to those categories already provided by TripAdvisor.
7th phase	Assessed the validity of the hypotheses proposed.

Data was gathered since 2006, when the first customer impression on social media was published, up to May 22, 2017. Seven hundred sixty-four evaluations were registered: some with only quantitative data (e.g., score from 1 to 5) and others with quantitative and qualitative data (e.g., descriptive reviews). However, to thoroughly analyze the assumptions of this study without compromising the sample quality, only 430 online reviews were considered from 2008 (Table 4). Of those discarded, 315 were blank reviews, while 19 had no specific quantitative features to be analyzed, as those consisted of short descriptions not identifying the service/product evaluated.

Table 4. Number of evaluations per online platform (units and %)

Source	Number of Evaluations	%
Facebook	146	34%
TripAdvisor	278	65%
Both	6	1%
Total	430	100%

The lodging attributes most discussed were identified from the ones most frequently mentioned within the reviews. After analyzing them, some were excluded because the sample size was insignificant (e.g., Wi-Fi and noises), while others were incorporated in more general attributes (e.g., bed quality and room security), as shown in Table 5. Moreover, these occurrence terms followed the study by Melo et al. (2017), where they analyzed rural lodgings in France and used the same categories and similar descriptions, except for the food category, which was not mentioned in their research.

Table 5. Features extracted from TripAdvisor and Facebook reviews

Features Based on Comments	Typology of the Features
Food/ Breakfast	Typical and general food; Variety
Decoration	Styles (e.g., rustic); Pillows and other adornments;
Peacefulness	Relaxing, quiet and calm; Nature
Exterior Place	Pool; Gardens; Animals; Landscape
Hosts	How hosts are (e.g., friendly)
Leisure Activities	Sports; Hiking; River Beach
Cleanliness*	Rooms and general spaces' cleanliness
Service*	Check-in; 24h service; Hospitality; Attention to customers' requests and needs
Location*	Cars, persons, and place accessibility and visibility; Road infrastructures;
Rooms*	Air conditioner; Cozy; Shower; Fridge; Bed quality; Rooms security; Comfort;

* These features are the same in TripAdvisor, where users can classify them from 1 to 5, but it is relevant to add them again since many consumers wrote about them.

Their sentiment polarity was computed to understand the sentiments drawn from qualitative features. Therefore, besides TripAdvisor standard categories' ranks, the sentiments from both Facebook's comments and TripAdvisor's reviews were extracted using Excel and quantified into positive (100%) or negative (0%), according to Jiménez-Zafra et al. (2016). All numerical features were included in SPSS for the t-student and Pearson correlation tests to assure statistical validity. All the information retrieved and feature extraction was made manually, according to the recommendations by Marcheggiani et al. (2014).

As the reviews' accommodations were pre-selected based on Schist Villages lodging and, consequently, not every person had his profile complete, there were missing values to deal with. From the 430 reviews, 284 were extracted from

TripAdvisor, while the remaining comments were obtained from Facebook. Since Facebook's profiles are undisclosed, this study only analyzed gender based on TripAdvisor's profiles, which account for more than half of the reviews. However, more than half of the reviewers do not post in their TripAdvisor profile their gender (n= 187) or their age (n=195), respectively. Nevertheless, from the ones who disclose both features, men, and people whose ages range from 35-49 (13%) and 50-64 (9%) are more likely to generate online content. Also, most reviewers travel as a couple (38%), followed by 29% as a family.

FINDINGS AND DISCUSSION

Next, the hypotheses drawn are evaluated based on the collected data.

H1: Of all TripAdvisor features, tourists value service the most.

Table 6 shows the average ranks obtained for each of the TripAdvisor features. Although all ranks shown are above 4.50, cleanliness is the feature that users overall granted the highest score. The next most valued features are "service" and "rooms". However, these last two are the only features that had negative scores within the sample (1 and 2 values), as shown in the following example for "service": "a tremendous lack of respect. I made the appointment via telephone where I received the confirmation by email, and when I arrived there, I had no reservation"; and the following example for "rooms": "there was no air conditioning as had been mentioned (…)". Both reviews were about the same accommodation unit, probably indicating the need to restructure such a unit to meet customers' expectations. Thus, overall, the hypothesis is not validated.

Table 6. TripAdvisor features mean

Ranking	TripAdvisor Features	Mean
1	Cleanliness	4.78
2	Service	4.73
3	Rooms	4.69
4	Sleep Quality	4.64
5	Value	4.58
6	Location	4.51

H2: Having friendly hosts is a relevant feature of rural tourism.

From the aspects extracted from the description of consumers' experience, it is interesting to note the frequency and diversity of opinions. The "hosts" feature only accounted for one negative review compared to the 289 positive ones (140 records were missing values for this feature). Consequently, it results in 67.2% positive reviews, the highest percentage among all features (Table 7). Therefore, the hypothesis is validated.

In textual reviews for rural tourism, people also quoted as positive the fact of being surrounded by nature where they can relax and stay out from the busy life in urban areas. Moreover, Portuguese gastronomy and a good and varied breakfast are other things tourists write favorably about. However, in the same reasoning, it can be noticed that the food topic is the most cited when people talk negatively about anything. The t-student test was computed to evaluate mean differences in total scores between the individuals that scored negatively and positively on the different variables. "Food/breakfast" and "rooms" are the only ones that have statistical evidence to state that the mean of those features is different from both groups (negative and positive) analyzed, according to p-value ≤ 0.05.

Table 7. Opinion about the additional features extracted

	Positive (%)	Negative (%)	Missing Value (%)
Food/Breakfast	39.3%	3.5%	57.2%
Decoration	25.8%	0.2%	74.0%
Peacefulness	46.0%	-	54.0%
Exterior Place	32.1%	0.7%	67.2%
Hosts	**67.2%**	0.2%	32.6%
Leisure Activities	21.2%	-	78.8%
Service	20.0%	0.2%	79.8%
Cleanliness	9.6%	0.2%	90.2%
Locations	13.0%	1.6%	85.4%
Rooms	14.9%	2.1%	83.0%

H3: People who wrote about leisure activities traveled with family.

Outdoor activities are one of the things that distinguish rural tourism from urban tourism. Thus, it is essential to study this issue. As a result, the feature mainly related to it is the one that describes who people travel with (e.g., family, couple, alone, business, and friends).

Because everyone who talked about these activities in their reviews only mentioned it positively, this hypothesis has the perspective of more who talk about leisure activities in rural areas. This way, crosstabs were made in the SPSS to understand if people who wrote about leisure activities traveled with family, friends, alone, in business, or as a couple. Of the 69 people who responded, 28 traveled by family, accounting for 40.6%. This result was followed by people who traveled as a couple (39.1%) and with friends (13.0%). Thus, H3 is supported.

H4: There is a difference between genders in the score granted.

There is no statistically significant difference between males and females and the total score since the p-value > 0.05. Therefore, the hypothesis is rejected. One of the reasons for this is that not only there exists few negative scores, but also the difference between the number of males and females is small within the sample. Nevertheless, H4 rejection implies that men and women have the same perception and satisfaction of the lodgings despite their personal tastes. For example, literature acknowledged that male tourists are more attracted to adventure and sports in their travel experience, while female tourists prefer to explore the place's culture and opportunities for family closeness (Meng & Uysal, 2008).

H5: Hosts of rural tourism units are replying to negative comments.

Statistically, there is no significant difference between "No response" and "Response" in total score since the p-value > 0.05. Thus, the hypothesis is not valid. The reason is that the sample is small, limiting hypothesis validation. However, if the set is divided into two groups, e.g., one containing all reviews with scores above 3, and another with reviews rated equal or below 3, then it is possible to verify that both ratios are less than 50%, implying that hosts do not reply independently of the total score. Further data is in demand for a more robust answer to such a question.

H6 a: The TripAdvisor member's duration positively correlates with the number of useful votes.

Pearson's r statistic for the correlation between the TripAdvisor member years and the number of useful votes is 0.305. Therefore, we can conclude that there is a weak positive correlation between member years and useful votes, meaning that this variable does not influence the credibility and trustworthiness of a review, as it might be seen.

H6 b: Tourists with more votes have, on average, more negative reviews.

In this hypothesis, the total score was transformed in the same two groups as in H5: score ≤ 3 and score > 3. Although statically, the difference of means in both groups is not significant (p-value > 0.05), the average of useful votes is higher in negative reviews than in positive reviews if it is considered that only 12 negative reviews are present, which is a limitation of this study. Nevertheless, results point out that when people plan their trip, they are more careful, interested, and trusting in reading negative reviews to make more judged decisions.

H7: The eWOM has been increasing in rural tourism.

The eWOM in Schist Villages accommodation has increased during the years, validating H7. This steady growth started after 2012, which agrees with the study made by Mauri et al. (2017), where they found a growth in online reviews after the date mentioned. The most significant difference in online reviews written by travelers is between 2014 and 2015, with more than 93 reviews. Since 2017 has data from less than half of the year, it is impossible to conclude this year. Most of these comments were retrieved from TripAdvisor, not Facebook, as the former platform is friendlier to use and evaluate and is associated with online booking platforms.

The average total score in the 430 online reviews analyzed is 4.77 in 5, which means that Schist Villages' accommodations are evaluated, on average, as excellent units, meeting customers' expectations with high satisfaction. Thus, the mode was 5, and the minimum and maximum score attributed was 1 and 5, respectively.

On the one hand, we can see that there are no negative reviews on the Facebook platform and just one with a mid-level classification, meaning that most people using Facebook to spread the word think positively about Schist Villages' lodging. On the other hand, TripAdvisor has many positive reviews and 12 reviews with negative and average scores. The reasons that may explain this difference between TripAdvisor and Facebook are: (1) TripAdvisor is a specifically designed hospitality reviews website, while Facebook is a generic online social network platform; (2) TripAdvisor has six features to classify from 1 to 5, making reviewers more aware about the criteria for evaluating accommodation units.

CONCLUSIONS AND IMPLICATIONS

Portugal and, particularly, the Schist Villages have not been a matter of study in terms of eWOM or rural tourist preferences prior to this research; thus, this study is the first aiming at better understanding it. Despite the limited sample, it can be concluded that eWOM has been increasing in rural tourism, a niche market, over the years, following a general tourism trend.

As TripAdvisor Insights (2015) suggested, properties can become more popular compared to others based on reviews' quality, quantity, and recency. It is like a vicious cycle because all these three aspects reflect on each other to continually attract more visitors and, consequently, more reviews on this website. Another way to increase the rating is to improve the lodger experience by using their feedback to respond to their needs. Thus, this study helped to collect reviewers' feedback that spent time in Schist Villages accommodation, motivating owners to improve and get to know their guests better and, consequently, convert the analyzed data into opportunities. Nevertheless, findings show that Schist Villages hold an excellent rating (4.77 from the reviews taken) and with few negative or average reviews.

The benefits of responding to online reviews were outlined in the positive and negative literature review. In terms of positive reviews, a response demonstrates that the reviewer's feedback is appreciated and attention is paid to it. In general, the consequences of replying are related to giving a competitive advantage to businesses by listening to customers and understanding their perceptions (TripAdvisor Insights, 2013). Schist Villages have to consider this and start responding to online reviews on TripAdvisor and Facebook pages in a timely and personalized manner to perform higher than their competitors and to increase their visibility and revenue while improving their reputation with a proper management response. For this purpose, suppliers could apply two solutions: hiring someone to do all this control or creating a system that could monitor and respond to comments.

Based on the collected dataset, we can split it into two types of consumer perceptions in this study: one related to the accommodation itself and the other connected to the surroundings. The results addressed a research gap, finding that content categories in comments differ at some point from those categorized in a rank.

From the results' insights, it was observed that rural reviewers evaluated positively the services provided by Schist Villages accommodation studied in the sample because more than half of the online reviews were favorable, judging the place with good and very good connotations. Therefore, it can be inferred that there is a high level of professionalism in providing rural experiences to customers. Despite the majority of positive posted comments and their featured evaluations provided by rural lodgings, it is necessary that managers of these tourism companies think about the reviews with low levels of satisfaction along with which features and why that happened. The most regular complaints were about rooms, service features, and food/breakfast.

In addition to these meaningful inferences from the study findings, some limitations must be considered. First, only user-generated content on Facebook and TripAdvisor was used. Although both platforms are well-known among potential rural tourists, others could be a matter of study, such as Booking.com or Toprural website. Second, the sample size was not extensive, considering that rural tourism is a niche in the market. Finally, it must be considered how difficult it is to deal with and identify motivational or emotional elements within text. For example, it was impossible to identify reactions to any feature within 19 comments.

It is recommended to expand the analysis to other types of establishments within TRA consideration, like rural hotels and in other rural regions of Portugal and other countries. In the same reasoning, it could be helpful to extend to other social media platforms to see if there are differences in how consumers perceive rural tourism. Last but not least, comparing different types of tourism with the same factors analyzed could reveal hints for suppliers.

ACKNOWLEDGMENT

This work was supported by national funds through FCT (Fundação para a Ciência e a Tecnologia) under the project - UIDB/04152/2020 - Centro de Investigação em Gestão de Informação (MagIC)/NOVA IMS and ISTAR-Iscte Projects: UIDB/04466/2020 and UIDP/04466/2020.

REFERENCES

Agapito, D., Mendes, J., & Oom do Valle, P. (2012). The Rural Village as an open door to nature-based tourism in Portugal: The Aldeia da Pedralva case. *Tourism (Zagreb)*, *60*(3), 325–338.

Albacete-Sáez, C., Fuentes-Fuentes, M., & Lloréns-Montes, F. (2007). Service quality measurement in rural accommodation. *Annals of Tourism Research*, *34*(1), 45–65. doi:10.1016/j.annals.2006.06.010

Aldeias do Xisto. (2015). *Relatório de Avaliação EEC PROVERE rede das Aldeias do Xisto* [EEC PROVERE Assessment Report Schist Villages network]. ADXTUR.

Aldeias do Xisto. (2023). *Aldeias.* Available at: https://aldeiasdoxisto.pt/aldeias

Alturas, B., & Oliveira, L. S. (2016). Consumers using Social Media: Impact on Companies' Reputation. Radical Marketing. In *Academy of Marketing Annual Conference Proceedings*. Newcastle Business School at Northumbria University.

Andraz, J. M., Norte, N. M., & Gonçalves, H. S. (2015). Effects of tourism on regional asymmetries: Empirical evidence for Portugal. *Tourism Management*, *50*, 257–267. doi:10.1016/j.tourman.2015.03.004

Avraham, E., & Ketter, E. (2017). Destination image repair while combatting crises: Tourism marketing in Africa. *Tourism Geographies*, *19*(5), 780–800. doi:10.1080/14616688.2017.1357140

Bandyopadhyay, S. (2016). Factors affecting the adoption of electronic word-of-mouth in the tourism industry. *International Journal of Business and Social Science*, *7*(1), 10–18.

Barreto, A. M. (2013). Do Users Look at Banner Ads on Facebook? *Journal of Research in Interactive Marketing*, *7*(2), 119–139. doi:10.1108/JRIM-Mar-2012-0013

Boukas, N. (2019). Rural tourism and residents' well-being in Cyprus: Towards a conceptualized framework of the appreciation of rural tourism for islands' sustainable development and competitiveness. *International Journal of Tourism Anthropology*, *7*(1), 60–86. doi:10.1504/IJTA.2019.098105

Brogan, C., & Smith, J. (2009). *Trust agents: Using the web to build influence, improve reputation, and earn trust*. Wiley.

Calheiros, A. C., Moro, S., & Rita, P. (2017). Sentiment Classification of Consumer-Generated Online Reviews Using Topic Modeling. *Journal of Hospitality Marketing & Management*, *26*(7), 675–693. doi:10.1080/19368623.2017.1310075

Chen, N., Dwyer, L., & Firth, T. (2014). Effect of dimensions of place attachment on residents' word-of-mouth behavior. *Tourism Geographies*, *16*(5), 826–843. doi :10.1080/14616688.2014.915877

Costa, C. A., & Chalip, L. (2005). Adventure sport tourism in rural revitalisation: An ethnographic evaluation. *European Sport Management Quarterly*, *5*(3), 257–279. doi:10.1080/16184740500190595

Costa, S. M., Moro, S., Rita, P., & Alturas, B. (2023). Customer experience through online reviews from TripAdvisor: The case of Orlando theme parks. *International Journal of Technology Marketing*, *17*(1), 48–77. doi:10.1504/IJTMKT.2023.127352

Coyl-Shepherd, D. D., & Hanlon, C. (2013). Family play and leisure activities: Correlates of parents' and children's socio-emotional well-being. *International Journal of Play*, *2*(3), 254–272. doi:10.1080/21594937.2013.855376

Dahka, Z. Y., Hajiheydari, N., & Rouhani, S. (2020). User response to e-WOM in social networks: How to predict a content influence in Twitter. *International Journal of Internet Marketing and Advertising*, *14*(1), 91–111. doi:10.1504/IJIMA.2020.106041

Dashper, K. (2014). *Rural Tourism: An International Perspective*. Cambridge Scholars Publishing.

Devesa, M., Laguna, M., & Palacios, A. (2010). The role of motivation in visitor satisfaction: Empirical evidence in rural tourism. *Tourism Management*, *31*(4), 547–552. doi:10.1016/j.tourman.2009.06.006

Drăgulănescu, I.-V., & Druţu, M. (2012). Rural Tourism for Local Economic Development. International. *Journal of Academic Research in Accounting, Finance and Management Sciences*, *2*(1), 196-203.

Duarte, M., Dias, Á., Sousa, B., & Pereira, L. (2023). Lifestyle Entrepreneurship as a Vehicle for Leisure and Sustainable Tourism. *International Journal of Environmental Research and Public Health, 20*(4), 3241. doi:10.3390/ijerph20043241 PMID:36833935

Duarte, P. (2010). Evolution of rural tourism in Portugal: A 25 years analysis. *e-Review of Tourism Research (eRTR), 8*(3), 41-56.

Ezeuduji, I. O. (2015). Strategic event-based rural tourism development for sub-Saharan Africa. *Current Issues in Tourism, 18*(3), 212–228. doi:10.1080/13683500.2013.787049

Figueiredo, E., Kastenholz, E., & Pinho, C. (2014). Living in a rural tourism destination - exploring the views of local communities. *Revista Portuguesa de Estudos Regionais, 36*, 3–12. doi:10.59072/rper.vi36.417

Fleischer, A., & Felsenstein, D. (2000). Support for rural tourism: Does it make a difference? *Annals of Tourism Research, 27*(4), 1007–1024. doi:10.1016/S0160-7383(99)00126-7

Frochot, I. (2005). A benefit segmentation of tourists in rural areas: A Scottish perspective. *Tourism Management, 26*(3), 335–346. doi:10.1016/j.tourman.2003.11.016

Gössling, S., & Lane, B. (2015). Rural tourism and the development of Internet-based accommodation booking platforms: A study in the advantages, dangers and implications of innovation. *Journal of Sustainable Tourism, 23*(8-9), 1386–1403. doi:10.1080/09669582.2014.909448

Gretzel, U., & Yoo, K. H. (2008). Use and impact of online travel reviews. In P. O'Connor, W. Hopken, & U. Gretzel (Eds.), *Information and communication technologies in tourism 2008, 35-46*. Springer. doi:10.1007/978-3-211-77280-5_4

Helm, R., & Mark, A. (2007). Implications from cue utilisation theory and signalling theory for firm reputation and the marketing of new products. *International Journal of Product Development, 4*(3/4), 396–411. doi:10.1504/IJPD.2007.012504

Heneghan, M., Caslin, B., Ryan, M., & O'Donoghue, C. (2016). *Rural Tourism: Rural Economy & Development Programme*. Teagasc.

Hennig-Thurau, T., Gwinner, K. P., Walsh, G., & Gremler, D. D. (2004). Electronic Word-of- Mouth via Consumer-Opinion Platforms: What Motivates Consumers to Articulate Themselves on the Internet? *Journal of Interactive Marketing, 18*(1), 38–52. doi:10.1002/dir.10073

Huete-Alcocer, N. (2017). A Literature review of word of mouth and electronic word of mouth: Implications for consumer behaviour. *Frontiers in Psychology*, *8*(July), 1256. Advance online publication. doi:10.3389/fpsyg.2017.01256 PMID:28790950

Jeong, M., & Jeon, M. M. (2008). Customer reviews of hotel experiences through consumer generated media (CGM). *Journal of Hospitality & Leisure Marketing*, *17*(1-2), 121–138. doi:10.1080/10507050801978265

Jiménez-Zafra, S., Martín-Valdivia, M., Martínez-Cámara, E., & Ureña-López, L. (2015). Combining resources to improve unsupervised sentiment analysis at aspect level. *Journal of Information Science*, *42*(2), 1–19. doi:10.1177/0165551515593686

Kastenholz, E. (2004). 'Management of demand' as a tool in sustainable tourist destination development. *Journal of Sustainable Tourism*, *12*(5), 388–408. doi:10.1080/09669580408667246

Kastenholz, E., Carneiro, M., Eusébio, C., & Figueiredo, E. (2013). Host-guest relationships in rural tourism: Evidence from two Portuguese villages. Anatolia. *Anatolia*, *24*(3), 367–380. doi:10.1080/13032917.2013.769016

Kastenholz, E., Carneiro, M., Marques, C., & Lima, J. (2012). Understanding and managing the rural tourism experience - The case of a historial village in Portugal. *Tourism Management Perspectives*, *4*(October), 207–214. doi:10.1016/j.tmp.2012.08.009

Kastenholz, E., Davis, D., & Paul, G. (1999). Segmenting tourism in rural areas: The case of north and central Portugal. *Journal of Travel Research*, *37*(4), 353–363. doi:10.1177/004728759903700405

Kavoura, A., & Bitsani, E. (2013). E-branding of rural tourism in Carinthia, Austria. *Tourism Review*, *61*(3), 289–312.

Kumar, S., Valeri, M., & Shekhar. (2022). Understanding the relationship among factors influencing rural tourism: A hierarchical approach. *Journal of Organizational Change Management*, *35*(2), 385–407. doi:10.1108/JOCM-01-2021-0006

Law, R., Leung, R., Lo, A., Leung, D., & Fong, L. H. N. (2015). Distribution channel in hospitality and tourism: Revisiting disintermediation from the perspectives of hotels and travel agencies. *International Journal of Contemporary Hospitality Management*, *27*(3), 431–452. doi:10.1108/IJCHM-11-2013-0498

Lee, H. A., Law, R., & Murphy, J. (2011). Helpful reviewers in TripAdvisor, an online travel community. *Journal of Travel & Tourism Marketing*, *28*(7), 675–688. doi:10.1080/10548408.2011.611739

Leung, D., Law, R., Hoof, H., & Buhalis, D. (2013). Social Media in Tourism and Hospitality: A Literature Review. *Journal of Travel & Tourism Marketing*, *30*(1-2), 3–22. doi:10.1080/10548408.2013.750919

Levy, S., Duan, W., & Boo, S. (2013). An analysis of one-star online reviews and responses in the Washington, D.C., lodging market. *Cornell Hospitality Quarterly*, *54*(1), 49–63. doi:10.1177/1938965512464513

Litvin, S., Goldsmith, R., & Pan, B. (2008). Electronic Word-of-Mouth in Hospitality and Tourism Management. *Tourism Management*, *29*(3), 458–468. doi:10.1016/j.tourman.2007.05.011

Madanaguli, A., Kaur, P., Mazzoleni, A., & Dhir, A. (2022). The innovation ecosystem in rural tourism and hospitality – a systematic review of innovation in rural tourism. *Journal of Knowledge Management*, *26*(7), 1732–1762. doi:10.1108/JKM-01-2021-0050

Madureira, L., & Alturas, B. (2022). Impact of Social Media Influencers on the Portuguese Tourism and Travel Industry in a Covid-19 Era. In Marketing and Smart Technologies. Proceedings of ICMarkTech 2021. Springer. doi:10.1007/978-981-16-9272-7_32

Marcheggiani, D., Täckström, O., Esuli, A., & Sebastiani, F. (2014). Hierarchical multi-label conditional random fields for aspect-oriented opinion mining. In *Advances in information retrieval* (pp. 273–285). Springer. doi:10.1007/978-3-319-06028-6_23

Marzo-Navarro, M., Pedraja-Iglesias, M., & Vinzón, L. (2015). Sustainability indicators of rural tourism from the perspective of the residents. *Tourism Geographies*, *17*(4), 586–602. doi:10.1080/14616688.2015.1062909

Mauri, A. G., & Minazzi, R. (2013). Web reviews influence on expectations and purchasing intentions of hotel potential customer. *International Journal of Hospitality Management*, *34*(September), 99–107. doi:10.1016/j.ijhm.2013.02.012

Melo, A., Hernández-Maestro, R., & Muñoz-Gallego, P. (2017). Service quality perceptions, online visibility, and business performance in rural lodging establishments. *Journal of Travel Research*, *56*(2), 250–262. doi:10.1177/0047287516635822

Meng, F., & Uysal, M. (2008). Effects of Gender Differences on Perceptions of Destination Attributes, Motivations, and Travel Values: An Examination of a Nature-Based Resort Destination. *Journal of Sustainable Tourism*, *16*(4), 445–466. doi:10.1080/09669580802154231

Meyers-Levy, J., & Sternthal, B. (1991). Gender differences in the use of message cues and judgments. *JMR, Journal of Marketing Research, 28*(February), 84–96. doi:10.1177/002224379102800107

Molera, L., & Albaladejo, P. (2007). Profiling segments of tourists in rural areas of south-eastern Spain. *Tourism Management, 28*(3), 757–767. doi:10.1016/j.tourman.2006.05.006

Moro, S., & Rita, P. (2018). Brand strategies in social media in hospitality and tourism. *International Journal of Contemporary Hospitality Management, 30*(1), 343–364. doi:10.1108/IJCHM-07-2016-0340

Moro, S., Rita, P., & Coelho, J. (2017a). Stripping customers' feedback on hotels through data mining: The case of Las Vegas Strip. *Tourism Management Perspectives, 23*(July), 41–52. doi:10.1016/j.tmp.2017.04.003

Moro, S., Rita, P., & Cortez, P. (2017b). A text mining approach to analyzing Annals literature. *Annals of Tourism Research, 66*(September), 208–210. doi:10.1016/j.annals.2017.07.011

Moro, S., Rita, P., & Oliveira, C. (2018). Factors influencing hotels' online prices. *Journal of Hospitality Marketing & Management, 27*(4), 443–464. doi:10.1080/19368623.2018.1395379

Munar, A., & Jacobsen, J. (2013). Trust and Involvement in Tourism Social Media and Web-Based Travel Information Sources. *Scandinavian Journal of Hospitality and Tourism, 13*(1), 1–19. doi:10.1080/15022250.2013.764511

Munar, A., & Jacobsen, J. (2014). Motivations for sharing tourism experiences through social media. *Tourism Management, 43*(August), 46–54. doi:10.1016/j.tourman.2014.01.012

OECD. (1994). *Tourism strategies and rural development: organisation for economic co-operation and development*. OECD.

Park, D., & Yoon, Y. (2009). Segmentation by motivation in rural tourism: A Korean case study. *Tourism Management, 30*(1), 99–108. doi:10.1016/j.tourman.2008.03.011

Phillips, W. J., Wolfe, K., Hodur, N., & Leistritz, F. L. (2013). Tourist word of mouth and revisit intentions to rural tourism destinations: A case of North Dakota, USA. *International Journal of Tourism Research, 15*(1), 93–104. doi:10.1002/jtr.879

Proença, S., & Souziakis, E. (2008). Tourism as an economic growth factor: A case study for Southern European countries. *Tourism Economics, 14*(4), 791–806. doi:10.5367/000000008786440175

Resnick, P., Zeckhauser, R., Friedman, E., & Kuwabara, K. (2000). Reputation systems. *Communications of the ACM*, *43*(12), 45–48. doi:10.1145/355112.355122

Ribeiro, M., & Marques, C. (2002). Rural tourism and the development of less favoured areas - Between rhetoric and practice. *International Journal of Tourism Research*, *4*(3), 211–220. doi:10.1002/jtr.377

Rosalina, P. D., Dupre, K., & Wang, Y. (2021). Rural tourism: A systematic literature review on definitions and challenges. *Journal of Hospitality and Tourism Management*, *47*(June), 134–149. doi:10.1016/j.jhtm.2021.03.001

Sparks, B., & Browning, V. (2010). Complaining in cyberspace: The motives and forms of hotel guests' complaints online. *Journal of Hospitality Marketing & Management*, *19*(7), 797–818. doi:10.1080/19368623.2010.508010

Thevenot, G. (2007). Blogging as a social media. *Tourism and Hospitality Research*, *7*(3-4), 287–289. doi:10.1057/palgrave.thr.6050062

Timothy, D. J. (2018). Geography: The substance of tourism. *Tourism Geographies*, *20*(1), 166–169. doi:10.1080/14616688.2017.1402948

Townsend, L., Wallace, C., Smart, A., & Norman, T. (2016). Building Virtual Bridges: How Rural Micro-Enterprises Develop Social Capital in Online and Face-to-Face Settings. *Sociologia Ruralis*, *56*(1), 29–47. doi:10.1111oru.12068

Trading Economics. (2016). *Portugal - Rural Population*. Available at: https://tradingeconomics.com/portugal/rural-population-percent-of-total-population-wb-data.html

TripAdvisor Insights. (2013). *How to optimize your attraction listing on TripAdvisor*. Available at: https://www.tripadvisor.com/TripAdvisorInsights/n710/how-optimize-your-attraction-listing-tripadvisor

TripAdvisor Insights. (2015). *All about your TripAdvisor bubble rating*. Available at: https://www.tripadvisor.com/TripAdvisorInsights/n2640/all-about-your-tripadvisor-bubble-rating

Tung, V., & Ritchie, J. (2011). Exploring the essence of memorable tourism experiences. *Annals of Tourism Research*, *38*(4), 1367–1386. doi:10.1016/j.annals.2011.03.009

Turismo de Portugal, I. P. (2015). *Turismo 2020: cinco princípios para uma ambição* [Tourism 2020: five principles for an ambition]. Available at: https://www.turismodeportugal.pt/Portugu%C3%AAs/turismodeportugal/newsletter/2015/Documents/TURISMO2020-5Principios.pdf

Valente, S., & Figueiredo, E. (2003). O turismo que existe não é aquele que se quer... [The tourism that exists is not what you want...]. 1º Encontro de Turismo em Espaços Rurais.

Vermeulen, L. E., & Seegers, D. (2009). Tried and tested: The impact of online hotel reviews on consumer consideration. *Tourism Management*, *30*(1), 123–127. doi:10.1016/j.tourman.2008.04.008

Walmsley, D. J. (2003). Rural tourism: A case of lifestyle-led opportunities. *The Australian Geographer*, *34*(1), 61–72. doi:10.1080/00049180320000066155

World Travel Awards. (2017). *World's Leading Destination 2017*. Available at: https://www.worldtravelawards.com/award-worlds-leading-destination-2017

Xiang, Z., & Gretzel, U. (2010). Role of social media in online travel information search. *Tourism Management*, *31*(2), 179–188. doi:10.1016/j.tourman.2009.02.016

Yang, J., Yang, R., Chen, M., Su, C., Zhi, Y., & Xi, J. (2021). Effects of rural revitalization on rural tourism. *Journal of Hospitality and Tourism Management*, *47*(June), 35–45. doi:10.1016/j.jhtm.2021.02.008

Zhou, L. (2014). Online rural destination images: Tourism and rurality. *Journal of Destination Marketing & Management*, *3*(4), 227–240. doi:10.1016/j.jdmm.2014.03.002

ADDITIONAL READING

Rita, P., & Moro, S. (2022). Tasting the Port wine cellar experience: What features please the most? *Journal of Wine Research*, *33*(2), 88–99. doi:10.1080/09571264.2022.2081140

Yu, T., Rita, P., Moro, S., & Oliveira, C. (2022). Insights from sentiment analysis to leverage local tourism business in restaurants. *International Journal of Culture, Tourism and Hospitality Research*, *16*(1), 321–336. doi:10.1108/IJCTHR-02-2021-0037

KEY TERMS AND DEFINITIONS

Cultural Tourism: Cultural tourism refers to the practice of traveling to destinations, events or attractions that have cultural and historical significance, in order to learn about and experience the unique cultural heritage of a particular place or community. This can include visiting museums, historical sites, festivals, and cultural events, as well as participating in activities that showcase local traditions, art, music, and cuisine.

Electronic Word-of-Mouth: Electronic word-of-mouth (eWOM) refers to the process of sharing opinions, recommendations, and other forms of information about products, services, or brands through electronic media, such as social media, online forums, and review websites. It involves the use of digital communication channels to spread information and influence the attitudes and behaviors of others, including potential customers, about a particular product or service.

Rural Tourism: Rural tourism refers to the practice of traveling to rural areas and engaging in activities that are related to the local culture, heritage, and environment. This can include visiting rural communities, participating in outdoor activities, exploring natural landscapes, and learning about the local agricultural practices and traditional ways of life. Rural tourism is often seen as a way to promote sustainable development and preserve local culture and natural resources, while also providing economic benefits to rural communities.

Social Media: Social media refers to digital platforms and tools that allow users to create, share, and exchange content, opinions, and information with others. Social media platforms enable users to connect with each other and engage in online communication, which can include text, images, videos, and other types of multimedia.

Chapter 5
Sustainable Development and Tourism

Julieta E. Salazar-Echeagaray
https://orcid.org/0000-0003-0689-532X
Universidad Autónoma de Sinaloa, Mexico

Teresa I. Salazar-Echeagaray
https://orcid.org/0000-0001-8785-6300
Universidad Autónoma de Sinaloa, Mexico

José G. Vargas-Hernandez
Tecnológico Superior de Jalisco, Mexico

ABSTRACT

This chapter exposes a theoretical methodology focused on the importance of sustainable development in the tourism sector. It remarks on the paramount importance of environmental care to protect the ecosystems in which the touristic centers are situated to promote the environmental care in the people who live there, the employees of the tourist centers, and the tourists.

INTRODUCTION

Human society needs the economic system so that its operation and its actions generate economic resources. In the same way, it requires leisure activities to have a better quality of life. These needs cause the increase of the tourism sector that supplies both, the need to generate income and sources of employment, as well as entertainment. Quarantine due to COVID-19 pandemic made us notice that people need leisure time. People have the desire to travel and visit places other than where

DOI: 10.4018/978-1-6684-7242-2.ch005

they live and work. At the same time, quarantine made us notice made us notice that we must take care of the environment, and the best option to achieve it is through sustainable development.

Social media was the main promoter of leisure during quarantine and became a source of opinions that influence the way people think. On these platforms, people can express their viewpoints, opinions, and be heard. The new generations have observed the need to protect the environment and shared their opinions with previous generations. Likewise, various researchers have conducted different studies regarding the need for the tourism industry to implement sustainable development.

This fact leads us to the need to strike a balance between tourism and sustainable development. Where society sees the economic benefit of tourism activity for tourism service providers. Including the recreation and the experience of the people who enjoy those services. And the care of the ecosystems where the tourist centers are located. And thus, achieve economic growth and development, job creation, leisure activities for tourists and care for the environment, so that current and future generations will be able to continue to enjoy the ecosystems where tourist services are located. This document will present the need to implement sustainable development in tourism in the United Mexican States. This is because the nation is rich in natural resources of flora and fauna, and they must be protected so that current and future generations can enjoy them.

BACKGROUND

The tourism sector has been a source of jobs and a generator of wealth since the industry began. The different nations offer everything from beach tourism, religious tourism, cultural activities tourism and adventure tourism, where people camp outdoors, climb mountains, visit exotic animals to name a few. There are different options for leisure such as travel packages with everything paid or you can hire a tour guide. Tourism activity is as old as human society.

Tourism has been motivated by migratory movements, war, trade, religion, and fashion to name a few. World tourism can be divided into stages: 1-Ancient Age (from 3000 BC to 476 AD); 2-Middle Ages (from 477 AD to 1453); 3-Modern Age (from 1454 to 1789) being this stage where the first lodgings appear and are named hotel; and 4-Contemporary Age (from 1790 to 3000) in the 16th century establishes the bases for modern tourism performance. In addition, the Contemporary Age is divided into a. Industrial Revolution (from 1790 to 1949) where industrialization increased the free time of workers causing a demand for holidays tourism activities, b. Second half of the 20th Century (from 1950 to 1999); and c. 21st Century (from

2000 to 3000). As can be seen we are in the last stage of the Contemporary Age. (Warrior, 2014).

Based on the activities conducted and tourist needs, tourism is classified as: exploratory travel, mandatory travel, pilgrimage travel, elite travel, and mass tourism. Exploratory tourism is the most archaic, with its first exponent being study trips, investigative exploration, and territorial trips conducted during the period from the fifteenth to the nineteenth century. Mandatory tourism arose from the need for merchants to sell their products in places other than their origin. Similarly, people who have been forced to migrate due to war are also considered mandatory tourists, and lastly, people who travel to obtain knowledge for their jobs are considered mandatory travelers.

Pilgrimage trips are made for religious issues of being in sacred places. Participating in important religious ceremonies or visiting places, where notable events or graves of transcendental figures in history were held. Elite tourism is the pleasure travel made by wealthy people. Who, at the time, were the pioneers of mass tourism in today's pleasure journeys. Mass tourism refers to pleasure trips made by people who have enough time to do it. They rise after the end of World War II and, showed a high diversification during the 1980s (Faraldo, 2014).

It is worth mentioning in a distinct way tourism culture. Cultural heritage is a resource used in the implementation of the design of a good tourism strategy. A respectful value must be given to the past by conjugating it with the future. So, that future generations can enjoy these spaces. Historic towns have a varied urban and architectural heritage that attracts people. By promoting culture, a link is made with the productive processes, generation of work sources, technological advances and producing economic well-being in the population where tourism activity is conducted (Troitiño, 2002).

Nowadays, the term tourism implies a trip, a process of organizing it, the promotion, and the provision of travel-related services from the origin point to destination (Quesada 2000: 8). Current tourists demand a complete service that includes transportation, lodging, and recreational activities. This requested demand has positively influenced in the tourist offer, generating services such as hairdressing, nursery, massages, and souvenir sales. Tourist activities were developing with a good market until the COVID-19 pandemic arrived, a fact that changed the perspective to the new normality.

MAIN FOCUS OF THE CHAPTER

At a global level, COVID-19 pandemic has had an impact in three ways: 1-production has been affected, 2-supply chain has been disrupted, and 3-financial markets and companies have been impacted (Sánchez, 2021: 8). Stopping the production and the supply chain to a complete standstill affected people's economy badly. However, an effect not contemplated at first impression was also observed: the decrease in contamination due to the low mobility of people when they were in confinement.

Tourism impacts the environment in two ways: in the transportation cycle and in the accommodation cycle. The first presents polluting emissions from tourist transport and noise pollution. While the second involves all the waste generated by tourists such as the use of air conditioning, energy consumption, water, drainage, and the interaction of the tourist on the streets when leaving the place of accommodation (Rivas, 2012). And the confinement showed that this impact decreased when the tourist activity was not conducted.

This decrease in pollution led service providers and tourist centers to observe the need to conduct their activities in a more environmentally friendly manner. And thus begin to carry out activities such as giving hotel guests the option of not changing the sheets daily, the use of different towels in the bedrooms and swimming pools due to the chemicals used in the latter, soap dispensers in the rooms to avoid the use of packaging, the recycling of waste and the use of transportation providers with a not more than five years old vehicles.

THEORETICAL BACKGROUND

Leisure travel tourism begins to appear in Greek civilization, the so-called cradle of democracy, with the aim of enriching itself in political, cultural, theater or bath activities. There are numerous travel stories, and they are a popular literary genre. Romans traveled because there was a period of prolonged peace, they had transport infrastructure such as roads in good condition and, as there was prosperity and wealth, there were high social classes that had free time to spend on tourist activities (Vogeler, 2018).

In the year 476 the Roman Empire falls causing a new stage in human history, the Middle Ages were characterized by famine, conflict, and death. During this stage, the roads disappeared. On some occasions the monasteries and castles flourished to avoid the barbarian invasions. Around the castles, towns, villages, and cities were founded where grazing, agriculture and trade achieved commercial prosperity giving rise to medieval cities. The universities of Oxford, Tolosa, Paris, Rome, Bologna, and Salamanca also arose, as well as cathedrals such as Westminster Abbey and Our Lady of Paris (Quesada, 2000).

After the Middle Ages, the stage of the Renaissance began and was motivated by the desire to discover new cultures and places, as well as expeditions to other continents. When the Renaissance ends, begins the industrial revolution where the

European countries develop. The industrial revolution continued after the Renaissance and with the development of steamships and trains managed to cover distances in less time and encouraged the establishment of tourist centers in cities located on the coasts that were preferred for health reasons (Martínez, 2022).

The term tourism arose from the movement of English tourists to Europe in the last decades of the eighteenth century and in the first decades of the nineteenth century (Quesada, 2000). Families of high social status sent their children away of their home to be educated and to travel in various cities to prepare them for their performance in adult life. In the present days tourism is a product of the lower middle class and the middle class that vacations away from their place of residence (Muñoz, 2003). This segment of tourism is the one with the highest volume that occurs in the world and for this reason they generate a large amount of economic spillover to the places where they visit.

Modern tourism is integrating rural areas as part of the tourist experience (Crosby, 2009). Tourists not only want to visit the areas where the beaches are, but also the cultural, religious, and various entertainment areas such as museums, aquariums, or zoos. Tourist's mentality is changing and therefore the providers of tourist services must adapt to their needs. Currently, the tourism sector is experiencing the greatest possible growth in its history (Ojeda, 2016). Before XXI century began, only complete families traveled, now you can see couples, groups of friends and even single people.

The present twenty-first century is profiling and deploying lines of tourism progress that will be deepening. Which are: the internationalization of entities, chains, tourism operators seeking to increase their market quotas and increase their dominant position in emerging markets. The economic importance of tourism as a sector of the economy in developed and developing nations. The attraction of exotic and new destinations for people with desires to live new experiences.

The latest offers for leisure are the places of sports attractions, and adventure. As well as spa, thermal waters, and major events such as the Olympics and world championships. The increased demands of people regarding the quality and price of services promote competition in the hotel market. And finally, the influence of the internet by offering largest amount of information of offers and cost reduction. With respect to the places, attractions, gastronomy, and entertainment offered by the tourist places to the diverse types of tourism. Some tourist spaces are already segmenting the market every day to cover the largest number of people who enjoy their services (Rivas, 2004).

People give more value to quality standards in both urban and natural areas. A demand for services that has taken importance in the twenty-first century is sustainable tourism. People are enjoying visiting natural places and performing activities such as hiking, nature observation, mountaineering, to name a few. This has caused the transformation of natural areas causing the need to restore such spaces to maintain the principles of sustainable development. The different nations have been implementing strategies intensively to maintain the natural state of those areas (Alonso-Sañudo, 2002).

Likewise, tourists have been observing the effect of global warming on the planet and its effects on the climate worldwide. They have also become more aware of the need to take care of the environment and how their actions are affecting the ecology of the place where they live and the places where they travel. Derived from this change in mentality, tourists are requesting that tourism service providers in their activities be more responsible with the care of the environment.

CONTEXTUAL FRAMEWORK

From the economic point of view, tourism is defined "as an act that involves displacement, which entails an expense of income in a place other than the one in which income originates, and in which one does not habitually reside" (Casanueva, 2012). People are recurring to vacations to de-stress from their daily lives, to discover novel places, to live new experiences, to try different gastronomy, to learn about diverse cultures and, in general, to disconnect from their daily life for a period generating new and pleasant memories.

Derived from the above, the requirement towards tourism service providers regarding both in terms of responsibility for the services they offer as in their quality. From this fact Corporate Social Responsibility is originated. Corporate Social Responsibility is a way in which companies seek a balance between achieving their financial, economic and development objectives, and the environmental impact that is caused by those activities (Ojeda, 2016).

Corporate Social Responsibility is the commitment that companies must support the benefit and development of the standard of living of communities, their workers, and the workers' families. That is why the decision-making of businesspeople directly affects society. Therefore, entrepreneurs must maintain an agreement within the three dimensions of sustainable development: the economic dimension, the social dimension, and the ecological dimension. Companies must seek society's welfare without neglecting the objective for which they were created (Martínez, 2005).

Corporate Social Responsibility has a relationship with sustainable development, especially in a more emphatic way with the inclusion of the pillar of social justice. So, if a company wants to apply sustainable development, it must respect the expectations of Corporate Social Responsibility (Wulf, 2018). Corporate Social Responsibility modernizes the concept of organizations and gives them an inclusive and broad dimension beyond the economic question, incorporating the three phases of sustainable development: social, economic, and environmental. Sustainable development is located as the purpose of achieving through an adequate establishment of a socially responsible business model (Reyno, 2007).

When a company choose an efficient and optimal management of resources and take in consideration local communities, it is called sustainable tourism. This fact is emphasized in conditioning and remodeling old building structures or traditional rural construction-houses systems. Turning them into small hotels where tourists enjoy different experiences from the traditional beach and sun tourism (Crosby, 2009). Sustainable tourism is a niche opportunity that is beginning to be exploited and could be a source of income for tourism providers.

Sustainable tourism is based on the theory of sustainable development. Sustainable development collaborates in sustained economic growth on the ability to internalize the social and ecological situations of democracy, justice, and equity. Sustainable development arises from the term sustainability that integrates the meanings of sustainable, involving the internalization of the economic process and the ecological circumstances and the durability of the economic process (Leff, 2004).

In recent years, the importance of caring for ecosystems and environment in different nations has been emphasized. The concern of human society is reflected in the year 1987 when, in the forum called Our Common Future or Brundtland Report, the World Commission on the Environment and Development (WCED) defines the concept of sustainable development as "development that meets the needs of the present generation without compromising the ability of future generations to meet their own needs." It is necessary that world society and governments accept and use sustainable development so that the planet remains in optimal conditions (CMMAD, 1987).

Nebel (1999) presents another definition of sustainable development as "the fact that a system or process can continue its operation indefinitely and correctly without exhausting its resources, that is, the production capacity must be conserved in accordance with the growth of the population" (p. 3). While Moyado (1995) suggests that sustainable development should "consider nature as a factory that needs to be maintained and perfected so that the efforts of management plans and derived technologies maximize their long-term productivity and minimize their negative effects" (p. 238).

Sustainable development is preceded by the following ideas: the current generation has the duty to preserve the balance between different environmental ecosystems to leave enough resources for future generations. The purpose of sustainable development will be achieved if the environmental, social, and economic dimensions are kept in balance through social cohesion, progress, economic growth, and respect for the environment (Fernández, 2011). Social, economic, and environmental dimensions must be coupled in the best way to achieve adequate and harmonious sustainable development.

People's movement for tourist activity has been related to the impact on natural resources and its relationship with economic development models (Maerk, 2000). Tourism is an example of the close correlation between the environment and the economic development, with its conflicts, tensions, and benefits. The impact of tourism on the environment is subject to tourists, their behavior, and the quality of tourism services. For this reason, it is important to adopt a global approach integrating economic development and environmental conservation (Bigné, 2000).

This fact leads us to the relationship between sustainable development and the sustainable tourism term. Sustainable tourism is known as all those activities of a tourist nature that are participatory and respectful of the social, cultural, and ecological environment, allowing a positive exchange of experiences between visitors and residents of the place to be enjoyed (Guerrero, 2014). That is, people dedicated to tourist activities must consider economic development and respect for the environment.

Ecotourism is a conservation movement where tourist trips are organized in a responsible way to conserve the environment and support the local community where they arrive. They are accompanied by ethical codes that promote recreation and education through study and observation of the place. The development of ecotourism generates resources to preserve nature, culture, and prosperity of the community where it takes place. Its main activities are sidereal observation, flora observation, geological observation, natural attractions observation, the rescue of fauna and flora, and hiking (Ivanova, 2012).

Society has noticed that tourism is a large industry with one of the most dynamic sectors of economic growth and by using resources it generates an environmental deterioration. It also causes social and cultural benefits and costs during the development of its process (Virgen, 2014). Caring for the environment and sustainable development dates to the United Nations Conference on the Environment in 1972, which originated the United Nations Environment Program the following year. The World Tourism Organization in 1980 presented the World Tourism Conference.

In 1987, the Brundtland Report-Our Common Future was published, defining sustainable development as "the one that meets the needs of the present without compromising the needs of future generations". In 1991, the Congress of the International Association of Scientific Experts in Tourism was held in Mahé, Seychelles. The following year in Rio de Janeiro, the United Nations Conference on Environment and Development, known as the Earth Summit, was presented. As a result of those works, Agenda 21 was created (Crespo et al., 2020).

In 1995, on April 27th and 28th, in Lanzarote, Canary Islands, the first meeting of the World Conference on Sustainable Tourism was held, promoted by the UNWTO, UNESCO and the United Nations Environment Program (UNEP). In this event, the Sustainable Tourism Charter was approved, where principles and objectives are adopted which promote the foundation of tourism development on sustainability criteria, with viability and positive economic contribution and with a social and ethical perspective for the benefit of the different communities. Likewise, it will be integrated into the human, natural and cultural environment, respecting the balance of tourist destinations. In the same way, international, national, regional, and local cooperation will be conducted (World Conference on Sustainable Tourism, 1995).

In 1999 the World Tourism Organization in its General Assembly issues the World Code of Ethics for Tourism. In that document, it can be observed in its article 3 that tourism is a factor of sustainable development. Focusing on the fact that people related to tourism development must protect natural resources and the environment. In addition, the authorities will encourage tourism development in a way that allows the saving of natural resources. In the same way, the activities will be programmed, and the infrastructure will be implemented toprotect the natural heritage that constitutes the different ecosystems (Organización Mundial del Turismo, 1999).

This Global Code of Ethics for Tourism has contributed to the implementation of environmental care activities in the various hotels at the international, national, regional, and local levels. The people who request the services of the tourism sector are interested in caring for the environment. And they request that the services they receive guarantee their compliance. If they observe that the lodging center or the service providers do not comply with the code, they stop requesting their services.

METHOD

The method used to analyze the results is the review of the various authors on the theories related to tourism such as the background with Guerrero, Vogeler, Quesada, and MartínezFaraldo, Troitiño. Reviewing tourism today by Muñoz, Crosby, and Ojeda Alonso-Sañudo. The examination of the definition of tourism by Casanueva. The review to corporate social responsibility by Wulf, and Reyno. The analysis to the impact of tourism on the environment exposed by Rivas, Maerk, and Bigné. The examination of ecotourism by Ivanova, Virgen and the World Tourism Organization and sustainable development by Leff.

The information of the World Commission on Environment and Development (WCED), World Conference on Sustainable Tourism (WCST), Nebel, and Fernández. In addition, data sources provided by the World Tourism Organization, the National Institute of Statistics and Geography (INEGI) and the Ministry of Tourism were consulted. Research on the subject is presented in the discussion section by the authors: Troitiño, Orgaz, Caro-González, Anton, Regalado, De Esteban, Meneses, Lalangui, Lorenzo, Aucancela and Martínez.

The revised documentation corresponding to the data presented in this document corresponds to the year two thousand twenty-two. For this reason, the temporal range of this document corresponds to the year two thousand twenty-two. They analyzed the documents submitted in the database of the World Tourism Organization presented in year two thousand twenty-two. The National Institute of Statistics and Geography (INEGI) database of the United States of Mexico for the year 2022 was also examined. Similarly, the database of the satellite account of tourism in Mexico in the year two thousand twenty-two was reviewed. Also, the Information and Monitoring Unit of the Tourism Secretariat during the year 2022 was analyzed.

ANALYSIS OF RESULTS

Worldwide, tourism began to reactivate and increase its influx of people with the lifting of restrictions by COVID-19. Based on a press release called: The recovery of tourism gains momentum after the reduction of restrictions and the increase in confidence issued by the World Tourism Organization, it is stated that international tourism increased by 182% during the period from January to March 2022. Worldwide, 117 million international arrivals were received compared to forty-one million in the first quarter of 2021. That is, it increased by seventy-six million during the first 3 months compared to the previous year, with March being the month in which international arrivals began to recover in greater numbers with forty-seven million people (World Tourism Organization, 2022).

During the first quarter of 2022 in Europe, an increase of 280% was received compared to the previous year of people from the same regions, noting the effect of the lifting of the quarantine by COVID-19. In the American continent the arrival of tourists doubled in an increase of 117%. In the Middle East, the arrival of tourists increased by 132% while in Africa the increase was 96% and, in the Pacific, and Asia the increase was 64% compared to the previous year (Organización Mundial del Turismo, 2022).

Southern Mediterranean Europe and the Caribbean are recovering faster with 75% compared to the year 2019 prior to the COVID-19 quarantine. The income generated by international tourism was 602,000 million dollars, an increase of 4% compared to 2020. The Middle East and Europe presented an increase in their income of 50% compared to pre-quarantine income levels (Organización Mundial del Turismo, 2022).

Experts from the World Tourism Organization conducted a survey to find out the latest confidence index of tourism professionals and found that there is optimism, seeing better prospects for 2022 by 83% of tourism professionals if they are softened or lifted. travel restrictions. Forty-eight percent of those surveyed see a return to the levels managed in 2019 and international tourist arrivals in 2022 are expected to reach between 55% and 70% compared to 2019. The outlook is optimistic for the sector tourism regarding the restrictions by COVID-19 (Organización Mundial del Turismo, 2022).

At the national level, the National Institute of Statistics and Geography (INEGI) in its press release NO. 539/21 dated September 24, 2021, called: Statistics regarding World Tourism Day (September 27) I publish relevant information. The Economic Censuses conducted during 2019 report that tourist activities were registered in 786,540 economic units, representing 16.4% of the total economic units at the national level. During that year, for everyone hundred pesos produced by the national economy, 8.7 correspond to tourism. Of everyone hundred pesos consumed by tourists within the country, 17.9% correspond to foreign visitors and 82.1% to national visitors. Tourism contributed 8.7% to the Gross Domestic Product (GDP) of the national economy. It is considered in tourism production that 11.4% corresponds to the production of goods and 88.6% to services (INEGI, 2021).

The tourism satellite account in Mexico (CSTM) identifies the consumption of services and goods made before and during their trip by national and foreign visitors within the country. As well as the tourist consumption of Mexicans abroad. This account shows that foreign consumption contributed 17.9% and national consumption was 82.1%. Paid occupied jobs (PTOR) are the labor inputs required to conduct the production of tourism services and goods, representing 5.8% of the national economy, that is, 2.3 million during the year 2019 (INEGI, 2021).

The Tourism Secretary, through its Information and Follow-up Unit, prepared the report on the Results of the Tourist Activity in June 2022. In this document, it presents interesting data on the behavior of the tourist activity in the country. During the period from January to June, a sample of seventy tourist centers was selected and they show the following information. Occupancy was 54.6% and +20.3 percentage points higher compared to 2021. International tourists corresponded to 27.6% of the total with an amount of ten million 494 thousand tourists, with respect to national tourists they corresponded to 72.4% of the total with an amount of twenty-seven million 574 tourists. In total, the arrival of tourists increased by +44.3% percentage points compared to the previous year (Secretaria de Turismo, 2022).

During the month of June when summer begins, the occupancy level of hotel rooms increased +11.7 percentage points compared to June 2021, showing an increase of 57.2% compared to the previous year. The arrival of tourists to the different tourist centers varied in the year 2022 with respect to the year 2021 in the month of June 27.1% with respect to foreign tourists and 12.8% with respect to national tourists. Showing an accumulated corresponding to the months of January to June of 58.8% foreign tourists and 39.4% national tourists (Secretaria de Turismo, 2022).

Regarding beach resorts, the percentage of hotel occupancy in June 2022 was 66.8%, a higher level of +10.4 percentage points compared to the same month during the previous year. Concerning the Gross Domestic Product (GDP), tourism registered an increase of 5.7% in the first quarter compared to the last quarter of 2021. Goods increased by 10.1% and services by 5.0%. Vaccination against COVID-19, the easing of travel restrictions and the revival of the economy have benefited the tourism sector (Secretaria de Turismo, 2022).

The industry dedicated to lodging represents a profitable and attractive business that must be consistent with sustainable development. Martínez (2020: 238) exposes awareness measures for tourist accommodation to become businesses committed to caring for ecology. Some of the measure are: 1-Participation in programs to reduce CO2 emissions such as planting trees; 2-A system to collect and manage the best use of water and compost in accommodations where it is most convenient for both staff and tourists; 3-Conduct a social awareness campaign among employees and customers; 4-Implementation of sustainable means of transport; 5-Carrying out energy audits in order to know the level of energy consumption, as well as involvement with environmental care to take appropriate measures in order to improve the situation; 6-The use of cleaning products free of toxic and chemical substances that can contaminate the environment; 7-The construction of establishments with adequate and efficient architecture such as the collection and reuse of rainwater, heating, adequate insulation and elements that allow the development of sustainable buildings that reduce energy consumption and are adapted to the environment, appropriate to the needs where they are located.

Rodríguez (2016:40) mentions the placement of posters in the rooms where the guest is asked to decide how often their towels are washed to maintain responsible use of water care. Pérez de las Heras (2003) indicates that when buildings constructions are conducted, must be avoid being made them nearby or on the natural courses of rivers, animal passages, that is, build in places that go unnoticed by animals. It is recommended that the tourist spaces are located close to each other. The use of sustainable designs in all the buildings used in the tourist complexes. Walking trails, paths or roads must be as camouflaged as possible in the terrain, as well as easy to clean and not excessively eroded. In other words, designing ecological buildings taking care of the flora and fauna of the place, as well as being useful for both visitors and the people who work there.

Companies that meet the requirements of being friendly to the environment and generate profits are diverse. SEMARNAT (2017) in a document called Sustainable Tourism exposes examples of companies friendly to the environment. As is the case of Ecoturixtlán (2023), a company located in the community of Ixtlán de Juárez, Oaxaca in the northern highlands of the state. It is considered for its biological diversity and has various activities for people who like adventure such as caving, zip line, rappelling, aerial games, horseback riding, climbing, temazcal, mountain biking, fishing, hiking, flora and fauna observation and cycling. Its main attractions are the cloud forest, the Cerro del Cuachirindoo viewpoint, the Mesófilo waterfall, the Grutas del Arco, the community forest industry park, and the 17th-century Santo Tomás temple. In addition, it has facilities for holding conventions, workshops, and meetings for 250 people.

A second company is Rancho La Bellota (2023): rural tourism located 65 km south of Tecate, Baja California. One of its attractions is receiving groups attended by the owners of the place with a maximum of sixteen people in six cabins completely isolated from civilization. The activities to do are hiking and horseback riding trails. Another place is called: Rancho La Acacia: ecotourism in protected natural areas. Its location is in the town of San Dionisio, south of La Paz, north of Los Cabos within the Sierra de la Laguna Biosphere Reserve. It is made up of ecotourism cabins equipped with ecological toilets, light from solar cells and made with local techniques such as the locked rod and palm roofs. They also offer a camping area (SEMARNAT, 2017).

The last company mentioned is the Hotel Taselotzin (2023) located in the town of Cuetzalan, Puebla. It offers the services of traditional cuisine, herbal medicine, massages, temazcal, handicrafts workshops, walks with observation of fauna and flora, guided tours in the Yohualichan archaeological zone, mountain bike rides and walks in caves and waterfalls of the zone. It is worth mentioning that it is a company founded by indigenous women from the area. These are some examples of companies that are generating jobs and wealth while maintaining care for the environment (SEMARNAT, 2017).

DISCUSSIONS

Historic towns are cultural heritage and a tourist resource that, by not having a renewable character, the resources that sustain them must be protected from a deterioration. The tourist activity presents a positive and negative interdependence by being a protagonist in the life of those locations. This fact causes the challenge of establishing an arrangement in the cities and establishing a regulation in the flow of visitors. The various international organizations call on state administrations to establish sustainable tourism and establish tourism control observatories to protect these spaces.

Tourism supported by sustainable development is necessary to establish and maintain a balance in the functioning of historical communities. The repercussions that must be integrated into tourism activity are diverse. To mentioning some, such as: cultural, social, environmental, and economic. This incorporation will be conducted through the inclusion of a current multifunctional reality in the reuse of these populations. However, it should be considered the fact that they are places with antiquity, and these make them fragile in their use. This fact emphasizes the need to implement sustainable development strategies. Activities should be implemented where the capacity of cities to receive visitors is observed so that cultural heritage is not damaged and isined for the benefit of future generations (Troitiño, 1998).

Sustainable development seeks to satisfy the present needs by preserving ecosystems for future generations. By applying it in tourism activity refers to the way in which this activity satisfies the need of the people who inhabit the tourist places and those who visit them. Tourism can be developed sustainably as is the case with community tourism. This is developed in a specific destination where residents plan activities and are the main beneficiaries of the economic spill. These activities help to preserve the cultural and natural heritage.

Community tourism is a way of implementing sustainable tourism by seeking to improve the socio-economic development of the community. And to preserve and promote the protection of the environment and of cultural and heritage resources. This type of tourism gains greater importance in underdeveloped destinations. These are the topics of generating wealth and preserving especially important resources in the survival of present and future communities. By implementing it, it generates economic resources. Also, it contributes to the social and economic development of the destination population and fosters respect and conservation towards the resources that the community possesses.

Society must be involved in the process of planning and managing activities through the preparation of a tourism development plan. In the same way, value the cultural aspects, gastronomy, folklore, historical heritage that they offer and the environmental resources that are offered. In addition, to raise awareness and educate the local population in the correct performance of tourism activity with axes such as: culture, care of the environment, foundation of companies, offering a quality service to the customer. And above all the desire to care for and protect the resources they possess for the benefit of the present and future inhabitants (Orgaz, 2013).

The study conducted in the city of Santiago de los Caballeros in the Dominican Republic obtained the opinion of experts dedicated to tourism activity. The main results obtained were that people have the will to improve their performance as tourist service providers. The destination has the potential to offer services for such a reason they should work on a common approach to correct the weaknesses. Such as low training, especially in foreign languages and in offering the services. Similarly, establish a better tourist transportation service more efficiently. In adittion, attract tourists who enjoy sport and mountain tourism. Residents should be involved in tourism activities and see the benefits of such actions in society (Caro-González, 2015).

The community's inhabitants of tourist centers should establish more cooperative relations with respect to the experts on the subject to work together. The consideration towards sustainable tourism processes depends on the demand of tourists who want socially responsible products in the care of ecosystems. Specific strategies required by rural context should be implemented for the benefit of environmental care. In addition, inhabitants of communities, people dedicated to tourism activities, public administration and tourists must be considered when implementing such strategies (Anton, 1996).

One of the most well-known and visited cultural destinations of Peru is the Historic Shrine of Machu Picchu, which has been affected by environmental deterioration. The authorities should consider decision-making such as ordering the territorial space, investigating, and recovering the heritage. In addition, educating, training, and spreading the benefits of sustainable development in the population. Also, monitoring and controlling the activities conducted in the said place. All this for the benefit of this space to tourism sustainability, its correct use in the present and in the future (Regalado, 2006).

Ecotourism, cultural tourism, and rural tourism have a favorable future. Only if they are monitored locally with the principles of sustainable development. Pre-planning with good practices should be conducted on a local, national, and global scale. This is to take care of the ecosystems little visited by tourists who remain on the planet and rescue the deterioration of the places that have been eroded by the participation of man in tourism activity (De Esteban, 2010).

Tourism planning involves a sensible and formal process involving the community, tourism service providers and authorities. This is because tourism facilitates a better quality of life in the population involved. By promoting the preservation of resources, the strengthening of the cultural identity of the community, the implementation of business activities based on the careful criteria imposed by sustainable development. The implementation of such indicators is inevitable to protect and preserve ecosystems for both present and future generations. (Meneses, 2013

Social networks have gained a lot of importance worldwide in the issue of externalizing opinions. In these spaces are presented the desires to preserve the different existing ecosystems. The number of organizations that practice responsible tourism, committed to the care of the environment and sustainable is increasing annually. Sustainable tourism is gaining strength at the business level because more tourists want to do activities without harming ecosystems. Hotels are the main representatives of tourism activities and have the responsibility to support the community where they are located. Regarding to the biodiversity of the location and the culture of the population. This is in line with sustainable development (Lalang, 2017).

The need to implement sustainable development in tourism is in response to the demands of tourists and the locations that receive them to protect and improve the opportunities of tourism activities in the future. It is necessary to manage resources to meet social and economic needs, respecting culture, ecological processes in communities and the diversity of the environment. Tourism activities generate jobs resulting in a better quality of life for people. Techniques should be implemented to contribute to more efficient analysis of the relationship between community development and sustainable tourism. (Lorenzo, 2014).

Tourism management contributes to adequate planning when developing tourism services and products. And implementing business and communication strategies. It is observed the importance of coordinating the different sectors involved in tourism activities to meet the targets and objectives required in tourist management. It is necessary to implement an action plan in the planning of activities. As well as implement a timetable of plans focused on strengthening the tourism sector, an adequate allocation of budget towards tourism activities.

Similarly, make a marketing and commercialization according to the products and services offered. People engaged in tourism activity agree that tourism activities contribute to sustainable development and should be supported in the correct use of renewable resources through sustainable mechanisms supported by triple sustainability. It encourages the development, formulation, and implementation of projects in support of tourism management to improve sustainable development in ecosystems (Aucancela, 2021).

Nature-focused tourism is of utmost importance as it is considered a starting an arrival point in the experimental processes of the tourism sector. This is due to the protection of cultural and natural heritage to maintain success in the tourism sector. Excursions in wonderful habitats rich in fauna and flora are popular in the tourism sector. This has resulted in private and public entities implementing sustainable development measures to preserve resources and recover those that have been lost. The care of the environment has caused a new valorization by proceeding to the enjoyment be with its preservation without modifying it or altering it.

This involves short- and medium-term efforts in planning, implementing subsistence policies to care for ecosystems and promoting tourism activities. The ability to succeed in the implementation of sustainable development in the tourism sector lies in observing, conducting prior studies of supply and demand, and implementing indicators that measure cultural and natural heritage. Implement strategic plans and policies in accordance with tourism activity. Also consider and recognize a promotion of tourism activities that preserve and protect ecosystems (Martínez, 2017).

The studies presented above are compared to each other to expose the similarity of thinking in which people involved in communities. Tourism service providers, authorities and tourists agree on the need to implement actions for the benefit of environmental care. Likewise, they agree to the adoption of the principles of sustainable development to preserve ecosystems for the benefit of present and future generations.

SOLUTIONS AND RECOMMENDATIONS

Caring for the environment is the utmost importance not only because of the needs of the current generation but also of future generations. For this reason, companies must implement activities that support the care of ecosystems. The process of global warming and the deterioration of ecosystems is causing problems of rains, monsoons, fires that are disrupting the economic and personal life of human society. It is necessary to implement measures to care for the environment to benefit current and future generations.

National and international organizations have been calling on governments and people to adopt a caring approach to the environment. During the COVID-19 quarantine, it was observed that since there was not much movement of people, contamination decreased at a rate that places that are normally highly contaminated were clearly seen for the first time. This fact showed us as a society the importance of caring for the environment and that the damage caused can be reversible.

The end of the quarantine marked that people resume their daily lives and among them tourist trips. The tourism sector is a major source of employment generation, as well as wealth in the economies. The COVID-19 quarantine taught us as a society to take care of the environment and to see the need for recreation in human beings. Both are important for the better development of human society. There are examples of companies that have been successful in providing their services and being friendly to the environment. Considering these cases, recommendations can be considered for companies related to tourism.

The companies that are related to tourism are the lodging centers, transportation, tourist servers and the people who enjoy the services offered. These people can make the following recommendations:

Optimize water care, recycling rainwater to irrigate green areas or store it for another use, reducing water consumption in showers and tubs, washing bedding and towels. Improve water management in swimming pools to be more efficient and avoid waste. In the bathrooms of the rooms, provide the bidet service for those guests who wish to use them. Use of biodegradable, contaminant-free and environmentally friendly cleaning implements. When activities such as hiking are conducted, promote reforestation as an activity for tourists, workers in tourist centers and the population in general. Implement environmentally friendly energy sources such as solar panels and wind power, as well as find ways to save energy consumption with the use of electrical appliances such as air conditioners by building and maintaining adequate spaces to have a more efficient way of using them. Implement constructions adapted to the needs of tourists and the ecosystem where they are located with materials from the region, with techniques that do not pollute the environment and to be as durable as possible. As well as that they are not an inconvenience in the passage of migratory animals or of the region. Use of lighting systems that turn themselves off when there are no people in the physical space, such as in rooms, conference rooms, restaurants, as well as energy-saving lights to increase energy conservation. Use environmentally friendly means of transportation such as electric cars and transportation services for a greater number of people who can travel comfortably. As well as providing transportation for employees to reduce pollution from the flow of people. Buy and encourage the consumption of food from producers who use environmentally friendly systems. Reduce food waste either by donating it to charities or if it is not possible to consume it, use it as compost in support of the green areas of the tourist center and ecosystem. Also launch actions to improve the management of waste in food preparation, such as the recycling of cooking oil. Establish protection for places where domestic and wild animals are cared for to make tourists and employees aware of the importance of caring for the various species. Institute nurseries in the facilities of tourist centers and in nearby towns to promote a culture of support for local fauna. Promote waste cleanup days in the streets, green areas, beaches by workers and communities near the lodging centers. Develop a culture of recycling glass, plastic, paper in the guests of the tourist centers, in the employees and in society in general and donate any profit from said action to the care of the flora and fauna of the region. Try to be more efficient in the use of energy used in elevators and escalators by using solar panels to power them. Develop ecological souvenirs such as those made of coconut, palm, and biodegradable fabrics to reduce the consumption of plastic and metal. Apply the greatest use of recycled paper for the various areas that need it in the lodging centers. In the gyms, implement bicycles that generate energy when people exercise and use this energy in the operation of the lodging center. In this way, energy is reused and care for the environment is encouraged in the guest. In the areas of care for minors, implement non-polluting activities such as pottery making where they can spend a long time sculpting, learning about this beautiful art and recycling materials. Implement the recycling of ceramic plates and cups in the decoration of lodging centers and even conduct activities with minors where they can design patterns and models where these wastes are used and thus are not wasted. And they understand how they can conduct the recycling process by themselves. Reuse furniture and electrical items in the decoration of lodging centers. Collaborate and establish relationships with private and public organizations dedicated to caring for the environment. Know and disclose the legal regulations of the care of the environment among the people who enjoy the services, the employees and society in general.

CONCLUSION

There is a need to strike a balance between tourism and sustainable development. The COVID-19 pandemic highlighted the importance of leisure activities and the need to protect the environment. Social media has become an influential platform for expressing opinions, including the need for sustainable development in the tourism industry. The United Mexican States, with its rich natural resources, must implement sustainable development in tourism to ensure economic growth, job creation, and the preservation of ecosystems for current and future generations.

Tourism industry has a long history, and it has been driven by various factors such as migration, war, trade, religion, and fashion. Different types of tourism exist, including exploratory travel, mandatory travel, pilgrimage travel, elite travel, and mass tourism. Cultural heritage is a resource that is used in the implementation of a good tourism strategy, and it is important to promote culture to create links with productive processes, generate work sources, and produce economic well-being. The current tourists demand complete services that include transportation, lodging, and recreational activities. However, the COVID-19 pandemic has had a significant impact on the tourism industry, and it has changed the perspective of the new normality.

The COVID-19 pandemic has had a significant impact on the tourism industry, and it has affected production, supply chains, and financial markets. However, it has also led to a decrease in pollution and environmental impact due to the low mobility of people during confinement. This decrease in pollution has made service providers and tourist centers realize the need to conduct their activities in a more environmentally friendly manner. As a result, they have begun to implement various activities, such as offering guests the option of not changing sheets daily, using different towels in bedrooms and swimming pools, using soap dispensers to avoid packaging, recycling waste, and using transportation providers with newer vehicles.

Tourism is an economic activity that involves displacement, generating income in places other than where it originates, and in which one does not habitually reside. The importance of responsible tourism and Corporate Social Responsibility has arisen as a way for companies to seek a balance between achieving their financial and economic objectives and the environmental impact that their activities cause. Sustainable tourism and sustainable development are closely related, and ecotourism has emerged as a conservation movement that promotes responsible tourism while preserving the environment and supporting local communities. The importance of caring for the environment and sustainable development has been recognized in various international conferences and agreements, and the Global Code of Ethics for Tourism has contributed to the implementation of environmental care activities in the tourism sector.

Tourists are becoming more aware of the impact of their actions on the environment, and they demand that tourism service providers be more responsible and respectful of the environment. There is a significant increase in international tourism during the first quarter of 2022 compared to the previous year, indicating a recovery in the tourism sector as a result of the lifting of COVID-19 restrictions.

It is important to promote sustainable tourism practices, with examples of businesses that are environmentally friendly and generate profits. Measures such as participation in programs to reduce CO2 emissions, the collection and management of water, carrying out social awareness campaigns, implementing sustainable means of transport, carrying out energy audits, and using cleaning products free of toxic and chemical substances are recommended for tourist accommodations to become businesses committed to caring for ecology. The analysis suggests that the tourism sector needs to take care of the flora and fauna of the place, as well as designing ecological buildings that are useful for both visitors and the people who work there.

It is also important that sustainable tourism arise and the need to protect cultural and natural heritage sites from deterioration. Sustainable tourism can help to balance the functioning of historical communities and promote socio-economic development, while preserving and promoting the protection of the environment and heritage resources. Community involvement and planning, education and training, and monitoring and control of tourism activities are important in achieving sustainable tourism. Furthermore, social networks, responsible tourism practices, and nature-focused tourism play important roles in promoting sustainable tourism. The discussions emphasize the importance of observing and implementing indicators that measure cultural and natural heritage, and implementing strategic plans and policies that support sustainable tourism activities while promoting the preservation and protection of ecosystems.

REFERENCES

Agüera, O. F. (2013). El Turismo Comunitario como Herramienta para el Desarrollo Sostenible de Destinos Subdesarrollados. Nómadas. *Journal of Social and Juridical Sciencies, 38*(2). https://www.theoria.eu/nomadas/38/francisco_orgaz.pdf

Alonso-Sañudo, A. I. (2002). La aplicación del concepto de turismo sostenible en los países desarrollados. In *Turismo sostenible* (pp. 25 – 36). Iepala Editorial, 2002.

Anton, C., Salvador, L., & Monné, R. (1996). Turismo rural, desarrollo local y preservación del ambiente. Elementos para un desarrollo sostenible del turismo en la zona de montaña Prades-Montsant, Cataluña. *Ería: Revista cuatrimestral de geografía, 41*, 227 – 238. Disponible en: https://dialnet.unirioja.es/servlet/articulo?codigo=34843

Aucancela, I., Betty, P., Velasco, S., & Víctor, M. (2021). Gestión turística como herramienta de desarrollo sostenible de la microcuenca del río Chimborazo, Cantón Riobamba. *Revista de Ciencias Sociales y Humanidades Chakiñan, 13*, 102-116. Disponible en: http://scielo.senescyt.gob.ec/pdf/rchakin/n13/2550-6722-rchakin-13-00102.pdf

Bigné Alcañiz, E., Font Aulet, X., & Andreu Simó, L. (2000). *Marketing de destinos turísticos. Análisis y estrategias de desarrollo.* Escuela Superior de Gestión Comercial y Marketing.

Caro-González, F., Javier, A., Guzmán, J., Alberto, O.-A., & Francisco, C.-V. Mario. (2015). Turismo, desarrollo sostenible y percepción de los stakeholders. Un estudio de caso en República Dominicana. *Revista de Economía del Caribe, 15*, 153 – 182. Disponible en: http://www.scielo.org.co/pdf/ecoca/n15/n15a06.pdf

Casanueva, R. C., & Gallego, A. A. (2012). *Empresas y organizaciones turísticas.* Pirámide.

Comisión Mundial para el Medio Ambiente y el Desarrollo (CMMAD). (1987). *Reporte de la Comisión Mundial del Medio Ambiente y Desarrollo: Nuestro Futuro Común.* Organización de las Naciones Unidas. Disponible en: http://www.un-documents.net/our-common-future.pdf

Conferencia Mundial de Turismo Sostenible (WCST). (1995). *Carta del turismo sostenible.* Programa Hombre y Biosfera. Centro Mundial del Patrimonio. Programa de las Naciones Unidas para el Medio Ambiente. Organización Mundial del Turismo. Consejo Científico Internacional para el Desarrollo de las Islas. Comisión de las Comunidades Europeas. Decenio Mundial para el Desarrollo Cultural. Gobierno de Canarias. Cabildo de Lanzarote. Secretaría General de Turismo. Ministerio de Obras Públicas, Transportes y Medio Ambiente. Universidad de la Laguna.

Crespo Jareño, J. A. (2020). *Orígenes e impactos del ecoturismo en la Revista Kalpana, Edición Especial, Número 18.* Disponible en https://dialnet.unirioja.es/descarga/articulo/7834208.pdf

Crosby, A., Prato, N., Solsona, J., & Gómez, O. A. (2009). *Re-inventando el turismo rural. Gestión y desarrollo.* Laertes.

De Esteban, Curiel, Javier, Antonovica, & Arta. (2010). El ecoturismo como modelo internacional de desarrollo sostenible del turismo cultural. *Revista Teoría y praxis, 6*(8), 43 – 53. Disponible en: http://risisbi.uqroo.mx/handle/20.500.12249/622

Ecoturixtlan. (2023). Recuperado de www.oaxaca-mio.com/ecoturixtlan.htm

Faraldo, J. M., & Rodríguez-López, C. (2014). *Introducción a la historia del turismo*. Alianza Editorial.

Fernández García, R. (2011). *La dimensión económica. Desarrollo sostenible*. Editorial Club Universitario.

Guerrero González, P., & Ramos Mendoza, J. R. (2014). *Introducción al turismo*. Grupo Editorial Patria.

INEGI (Instituto Nacional de Estadística y Geografía). (2021). *Comunicado de prensa Numero 539/21 con fecha 24 de septiembre de 2021*. Estadísticas a propósito del día mundial del turismo. Disponible en: https://www.inegi.org.mx/contenidos/saladeprensa/aproposito/2021/EAP_Turismo21.pdf

Ivanova, A., & Ibáñez, R. (2012). *Medio ambiente y política turística en México. Tomo I: Ecología, biodiversidad y desarrollo turístico*. D.F. Secretaría de Medio Ambiente y Recursos Naturales. Instituto Nacional de Ecología. Universidad Autónoma de Baja California Sur.

Lalangui, J., Espinoza Carrión, C. R., & Pérez Espinoza, M. J. (2017). Turismo sostenible, un aporte a la responsabilidad social empresarial: Sus inicios, características y desarrollo. *Universidad y Sociedad, 9*(1), 148-153. Disponible en: http://scielo.sld.cu/pdf/rus/v9n1/rus21117.pdf

Leff, E. (2004). *Saber ambiental. Sustenibilidad, racionalidad, complejidad, poder*. Siglo XXI Editores.

Lorenzo Linares, H., & Morales Garrido, G. (2014). Del desarrollo turístico sostenible al desarrollo local. Su comportamiento complejo. *Pasos (El Sauzal), 12*(2), 453–466. doi:10.25145/j.pasos.2014.12.033

Maerk, J., y Boxill, I. (2000). *CDMX, México*. Turismo en el Caribe. Plaza y Valdés, S.A. de C.V.

Martínez Herrera, H. (2005). *El marco ético de la responsabilidad social empresarial*. Editorial Pontificia Universidad Javeriana.

Martínez Quintana, V. (2017). El turismo de naturaleza: Un producto turístico sostenible. *Arbor, 193*(785), a396. doi:10.3989/arbor.2017.785n3002

Martínez Salvador, S. (2020). *Dirección de alojamientos turísticos*. Editorial Paraninfo.

Martínez Salvador, S. (2022). *Animación Turística*. Editorial Parainfo.

Meneses, T. Á. (2013). La planificación turística: Un aspecto clave para el desarrollo sostenible y regional de Boyacá. Revista de Investigación. *Desarrollo e Innovación*, *3*(2), 101–110. doi:10.19053/20278306.2169

Moyano Bonilla, C. (1995). Derecho a un Medio Ambiente Sano. *Boletín del Instituto de Investigaciones Jurídicas de la UNAM, 84*. Disponible en http://biblio.juridicas.unam.mx/revista/DerechoComparado/indice.htm?n=82

Muñoz, F. (2003). *El turismo explicado con claridad*. Amertown International.

Nebel, B. J., & Wright, R. T. (1999). *Ciencias ambientales: ecología y desarrollo sostenible*. D.F. Pearson Educación.

Ojeda García, C., & Mármol Sinclair, P. (2016). *Marketing Turístico*. Parainfo.

Organización Mundial del Turismo. (1999). *Código Ético Mundial para el Turismo. Por un turismo responsable*. Asamblea General de la OMT. Disponible en https://webunwto.s3.eu-west-1.amazonaws.com/s3fs-public/2019-10/gcetpassportglobalcodees.pdf

Organización Mundial del Turismo. (2022). *Comunicado de prensa. La recuperación del turismo gana impulso tras la disminución de las restricciones y el aumento de la confianza*. Disponible en: https://webunwto.s3.eu-west-1.amazonaws.com/s3fs-public/2022-06/220606-unwto-barometer-es.pdf?VersionId=qi2kpKlUpSD4j_AJQilm9kIlpZ_tiHMt

Pérez de las Heras, M. (2003). *La guía del ecoturismo. O como conservar la naturaleza a través del turismo*. Mundi-Prensa Libros, S.A.

Quesada Castro, R. (2000). *Elementos del turismo*. Editorial Universidad Estatal a Distancia.

Rancho La Bellota. (2023). Recuperado de: www.rancholabellota.com

Regalado Pezúa, O., & Arias Valencia, J. (2006). Desarrollo Sostenible en Turismo: Una propuesta para Machu Picchu. *Journal of Economics, Finance and Administrative Science, 11*(20), 63-73. Disponible en: https://www.redalyc.org/pdf/3607/360735259003.pdf

Reyno Momberg, M. (2007). *Responsabilidad Social Empresarial (RSE) como Ventaja Competitiva*. Universidad Técnica Federico Santa María.

Rivas, García, & Jesús. (2004). *Estructura y Economía del Mercado Turístico*. Septem Ediciones, S. L.

Rivas García, J., & Magadán Díaz, M. (2012). *Planificación turística y desarrollo sostenible*. Septem Ediciones.

Rodríguez Sánchez-Escalonilla, N. (2016). *Organización y prestación del servicio de recepción en alojamientos UF0052*. Ediciones Paraninfo, S.A.

Sánchez Pérez, M., Terán, Yépez, E., Marín-Carrillo, M. B., Marín-Carrillo, G. M. (2021). *La COVID-19 y el sector turístico en España: Impacto sobre el comportamiento del consumidor turístico*. Editorial Universidad de Almería.

Secretaría de Turismo. (2022a). *Unidad de Información y Seguimiento*. Disponible en https://www.datatur.sectur.gob.mx/SitePages/versionesRAT.aspx

Secretaría de Turismo. (2022b). *Resultados de la Actividad Turística junio 2022*. Disponible en: https://datatur.sectur.gob.mx/RAT/RAT-2022-06(ES).pdf

TaselotzinH. (2023). Recuperado de: http://taselotzin.mex.tl/frameset.php?url=/intro.html

Troitiño, V., & Miguel, Á. (1998). Turismo y desarrollo sostenible en ciudades históricas. Ería, Revista cuatrimestral de geografía. *Ejemplar dedicado a: El turismo en las ciudades históricas, 47*, 211-227.

Troitiño, V., & Miguel, Á. (2002). *El patrimonio arquitectónico y urbanístico como recurso. In La función social del patrimonio histórico: El turismo cultural*. Ediciones de la Universidad de Castilla-La Mancha.

Virgen Aguilar, C. R. (2014). *Turismo y desarrollo sustentable. Un acercamiento al estudio del turismo*. Asociación Mexicana de Centros de Enseñanza Superior en Turismo y Gastronomía, A.C., Universidad de Guadalajara. Universidade Federal do Parná.

Vogeler Ruiz, C., & Hernández Armand, E. (2018). *Introducción al turismo. Análisis y estructura. Editorial Universitaria*. Ramón Areces.

Wulf Betencourt, E. (2018). *Responsabilidad Social Empresarial. Un desafío corporativo*. Editorial Universidad de la Serena.

ADDITIONAL READING

Becken, S. (2014). *Tourism and water: Interactions, impacts and challenges*. Channel View Publications.

Bramwell, B. (2011). Planning for sustainable tourism: A case study of a coastal resort in Thailand. *Journal of Sustainable Tourism, 19*(1), 69–88.

Buhalis, D., & Darcy, S. (Eds.). (2011). *Accessible tourism: Concepts and issues*. Channel View Publications.

Fennell, D. A., & Dowling, R. K. (2019). *Ecotourism and environmental sustainability: Principles and practice*. Routledge.

Gössling, S., Scott, D., & Hall, C. M. (2013). *Tourism and water: Interactions, impacts and challenges*. Channel View Publications.

Hall, C. M., & Lew, A. A. (Eds.). (2018). *Sustainable tourism: A comprehensive guide*. Routledge.

Lindberg, K., & Hawkins, D. E. (Eds.). (2019). *Sustainable tourism futures: Perspectives on systems, restructuring and innovations*. Routledge.

Mowforth, M., & Munt, I. (2015). *Tourism and sustainability: Development, globalisation and new tourism in the third world*. Routledge. doi:10.4324/9781315795348

Telfer, D. J., & Sharpley, R. (2015). *Tourism and development in the developing world*. Routledge. doi:10.4324/9781315686196

Weaver, D. (2014). *Sustainable tourism: Theory and practice*. Routledge.

KEY TERMS AND DEFINITIONS

Confidence Index: A measure of the level of confidence or optimism among professionals in a particular industry or market.

Ecosystems: Communities of living and nonliving things that interact with each other in a particular environment.

Influx of People: The number of people entering a country or region during a particular period of time.

Lower Middle Class and Middle-Class Tourism: The segment of tourism that involves individuals from the lower middle class and middle class who take vacations away from their place of residence.

Responsible Tourism: A form of sustainable tourism that involves responsible behavior and practices by tourists and tourism service providers, with a focus on reducing negative impacts on the environment, society, and culture.

Rural Tourism: A form of tourism that involves visiting rural areas for their natural, cultural, and/or historical attractions.

Sustainable Tourism: A form of tourism that takes into account the economic, social, and environmental impacts of tourism on the host community, and seeks to minimize negative impacts while maximizing positive impacts.

Tourism Development Plan: A formal and strategic planning process that involves the community, tourism service providers, and authorities in the development and management of tourism activities.

Tourism Satellite Account: A system for measuring the economic impact of tourism in a particular country or region, including the consumption of goods and services by tourists.

Triple Sustainability: A framework that integrates economic, social, and environmental sustainability in tourism development.

Chapter 6
Methodologies in Dark Tourism Issues:
Modelling the Dark Experience

Maximiliano Emanuel Korstanje
University of Palermo, Buenos Aires, Argentina

ABSTRACT

The growth of dark tourism studies is an unquestionable reality. Over recent years, a burgeoning number of publications have taken dark tourism as their main object of study. However, the evolution of literature which is based on dark experiences has been subject to great controversy. The present chapter reviews the strengths and weaknesses of dark tourism research today as well as the future agenda for the next years. The authors have identified five clear clusters that explain very well not only the dark tourists´ motivations but also the phenomenology of the dark experience.

INTRODUCTION

Dark tourism is doubtless one of the global themes that have captivated international researchers worldwide. The concept debates the nature of tourism confronting our already-existing beliefs. Tourism should be understood -at least in the literature- as a result of the combination between two factors, movement and leisure. In this vein, the rise of tourism combined a strange "British taste" for pleasure as well as an apollonian sense of beautyness that led travelers to enjoy the visited landscapes (Ousby, 1990). In consonance with this, some authors alert that the turn of the twentieth century has facilitated the emergence of new morbid forms of tourism consumption which include spaces of death or war, suffering, slumming or mourning

DOI: 10.4018/978-1-6684-7242-2.ch006

(Freire Medeiros 2014; Korstanje 2016; Mionel 2019). Dark tourism defies our concept of leisure putting the problem of morbid taste into the tapestry. Having said this, dark tourism offers a more than interesting spectrum for applied research in the years to come but with no few methodological problems (Bowman & Pezzulo 2009; Martini & Buda 2020). At first glimpse, there is a great dispersion of definitions as well as methodologies in dark tourism studies (Hooper & Lennon 2016). Secondly and most importantly, applied research is mainly based on what A. Franklin dubbed as "tourist-centricity" which means an obsession to take tourists as the only valid source of information or knowledge. For that, dark tourism studies are conducted on visitors who express or manifest their so-called motivations for visiting these places (Seraphin & Korstanje 2021). Methodologically speaking, centering the research on the questionnaire application has two fundamental problems. On one hand, interviewees often are unfamiliar with their real motivations and on another, they can lie to protect their interests. This behooves us to consider new emerging methodologies for dark tourism fields (Korstanje 2020).

One of the main problems of dark tourism research, as above referenced, seems to be the lack of homogeneity in discussing shared definitions and methodologies to study the object. As Hooper & Lennon (2016) put it, plenty of scholars have been captivated to study dark tourism but this process was chaotically guided. There are today many definitions of the same term, or simply many synonymous for the same phenomenon. As a result of this, dark tourism studies navigate in a sea of great fragmentation. As Korstanje noted, three academic waves debate -these days- the nature and future of dark tourism (Korstanje 2020; Korstanje & Olsen, 2020). The anthropology school define dark tourism as a type of mediation between visitors´ life and the "significant Other´s" death (Stone 2012; Stone & Sharpley 2008), The heritage school suggests that tourism should be understood as a sacred pilgrimage that helps ultimately to domesticate the future (Cohen 2011; Collins-Kreiner 2016; Strange & Kempa, 2003); and of course, the critical turn, a new emerging academic perspective that deals with the idea dark tourism acts as an ideological discourse disposed of for affirming the center-periphery dependency in a post-colonial time (Bowman & Pezzulo, 2009; Korstanje 2016; Tzanelli 2016). Having said this, these lines of argumentation may converge in the same common-shared point but often are reluctant to bolster a fluid dialogue (Ashworth & Isaac, 2015). Although the number of studies focusing on dark tourism has been notably triplicated (Shekhar & Valeri, 2022; Tarifa-Fernandez, Carmona-Moreno & Sanchez-Fernandez, 2022), less attention was paid to the methodological issues as well as the epistemological problems to approach these types of objects (Wight 2006). To fill this gap, the present book chapter explores the strengths and weaknesses of dark tourism research in our days as well as the epistemological dilemmas for the not-so-long distant future. The goals of the present piece are threefold. First and foremost, we offer a short description of preliminary definitions of dark tourism and its compatibilities or differences with associated terms such as mourning tourism, thana tourism and so forth. Secondly, we explore what experts dubbed *the phenomenology of dark experience.* This experience is based on the contradiction of what tourists overtly said (plausibly in questionnaires) and their real motivations which are inexpugnable to the researcher´s eye. Third, we give a snapshot of some promising managerial points which should be followed by policymakers to revitalize or sanitize affected destinations. Last but not least, there is a significant relationship between niche tourism and dark tourism. Niche tourism is a term originally coined in contraposition to mass tourism. Niche tourism is reserved for a small number of people or consumers. As Peter Tarlow observes, dark tourism is part of niche tourism simply because the memory of victims should be protected. Mass tourism is seen as an undesired phenomenon that affects the site and the memory of victims, in many cases, mass tourism is not preferred or accepted in dark tourism sites (Tarlow 2007). However, as we will see in the following lines, not always this is the case. Stone (2006) has shed light on this point offering a conceptual spectrum which ranges from the lightest to the darkest forms of dark tourism. Stone suggests that some sites are open to mass tourism and structured in higher investment in infrastructure, while others are oath to welcome mass tours. This occurs because each site is structured according to different levels of loss and trauma.

BACKGROUND: THE EVOLUTION OF DARK TOURISM

There is no clear consensus among scholars on whether dark tourism is a modern phenomenon or simply it was practised in the pastime. Some studies suggest that ancient Romans visited cemeteries or spaces of disasters in ancient history (Hartmann et al. 2018). Contrariwise, others allude to dark tourism as an experience that surfaced in modernity because of the advance of the secularizing process. This means that the fear of death led invariably us towards a negation of death. This sentiment negation or misunderstanding wakes up dormant anxieties today that are placated through different rites. Dark tourism, above all, offers a type of mediation (if not sublimation) through the "Other´s death" that ultimately helps visitors to contemplate their finitude (Stone & Sharpley, 2008). What is more important, dark tourism is gradually situating as a leading sub-field just after the 2000s. Today it comprises one of the most consulted topics and matters of debate in conferences, and academic events while transpiring the content of leading tourism-related journals (Shekhar & Valeri 2022; Ogretmenoglu, Mavric & Dincer, 2022). At a closer look, the specialized literature is growing yearly given the interest within and outside Academia. The proliferation of natural disasters, environmental degradation or the radicalization of political violence has generated further attraction for these types of spectacles (Korstanje 2020).

A Short Companion: Issues, Controversies, and Problems

Far from being new, the fascination for death is been packaged and consumed by a global audience through the articulation of what Foley and Lennon (1996) dubbed as "morbid taste".

Nowadays, dark tourism is seen as an emerging niche or subfield of research that is mainly marked by the visit of tourists to spaces of mass death or macabre spectacles. Studied from different angles, dark tourism is situated as a leading object of study in academia (Stone 2013; Olsen & Korstanje 2019).

Hooper and Lennon (2016) has called attention to the dispersion of knowledge production as a direct consequence of the attention paid by scholars and journalists to dark tourism issues. What is more important, excess attention is somehow creating a paradoxical condition for this subfield. On one hand, scholars who operate under the influence of different disciplines and methodological paradigms develop their definitions of dark tourism. On another, terms like thana tourism, dark tourism or mourning tourism are simply lumped together without any critical reasoning.

The Three Stages of Dark Tourism Studies

In time, the evolution of literature has gone through three different stages: the thanaptopic tradition, the dark heritage tradition, and the critical turn.

The thanaptopic tradition put efforts into deciphering the role of thanaptosis as the cornerstone of dark tourism sites. Thanaptosis is defined as a process where visitors contemplate their lives through the "significant Other´s death". Per this academic platform, thanaptosis results from the advance of a much deep secularizing process that recedes religiosity to the private sphere (Seaton, 1996, 1998; Stone, 2013). Secondly, the heritage-based tradition focuses on dark tourism as a type of sacred pilgrimage that ultimately helps visitors to understand traumatic events (Biran, Poria & Oren, 2019). The critical turn ignites a hot debate revolving around the connection between dark consumption (if not heritage) and colonialism. Having said this, dark tourism reaffirms long-dormant discourses oriented to suppress the autonomy of a local periphery respecting the colonial center (Tzanelli, 2016; Korstanje, 2016). Of course, each tradition has created and developed different methodological approaches and bases which are far from being in dialogue. It is safe to say dark tourism has received broad coverage in the media as well as the academic circles which crystalizes their advances in leading publications (Shekhar & Valeri, 2022). This changes forever the morphology of holidays (Sinclair, 2018). Classic mass tours were originally organized revolving around an Apollonian sense of beauty. This means that our parents travelled to beaches, and paradisiacal destinations to spend their holidays. Now, this has been replaced by a "morbid form of consumption" which recycles and commoditizes death as a main criterion of tourist attraction (Korstanje, 2016). This point is vital to understand how tourism research shifts in dark tourism issues. To wit, the visitor´s experience is previously determined by a disaster or a catastrophe where mass death takes place. Visitors are psychologically moved to get authentic experiences that sublimate their own lives, helping them understand their finitude (Miles, 2014). Under some conditions, this morbid curiosity leads to a rapid acceleration of recovery timeframe just after a disaster takes a hit.

There is valuable evidence that shows how local communities which embrace dark tourism recover faster in a post-disaster context (Isaac, Cakmak, & Butler, 2019). Joy Sather Wagstaff (2016) has enthusiastically debated to what extent disasters generate positive emotions like empathy and reciprocity in the audience. This visitor´s engagement is grounded in the fact that all men are equal before death.

Last but not least, dark tourism research evinces some methodological problems which need to be discussed in the next section. These setbacks include a lack of coherent definitions or what has been termed as *knowledge fragmentation,* associated with tourist-centricity, a leading conceptual background that put the tourist as the only valid source of knowledge. At the same time, scholars have not reached a consensus to find what are the real motivations behind dark tourism. While some studies emphasize the anthropological attempt to understand death or pedagogic goals (Cohen, 2011; Collins-Kreiner, 2016), others allude to pathological drives (Korstanje, 2016; Tzanelli, 2016).

The Dark Experience: New Problems and Recommendations

What are the factors that motivate tourists to visit dark tourist places, or what is more important how the dark experience is formed are two key questions scholars have formulated from the outset.

In this respect, dark tourism represents a disturbing (if not new) form of travel behavior that interrogates furtherly postmodern cultures on novel forms of relations with death (Seaton, 1996). The motivations of dark tourists take a wide spectrum ranging from lightest to darkest subtypes (Stone, 2006; Light, 2017). As Stone (2006) eloquently observes, the dark tourists´ experiences are variable and very hard to grasp. For that reason, the typification of the different spectrums should be based on palpable variables such as the dark site formation or the game between supply and demand. Since dark tourists move for the curiosity of macabre spectacles as well as the contemplation of the "Other´s pain" no less true seems to be that tourists have successfully adjusted their preferences to the rise of new postmodern landscapes marked by disasters (Strange & Kempa, 2003; Farmaki, 2013). In consonance with this, Rajasekaram, Hewege, and Perera (2022) call attention on the lack of understanding in the literature about the dark tourist experience. At least in applied research, the concept is very hard to operationalize, or what is more important, to measure in empirical bases. Although Stone has given some explanations on the spectrum of dark tourism, further qualitative research is at least needed. The dark tourism experience has an emotional impact that often exceeds rational thinking. At first glimpse, the dark spectrum -offered by Stone- facilitates some things to decipher the interplay between emotionality and cognition in dark sites. However, some reports have alerted on the importance of emotionality in darker experiences. The obtained outcome evinces how dark consumers subject to exaggerated stimulation of darkness are affected in their behavioral intentions (Lv et al., 2022a, 2022b).

Starting from the premise that dark tourism consumption is a behavioral phenomenon some voices have claimed that the advances in literature are not suffice to say we get conclusive evidence that explains dark tourists´ motivations (Cave & Buda, 2018; Jordan & Prayag, 2022; Podoshen, 2013).

Academic Schools in Dark Tourism Studies

It is not otiose to say that grouping all studies into conceptual families or academic platforms is a simplistic practice. Anyway, it assists readers in understanding the differences and commonalities among authors and their gravitations in other fields. Whatever the case may be, we have made a classification of reviewed literature according to five clear-cut bases. Those studies which focus on morbid curiosity as the key factor that explains dark tourism, are particularly organized around the idea that the visit to post-disaster sites is explained by a natural human curiosity. Even though dark tourism defies all conventional definitions of leisure practices, it is true to mention that the curiosity for death is cemented not only in a human drive but also in the force of a much deep secularizing process which is proper of postmodernity.

In this way, the mass media, widely supported by digital technologies, commoditizes disasters to be consumed 24/7 worldwide. As a result of this, dark tourism not only helps culture to understand traumatic event but also enhance resiliency as well as the acceleration of recovery timeframes (Roberts & Stone, 2014; Light, 2017; Robinson, 2015). A second family of theories is given by what specialists dubbed "nostalgia". Having said this, nostalgia becomes a type of host-guest relations where both try to domesticate the future. Nostalgia opens a door between the negated present which is traumatic for survivors and the future (Tarlow, 2007; White & Frew, 2013). A third family of theories leads us to pull and push factors (or the interplay between supply and demand).

This economic-centered academic platform speaks to us of dark tourism as the convergence between what tourists look at and what the site ultimately offers. Dark tourism cannot prosper if there is no supply on the site. Many cases witness how dark tourists are unwelcomed whenever locals are not willing to develop dark tourism consumption in the site. At the same time, the opposite is equally valid, some sites are fabricated in memory of the victims though they are not located where events happened (Cohen, 2011). For that, it is important to mention that a dark market mediates not only in the tourists´ expectations but also in their experiences.

The collected evidence suggests two important things. On one hand, dark consumption is collectively negotiated between the agency and the structure. On another, each visitor selects a site according to its emotional disposition or its biography (Farmaki, 2013; Robinson, 2015). The question of whether history plays a leading role in dark tourists´ experience has been studied by a fourth family of studies which emphasizes on significance and importance of authenticity and dissonance heritage. Dark tourism also offers a historical explanation of traumatic events as well as the next steps to avoid them in the future (Cohen, 2011; Walby & Piche, 2011).

A last set of publications, known as the "critical turn" holds a polemic but not for this less interesting thesis; dark tourism acts as a mechanism of indoctrination culturally orchestrated by European powers to affirm a post-colonial dependency. Far from being emphatic with the "Other´s suffering" dark tourists are in quest of revitalizing their institutions. Although they are motivated to consume spaces of mass death derived from genocide or ethnic cleansing in the periphery, dark tourists are unfamiliar with the crimes committed by European powers during the colonial expansion. After all, (dark) heritage represents a fabricated instrument of control successfully fabricated to keep lay citizens under control (Tzanelli, 2016; Sather Wagstaff, 2016; Kostanje, 2016).

METHODOLOGIES IN DARK TOURISM

As stated in the introductory section, dark tourism research has notably gained traction in these years (Mowatt & Chancellor, 2011). Dark tourism studies include quantitative research formed by applied questionnaires on visitors, as well as qualitative methodologies such as ethnography, netnography or even critical analysis. It is not simplistic to say that dark tourism studies show strengths and weaknesses.

Among the strengths, dark tourism studies have shed light on the academic career of many scholars as well as policy-makers in developing more resilient destinations which successfully adapted to disasters. This body of knowledge not only incorporated notions elaborated in other disciplines but also offers a fertile ground to understand the complex effects of death in culture (Isaac, 2021; Light, 2017; Light & Ivanova, 2022; Raine, 2013).

However, this body of research has some methodological inconsistencies which need to be cleared. Part of these limitations starts from what other voices have alerted as knowledge fragmentation, which means a wide diversification of methods and definitions which operate from incompatible epistemologies. When this happens, a shared definition is almost impossible (Hooper & Lennon, 2016). At the same time, the term dark obscures more than it clarifies. The term dark leads some scholars to think that visitors are influenced by perverse drives (Bowman & Pezzulo, 2009). Secondly, dark tourist studies are centered on questionnaires or wider samples conducted on tourists, in that case, other agents are pushed to the margins of the game. Methodologically speaking, interviewees are frequently unfamiliar with their psychological world negating their innermost emotions, or simply lying to care about their interests. To put this bluntly, what would respond a gangster do if we ask: what is your job or your profession? For sure, he will reply “I am a respectable businessman!”. Needless to re-question, is he a real businessman? (Korstanje, 2016; Tzanelli, 2016).

In terms of A. Franklin, the tourist-centricity in applied research hides the correlation among variables. The tourist-centricity -as a leading tourism academic paradigm- rests on the premise tourists are the only valid source to extract reliable knowledge affordable for fieldworkers but sometimes this is a big mistake. In Stone´s words, there is a pseudo-relation between variables which should be at least reconsidered (Stone, 2006). Third, the excessive attention paid to the theme has created a type of dispersion where countless definitions are put together (Hooper & Lennon, 2006). Fourth, dark tourism unless ethically regulated may lead to distorted or biased interpretations of pastimes paving the way for the commoditization of disasters. Hence, the conditions that facilitated the disaster -far from being fixed- are replicated. Whenever the preconditions of the disaster are forgotten the disaster repeats in the future (Potts, 2012; Korstanje, 2016).

Managerial Perspective in Dark Tourism With Focus in Nice Tourism

Dark tourism offers a fertile ground to accelerate the recovery timeframe in sites or destinations affected by political violence or natural disasters. Dark tourism acts not only as a catalyst to help locals in the recovery stage but also attracts further investors. At the same time, those destinations that recovered through the stimulation of dark tourism have shown more capacity of resilience to face similar disasters. This happens because the act of rememorizing gives a lesson to survivors and visitors generating reciprocity and synergy among stakeholders. Among the positive aspects of dark tourism, the lines of action should include:

- Dark tourism provides advantages for the local community to find and share their own culture. This includes a history of ghosts, disasters, or the history of death.
- Dark tourism or disaster tourism generates genuine profits and employment which are mainly canalized to fast the recovery timeframe. This includes road improvements, water supply, and the restoration of communications, or energy systems.
- Visitors are assisted to develop empathy with the visited sites or engaged with the story. Children can help to empower their understanding of history.
- There are some social benefits derived from dark tourism such as the organization of volunteer tourism while generating more solidarity among involved stakeholders.
- Dark tourism offers some practical benefits in degrowth tourism or the adoption of more sustainable practices to avoid a similar disaster or tragedy in the future.
- Positive spillover effects can involve neighboring cities or nearby destinations.

CONCLUSION

To hear we have debated the ebbs and flows as well as the main limitations of dark tourism studies. Despite its burgeoning growth, the literature stands at a halt. Part of the problems of specialized literature consists of the lack of shared definitions or methodologies to study dark tourism as a main object.

Secondly, it is associated with an over-valorization of tourists as the only source of valid knowledge. As a behavioral issue culturally linked to the phenomenology of experience, dark tourism shows many roots. Over the years, scholars of all pundits have discussed the nature of dark tourism motivations. Some studies have systematically encroached on the field of morbid taste and curiosity. The idea of curiosity invariably leads us to the mediation theory, which punctuates dark tourism acts as a bridge that helps us contemplate our finitude through the "Other´s death". At the same time, a second family of theories, which is based on the role played by nostalgia, explores dark tourism as an anthropological attempt to domesticate death and the future.

In parallel, an economic-centered paradigm explains the phenomenon from the lens of push-and-pull factors. Per this academic platform, there is a type of dark market conditioned by the supply-demand game. Fourthly, the dark experience is formed by a set of different pedagogic instruments deployed by society to learn the lessons behind traumatic events.

Last but not least, a critical standpoint based on the intersection between politics and colonial exploitation has recently taken a room in the constellations of dark tourism. Each theory has shed light on a specific spectre in dark tourism consumption leaving some open points which should be reviewed in the next layouts.

REFERENCES

Ashworth, G. J., & Isaac, R. K. (2015). Have we illuminated the dark? Shifting perspectives on 'dark' tourism. *Tourism Recreation Research*, *40*(3), 316–325. doi:10.1080/02508281.2015.1075726

Biran, A., Poria, Y., & Oren, G. (2011). Sought experiences at (dark) heritage sites. *Annals of Tourism Research*, *38*(3), 820–841. doi:10.1016/j.annals.2010.12.001

Bowman, M. S., & Pezzullo, P. C. (2009). What's so 'dark'about 'dark tourism'?: Death, tours, and performance. *Tourist Studies*, *9*(3), 187–202. doi:10.1177/1468797610382699

Cave, J., & Buda, D. (2018). Souvenirs in dark tourism: Emotions and symbols. Palgrave Macmillan UK.

Cohen, E. H. (2011). Educational dark tourism at an in populo site: The Holocaust Museum in Jerusalem. *Annals of Tourism Research*, *38*(1), 193–209. doi:10.1016/j.annals.2010.08.003

Collins-Kreiner, N. (2016). Dark tourism as/is pilgrimage. *Current Issues in Tourism*, *19*(12), 1185–1189. doi:10.1080/13683500.2015.1078299

Farmaki, A. (2013). Dark tourism revisited: A supply/demand conceptualisation. *International Journal of Culture, Tourism and Hospitality Research*, *7*(3), 281–292. doi:10.1108/IJCTHR-05-2012-0030

Foley, M., & Lennon, J. J. (1996). JFK and dark tourism: A fascination with assassination. *International Journal of Heritage Studies*, *2*(4), 198–211. doi:10.1080/13527259608722175

Freire-Medeiros, B. (2014). *Touring poverty*. Routledge. doi:10.4324/9780203840719

Hartmann, R., Lennon, J., Reynolds, D. P., Rice, A., Rosenbaum, A. T., & Stone, P. R. (2018). The history of dark tourism. *Journal of Tourism History*, *10*(3), 269–295. doi:10.1080/1755182X.2018.1545394

Hooper, G., & Lennon, J. (2016). *Dark tourism*. Taylor & Francis doi:10.4324/9781315575865

Isaac, R. K. (2021). Editorial special issue in Dark Tourism. *Journal of Heritage Tourism, 16*(4), 363–366. doi:10.1080/1743873X.2021.1920963

Isaac, R. K., Çakmak, E., & Butler, R. (Eds.). (2019). *Tourism and hospitality in conflict-ridden destinations*. Routledge. doi:10.4324/9780429463235

Jordan, E. J., & Prayag, G. (2022). Residents' cognitive appraisals, emotions, and coping strategies at local dark tourism sites. *Journal of Travel Research, 61*(4), 887–902. doi:10.1177/00472875211004761

Korstanje, M. E. (2016). *The rise of thana-capitalism and tourism*. Routledge. doi:10.4324/9781315457482

Korstanje, M. E. (2020). The dark tourist: Consuming dark spaces in the periphery. In M. Korstanje & H. Seraphin (Eds.), *Tourism, terrorism and security*. Emerald Group. doi:10.1108/978-1-83867-905-720201009

Korstanje, M. E., & Olsen, D. H. (2020). Negotiating the intersections between dark tourism and pilgrimage. In D. Olsen & M. Korstanje (Eds.), *Dark tourism and pilgrimage* (pp. 1–15). CABI. doi:10.1079/9781789241877.0001

Light, D. (2017). Progress in dark tourism and thanatourism research: An uneasy relationship with heritage tourism. *Tourism Management, 61*, 275–301. doi:10.1016/j.tourman.2017.01.011

Light, D., & Ivanova, P. (2022). Thanatopsis and mortality mediation within "lightest" dark tourism. *Tourism Review, 77*(2), 622–635. doi:10.1108/TR-03-2021-0106

Lv, X., Lu, R., Xu, S., Sun, J., & Yang, Y. (2022b). Exploring visual embodiment effect in dark tourism: The influence of visual darkness on dark experience. *Tourism Management, 89*, 104438. doi:10.1016/j.tourman.2021.104438

Lv, X., Luo, H., Xu, S., Sun, J., Lu, R., & Hu, Y. (2022a). Dark tourism spectrum: Visual expression of dark experience. *Tourism Management, 93*, 104580. doi:10.1016/j.tourman.2022.104580

Martini, A., & Buda, D. M. (2020). Dark tourism and affect: Framing places of death and disaster. *Current Issues in Tourism, 23*(6), 679–692. doi:10.1080/13683500.2018.1518972

Miles, S. (2014). Battlefield sites as dark tourism attractions: An analysis of experience. *Journal of Heritage Tourism, 9*(2), 134–147. doi:10.1080/1743873X.2013.871017

Mionel, V. (2019). Dark tourism and thanatourism: Distinct tourism typologies or simple analytical tools? *Tourism: An International Interdisciplinary Journal, 67*(4), 423–437.

Mowatt, R. A., & Chancellor, C. H. (2011). Visiting death and life: Dark tourism and slave castles. *Annals of Tourism Research, 38*(4), 1410–1434. doi:10.1016/j.annals.2011.03.012

Ogretmenoglu, M., Mavric, B., & Dincer, F. I. (2022). Using a bibliometric approach to shed light on dark tourism. *Podium (São Paulo), 11*(2), 328–352. doi:10.5585/podium.v11i2.19902

Olsen, D. H., & Korstanje, M. E. (Eds.). (2019). *Dark tourism and pilgrimage*. CABI.

Ousby, I. (1990). *The Englishman's England: taste, travel and the rise of tourism*. Cambridge University Press.

Podoshen, J. S. (2013). Dark tourism motivations: Simulation, emotional contagion and topographic comparison. *Tourism Management, 35*, 263–271. doi:10.1016/j.tourman.2012.08.002

Potts, T. J. (2012). 'Dark tourism'and the 'kitschification'of 9/11. *Tourist Studies, 12*(3), 232–249. doi:10.1177/1468797612461083

Raine, R. (2013). A dark tourist spectrum. *International Journal of Culture, Tourism and Hospitality Research, 7*(3), 242–256. doi:10.1108/IJCTHR-05-2012-0037

Rajasekaram, K., Hewege, C. R., & Perera, C. R. (2022). "Tourists' experience" in dark tourism: A systematic literature review and future research directions. *Asia Pacific Journal of Tourism Research, 27*(2), 206–224. doi:10.1080/10941665.2022.2046118

Roberts, C., & Stone, P. (2014). *Dark tourism and dark heritage: Emergent themes, issues and consequences* (I. Convery, G. Corsane, & P. Davis, Eds.). Newcastle University Press.

Robinson, N. (2015). *Dark tourism motivations: an investigation into the motivations of visitors to sites associated with dark tourism* [Doctoral dissertation]. University of Salford.

Sather-Wagstaff, J. (2016). *Heritage that hurts: Tourists in the memoryscapes of September 11*. Routledge. doi:10.4324/9781315427539

Seaton, A. V. (1996). Guided by the dark: From thanatopsis to thanatourism. *International Journal of Heritage Studies, 2*(4), 234–244. doi:10.1080/13527259608722178

Seaton, T. (2018). Dark tourism history. In *The Palgrave handbook of dark tourism studies* (pp. 1–2). Palgrave Macmillan. doi:10.1057/978-1-137-47566-4

Seraphin, H., & Korstanje, M. E. (2021). *Dark tourism tribes: social capital as a variable. In Consumer Tribes in Tourism: Contemporary Perspectives on Special-Interest Tourism*. Springer-Nature.

Shekhar, S., & Valeri, M. (2022). Evolving Themes in Dark Tourism Research: A Review Study. *Tourism (Zagreb), 70*(4), 624–641. doi:10.37741/t.70.4.6

Sinclair, D. (2018). What justification is there for including the mass suicide of Jonestown as part of a Guyana dark tourism narrative in 2025? *Worldwide Hospitality and Tourism Themes, 10*(5), 592–604. doi:10.1108/WHATT-05-2018-0035

Stone, P. (2013). Dark tourism scholarship: A critical review. *International Journal of Culture, Tourism and Hospitality Research, 7*(3), 307–318. doi:10.1108/IJCTHR-06-2013-0039

Stone, P., & Sharpley, R. (2008). Consuming dark tourism: A thanatological perspective. *Annals of Tourism Research, 35*(2), 574–595. doi:10.1016/j.annals.2008.02.003

Stone, P. R. (2006). A dark tourism spectrum: Towards a typology of death and macabre related tourist sites, attractions and exhibitions. *Tourism: An International Interdisciplinary Journal, 54*(2), 145–160.

Stone, P. R. (2012). Dark tourism and significant other death: Towards a model of mortality mediation. *Annals of Tourism Research, 39*(3), 1565–1587. doi:10.1016/j.annals.2012.04.007

Strange, C., & Kempa, M. (2003). Shades of dark tourism: Alcatraz and Robben Island. *Annals of Tourism Research, 30*(2), 386–405. doi:10.1016/S0160-7383(02)00102-0

Tarifa-Fernández, J., Carmona-Moreno, E., & Sánchez-Fernández, R. (2022). An attempt to clarify what deserves to remain dark: A long look back. *Tourism and Hospitality Research*. doi:10.1177/14673584221110358

Tarlow, P. E. (2007). Dark tourism: The appealing 'dark'side of tourism and more. In M. Novelli (Ed.), *Niche tourism* (pp. 47–58). Routledge.

Tzanelli, R. (2016). *Thanatourism and cinematic representations of risk: Screening the end of tourism*. Routledge. doi:10.4324/9781315624105

Walby, K., & Piché, J. (2011). The polysemy of punishment memorialization: Dark tourism and Ontario's penal history museums. *Punishment & Society, 13*(4), 451–472. doi:10.1177/1462474511414784

White, L., & Frew, E. (2013). *Dark tourism and place identity: Managing and interpreting dark places*. Routledge. doi:10.4324/9780203134900

Wight, A. C. (2006). Philosophical and methodological praxes in dark tourism: Controversy, contention and the evolving paradigm. *Journal of Vacation Marketing, 12*(2), 119–129. doi:10.1177/1356766706062151

ADDITIONAL READING

Alexander, E. (2015). *The Dark Tourism of the Bosnian Screen* (Vol. 4). Andrews UK Limited.

Ashworth, G. J., & Isaac, R. K. (2015). Have we illuminated the dark? Shifting perspectives on 'dark' tourism. *Tourism Recreation Research, 40*(3), 316–325. do i:10.1080/02508281.2015.1075726

Cai, Y., Li, G., Liu, C., & Wen, L. (2022). Post-pandemic dark tourism in former epicenters. *Tourism Economics, 28*(1), 175–199. doi:10.1177/13548166211034639

Handayani, B., Ivanov, S. H., & Korstanje, M. E. (2017). Chapter—Smart Tourism for Dark Sites: The Sacred Site of the Dead, Trunyan Cemetery. Gazing at Death: Dark Tourism as an Emergent Horizon of Research. Nova Science Publishers, 15-42.

Handayani, B., Seraphin, H., & Korstanje, M. (2019). Dark Tourism in the Philippines Islands. In Special Interest Tourism in Southeast Asia: Emerging Research and Opportunities (pp. 23-42). IGI Global. doi:10.4018/978-1-5225-7393-7.ch002

Hartmann, R. (2014). Dark tourism, thanatourism, and dissonance in heritage tourism management: New directions in contemporary tourism research. *Journal of Heritage Tourism, 9*(2), 166–182. doi:10.1080/1743873X.2013.807266

Kennell, J., & Powell, R. (2022). Dark tourism and World Heritage Sites: a Delphi study of stakeholder perceptions of the development of dark tourism products. In *Dark Tourism Studies* (pp. 9–23). Routledge. doi:10.4324/9781003266723-2

Kidron, C. A. (2013). Being there together: Dark family tourism and the emotive experience of co-presence in the holocaust past. *Annals of Tourism Research, 41*, 175–194. doi:10.1016/j.annals.2012.12.009

Korstanje, M., & George, B. (Eds.). (2017). *Virtual traumascapes and exploring the roots of dark tourism*. IGI global.

Korstanje, M. E. (2011). Detaching the elementary forms of dark-tourism. *Anatolia*, *22*(3), 424–427. doi:10.1080/13032917.2011.620800

Korstanje, M. E., Séraphin, H., & Maingi, S. W. (Eds.). (2022). *Tourism Through Troubled Times: Challenges and Opportunities of the Tourism Industry in 21st Century*. Emerald. doi:10.1108/9781803823119

Robb, E. M. (2009). Violence and recreation: Vacationing in the realm of dark tourism. *Anthropology and Humanism*, *34*(1), 51–60. doi:10.1111/j.1548-1409.2009.01023.x

Seraphin, H., & Gowreesunkar, V. (2018). On the use of qualitative comparative analysis to identify the bright spots in dark tourism. In *Virtual Traumascapes and Exploring the Roots of Dark Tourism* (pp. 67–83). IGI Global. doi:10.4018/978-1-5225-2750-3.ch004

Seraphin, H., & Korstanje, M. E. (2021). Dark tourism tribes: social capital as a variable. *Consumer Tribes in Tourism: Contemporary Perspectives on Special-Interest Tourism*, 83-99.

Sharma, N. (2020). Dark tourism and moral disengagement in liminal spaces. *Tourism Geographies*, *22*(2), 273–297. doi:10.1080/14616688.2020.1713877

Stone, P. R. (2012). Dark tourism and significant other death: Towards a model of mortality mediation. *Annals of Tourism Research*, *39*(3), 1565–1587. doi:10.1016/j.annals.2012.04.007

Tzanelli, R., & Korstanje, M. E. (2016). Tourism in the European economic crisis: Mediatised worldmaking and new tourist imaginaries in Greece. *Tourist Studies*, *16*(3), 296–314. doi:10.1177/1468797616648542

Walter, T. (2009). Dark tourism: Mediating between the dead and the living. *The darker side of travel: The theory and practice of dark tourism*, 39-55.

Weaver, D., Tang, C., Shi, F., Huang, M. F., Burns, K., & Sheng, A. (2018). Dark tourism, emotions, and postexperience visitor effects in a sensitive geopolitical context: A Chinese case study. *Journal of Travel Research*, *57*(6), 824–838. doi:10.1177/0047287517720119

Wight, A. C. (2006). Philosophical and methodological praxes in dark tourism: Controversy, contention and the evolving paradigm. *Journal of Vacation Marketing*, *12*(2), 119–129. doi:10.1177/1356766706062151

KEW TERMS AND DEFINITIONS

Adaptation: It is defined as the capacity to recover in context of vulnerability.

Dark Tourism: It is a type of new niche tourism that fosters the visit to spaces mainly marked by disasters or mass death.

Disasters: The term refers to a state of exemption generated by chaos, violence, or a state of crisis.

Niche Tourism: It represents a new segment of tourism consumption based on what the consumers see, or their experience. The term suggests that the market is a small size.

Resiliency: It is a term coined by Victor Frankl to denote the mobilization of resources and human energy to recover from adverse situations or disasters.

Chapter 7
Dark Tourism:
A Novel Trending Sector in Tourism – A Study in the Indian Subcontinent

Oindrila Chakraborty
J.D. Birla Institute, India

ABSTRACT

The chapter delves into a relatively newer segment in tourism arena: dark tourism. Dark tourism is a phenomenon that embodies the dramatic side of society including some horrific experiences. It is a complex fusion of history and heritage, tourism, and catastrophes. It has a tonne of promises and could definitely improve the economic progress of any country. It has the most vibrant youthful tourists. Due to its own attitudes and views, the Indian subcontinent lags in recognising this gold mine and promoting it, though it has always been observed as a spiritual hub and capital of occultism, hovering around the esoteric, supernatural beliefs and practices. If the Indian subcontinent adopts the global perspective on dark tourism and develops the necessary rules, infrastructure, and defences to deal with controversies and political concerns, it will see a significant increase in its domestic and international tourism-related earnings through exploration of this vast domain of possibilities.

INTRODUCTION

Traveling is becoming increasingly popular, leading to a demand for fresh, less-commercialized places to explore. *Dark Tourism* is a niche sector that offers adventures away from safe and sound tourism (Bissell, 2009). As per Seaton in 1996, "travel to a location fully, or largely, motivated by the desire for actual or symbolic brushes with

DOI: 10.4018/978-1-6684-7242-2.ch007

death," is what the terms "*Dark Tourism*" or "*Thana Tourism*" refers to. The word "*Thanatos*" comes from ancient Greek and signifies "*death*" in English. Thanatos, according to Gerard Corsane (2005), is 'the embodiment of death, more specifically violent death.' It was described as 'the phenomenon which covers the presentation and consumption (by tourists) of genuine and commodified death and disaster sites' by Foley and Lennon in 1996. Darkness is defined as the absence of light, the existence of evil, devastation, and grief, among other things (Meriam-Webster).

The global tourism landscape has been shifting towards niche tourism or concept tourism due to the complexity and diversity of tourism, which involves a wide range of individuals searching for novel and distinctive experiences. (Seabra, Abrantes, & Karstenholz, 2014; in Fonseca, Seabra, & Silva, 2016, p. 1). The idea of enjoyable entertainment in attractive settings is evolving and widening to include new market demands that are more sophisticated and even strange (Wight, 2004). Dark tourism finds its way in this gamut of concept tourism to attract people, interested in occultism.

The idea of 'Dark Tourism' and 'Thantourism' has given researchers in this field of tourism, the chance to investigate fresh problems and expand the parameters of their field (Ashworth & Issac, 2015; in Light, 2017). In the last ten years, dark tourism (and, to a lesser extent, Thanatourism) have established themselves as mainstream research topics in tourism studies and tourist management.

LITERATURE REVIEW ON THE DARK TOURISM

Dark Tourism, also known as Thanatourism or Grief Tourism, is the act of visiting places associated with death, tragedy, or suffering. In a recent review of dark tourism research, Stone and Sharpley (2021) found that scholars have examined various aspects of dark tourism, including its motivations, impacts, and management. One area of focus in dark tourism research is the motivations behind tourists' desire to visit sites associated with death and tragedy. Stone and Sharpley (2021) noted that scholars have identified a range of motivations, including a desire for education and learning, a search for authenticity, and a fascination with death and the macabre. Additionally, researchers have found that visitors often seek to pay their respects to victims or to commemorate historical events.

Another area of focus in the literature is the impact of dark tourism on visitors and local communities. Stone and Sharpley (2021) noted while some scholars have argued that dark tourism can have positive effects, like promoting historical understanding and empathy, while others have raised concerns about potential negative impacts, like the commodification of tragedy or the exploitation of local communities.

Finally, researchers have also examined the management of dark tourism sites. Stone and Sharpley (2021) noted that effective management is critical to balancing the needs of visitors and local communities, and that strategies such as interpretation, education, and community involvement can be effective in promoting responsible tourism practices.

Overall, the literature suggests that dark tourism is a complex and multifaceted phenomenon that requires careful consideration and management. Foley and Lennon (2021) provided an overview of dark tourism, including definitions, typologies, and visitor motivations. Stone (2021) explored the relationship between dark tourism and place identity, and how visitor experiences at dark tourism sites can shape perceptions of contested places. The author drew on case studies of sites associated with genocide, war, and terrorism to illustrate the ways in which visitors' emotional and sensory experiences can influence their understanding of place. Sharpley (2020) explored the recent impact of the COVID-19 pandemic on dark tourism, and the challenges faced by destinations in managing the risks associated with Thanatourism during a public health crisis. Yankholmes and Hoque (2020) provided a critical review of the literature on the socio-cultural impacts of dark tourism. The authors analysed the ways in which dark tourism can affect host communities, including issues related to commodification, exploitation, and cultural heritage. Kastenholz; Carneiro; Marques and Oliveira (2020) explored the relationship between dark tourism and well-being, and how emotional experiences and cultural learning can contribute to positive visitor outcomes.

MAIN FOCUS OF THE CHAPTER

The focus of this chapter is to present pertinent conceptual and primary research-based findings, which will help to develop a fresh and comprehensive view of the dark tourist industry with special reference to Indian subcontinents. The study would investigate the conceptual relevance of the darker tourism with a telephonic interview and laconic consumer survey to check the preference and diverse aspects of Indian Subcontinental tourists, who are primarily interested in Dark Tourism and want to be part of it on regular basis. The chapter would primarily deal with a host of research questions:

1. What are the fundamental concepts of Dark Tourism?
2. What are the types of Dark Tourism commonly followed worldwide?
3. What is the classification of tourists in this niche domain?
4. What are the purposes of Dark Tourism?
5. What are the awareness levels and trends in Dark Tourism, especially in Indian Subcontinent?
6. What is the future of Dark Tourism as a Concept tourism?

7. What are the challenges, issues, and problems in the field of Dark Tourism?
8. What are the areas in Indian subcontinents falling under the category?
9. Is there any change in consumption pattern of dark tourism after pandemic?

DESIGN OF THE CHAPTER

The chapter has been designed with the quintessential purpose of finding the basic information, connected with the research questions. With that intention, the chapter is segregated in two types of quest for information: 1)Literature Review to elaborate the framework of fundamental concept, types, purposes, challenges and constraints of dark tourism;2) A telephonic interview and Primary data based concise survey (a telephonic conversation [an unstructured interview with the help of phone calls, WhatsApp voice call/ video call/ chat] to reveal insightful information) in phase one and an online survey in the next phase to map the awareness and related aspects of dark tourism from the interested respondents from Indian subcontinents (India, Bangladesh, Nepal, Bhutan). The survey has been done through a 'chain referral system' or a 'snowball' sampling technique to increase the engagement of the respondents. The survey was incorporated in two phases, as already highlighted: First one to check the awareness of the respondents about the concept of dark tourism (with the 254 respondents) and the next phase (with the 202 respondents), only to consider those, who are well aware about the concept (sometimes with different synonyms, even in local language or vernacular). The respondents are from various academic (under graduate, post graduate, Research Scholar) and economic backgrounds (students, service, business, professionals like lawyer, doctor, tax consultants, housewife etc.). None of the respondents is below college level (just to ensure they are not influenced by superstitions of any kind due to ignorance).

Table 1 and Table 2 are showing the cross tabulation of sample distribution of respondents for phase-1 and phase-2. Table 1(A) shows the sample profile with regards to educational qualifications of the respondents across the borders and Table 1 (B) shows the levels of awareness within that community of respondents. Table 2 Shows the sample distribution of the Respondents who know some information about Dark Tourism in Indian Subcontinents. Table 3 shows only the cross tabulation for the Knowledgeable Respondents in both the phases.

Table 1(A). To show the educational background of sample distribution in Indian subcontinents with the different awareness levels about dark tourism (Phase 1)

Serial No.	Country	Total Number of Respondents	Undergraduate	Post graduate	Research Scholar
1	India	78	31 (40%)	33 (42%)	14 (18%)
2	Nepal	59	23 (39%)	16 (27%)	20 (34%)
3	Bhutan	52	21 (40%)	26 (50%)	5 (10%)
4	Bangladesh	65	27 (42%)	30 (46%)	8 (12%)
Total		254	102 (40%)	105 (41%)	47 (18%)

Table 1(B). To show the sample distribution in Indian subcontinents with the different awareness levels about dark tourism (Phase 1)

Serial No.	Country	Total Number of Respondents	Aware of Dark Tourism	Ignorant of Dark Tourism	Level of Awareness (in Percentage)
1	India	78	70	8	90
2	Nepal	59	47	12	80
3	Bhutan	52	32	20	62
4	Bangladesh	65	53	12	82
Total		254	202	52	80

Table 2. To show the sample distribution of the respondents knowledgeable of dark tourism in Indian subcontinents (Phase 1 and Phase 2)

Serial No.	Country	Total Number of Respondents	Female	Male
1	India	70	32	38
2	Nepal	47	25	22
3	Bhutan	32	19	13
4	Bangladesh	53	35	18
Total		202	111	91

Table 3. To show the sample distribution in Indian subcontinents of the knowledgeable respondents age wise and gender wise (Phase 1 and Phase 2)

Age Group	Female(No. of Respondents)	Male(No. of Respondents)	Total
18-28	32	29	61
29-39	41	26	67
40-50	28	27	55
<50	10	9	19
Total	111	91	202

FUNDAMENTAL CONCEPTS OF DARK TOURISM

In the context of tourism, the phrase "dark" alludes to a sense of ostensibly unsettling activities and morbid goods and experiences (Stone, 2006, p. 146). An innate interest about mortality and the shadowy sides of humanity is responsible for the attention paid to incidents of death, misery, and cruelty as well as the emergence of gloomy tourism destinations (Foley, 2009; in Fonseca et al., 2016, p. 1). The curiosity in death among humans dates back to the era of pilgrimages (Titta, 2010). Death is undoubtedly one of those topics that causes uncertainty and anxiety, making it important to keep it out of popular culture and treat the subject as taboo.

Berger's key essay from 1967 claimed that death was a necessary aspect of the human experience and that people needed to learn coping skills to deal with it. As evidenced by Harrrison's (2003) logic of absorption carried out through graves, images, literature, architecture, and monuments, as well as Lee's (2002) concept of disenchantment of death, which advocates death as 'coming out of the closet to redefine our assumptions of life,' commentators are now challenging death taboos by investigating contexts where the dead share world with the living. 'We never live through our own death, but we do live through the deaths of others, and their death gives significance to our success'; as per Bauman (1990,1992) about survival amongst the idea of sharing space with the dead.

According to Seaton (1996), the connection between death and tourism, or 'Thanatourism,' grew stronger from the eighteenth century onward, although 'Dark Tourism' has only more lately come under increased academic scrutiny (Byran & Hyde, 2013; Sharpley & Stone, 2009; Stone, 2013). The social sciences are becoming more interested in death and dying-related topics, as seen by the multidisciplinary study focus known as 'Death Studies,' which is becoming more prominent (Light, 2017). The University of Central Lancashire (UK) has founded the *Institute of Dark Tourism Research* (IDTR)(Light, 2017). Several nations have attempted to include black tourism as a product into their tourism industries when it was recognised as a phenomenon (Blom, 2000). Stone (2013) provided a critical review of existing literature on dark tourism, examining the key themes, debates, and gaps in the field. According to Stone (2013), the political movement of West Belfast has played a significant role in the development of dark tourism. Stone argued that the development of dark tourism in West Belfast is closely linked to the political movement and the struggle for national liberation. Overall, it highlights the importance of understanding the political and cultural context in which dark tourism develops and links to the promotion of a community's identity and heritage. Skinner (2016) examined the significance of movement and walking in the context of dark tourism in West Belfast. He argued that the physical experience of walking through these spaces is a key aspect of the dark tourism experience. Carrigan (2014) explored the intersections between dark tourism and postcolonial studies and argued that dark tourism could be seen as a form of Neocolonialism. Korstanje (2016) developed his concept of thana-capitalism and explored its implications for tourism, arguing that thana-capitalism is a response to the perceived existential threat of modernity. Korstanje (2020) further focused on the consumption of dark spaces in the periphery, arguing that dark tourism can be seen as a response to the perceived threats of globalization and modernity. He also investigated the concept of "Thana-capitalism," which he saw as a driving force behind the growth of dark tourism. Tzanelli (2016) explored the representation of risk in Thanatourism through cinema. She argued that these representations can help to shape our perceptions of risk and safety in tourism. Sather-Wagstaff (2017) explored the heritage of death in relation to dark tourism in the book "*Heritage of Death*". Dark tourism is an important part of a nation's income and reputation, so new places, attractions, and exhibitions are being opened to accommodate those who wish to observe genuine or reproduced death (Sharpley & Stone, 2009).

Dark tourism is known by a variety of names, including *'Thanatourism'*(Dann, 1994 ; Seaton 1996*), 'Milkingthe Macabre'*, *'the Dark Side of Tourism'*(Dann, 1998; in Dann& Seaton, 2001), *'Tragedy Tourism'*, and *'Mea Culpa Tourism'*(Richter, 1999 ; in Dann & Seaton, 2001), *'Black-Spot Tourism'* (Rojek 1993), *'Morbid Tourism'* (Blom, 2000), *'GriefTourism'* (O' Neill,2002), *'Atrocity Tourism'*(Ashworth,2002), *'Phoenix Tourism'* (Causevic & Lynch,2011), *'Natural Disaster Tourism'* (Miller,2008 ; in Ashworth & Isaac, 2015), *'Disaster Tourism'*, *'Hot-Spot Tourism'* (Pelton, 2003;in Ashworth and Isaac, 2015), *'Holocaust Tourism', 'Prison Tourism'* (Strange & Kempa, 2003; in Fonseca, Seabra& Silva, 2016), *'Cemetery Tourism'* (Abranja,2012; in Fonseca, Seabra & Silva, 2016), *'Ghost Tourism'*(Davies, 2007; in Fonseca, Seabra & Silva, 2016), *'Battlefield Tourism'* (Henderson, 1997; in Seaton, 2012), *'Conflict Tourism'* (Ryan & Kohli, 2006). It is first and foremost vital to mention how dark tourism as a significant commodity came into existence in order to get better grasp of it. They then clarified their claim that black tourism is a sign of postmodernity (Lennon & Foley, 2000). Time-space compression and gloomy tourism destinations undermine modernity's development, logic, and order, leading to time-space compression.

The emerging field of dark tourism often uses two fundamental bases for analysis. Seaton investigates 'Dark Tourism' or 'Thanatourism' as a behavioural phenomenon, testifying to the presence of the individuals. Thus, thanatourism is a type of tourism consumption according to Seaton (2004). Seaton and Lennon (2004; in Farmaki; 2013), identified two primary drivers:

Schadenfreude: It is known as the enjoyment of others' suffering.

Thanatopsis: It is the contemplation of death.

BACKGROUND OF DARK TOURISM

Dark tourism is not a recent phenomenon; in fact, it may be one of tourism's oldest subtypes (Bissell, 2009). Violence and death have always been popular forms of entertainment (Seaton & Lennon, 2004). According to Seaton's definition from 1996 the amphitheatres in ancient Rome were a well-known site for dark tourism. Gladiators amused the crowd by engaging in combat, where they often suffered grave injuries or brutal deaths. Venatio, the hunting and killing of wild animals, was practised along with the execution of criminals (Dunkle, 2008). Ancient pilgrimages, according to Lennon and Foley (2002), can be considered the earliest example of dark tourism because many pilgrimage locations were primarily associated with the misery or demise of individuals or groups. Dark tourism was therefore practised before "modern" tourism emerged (Seaton & Lennon, 2004).

TYPES OF DARK TOURISM

There are only five conceivable categories of dark trip activity, according to Seaton (1996):

1. to see public death re-enactments; 2. to visit the scene of a single death or a number of deaths; 3. to visit memorial or interment sites; 4. to see symbolic death representations; and 5. to see public death re-enactments.

There are several subcategories of dark tourism, each of which describes a particular type of travel within the main topic. Examples include War Tourism, Grief Tourism, Slum Tourism, Suicide Tourism, and Doomsday Tourism (Kendle, 2008). According to Cooke and Dickson (2006), even the traveller themselves might be divided into many types of dark tourists. The authors provided a classification for the dark tourist based on five categories:

- Opaque (the degree of morbidity attached to the experience/ memorial/play is not extensively dark in nature).
- Dark (the degree of morbidity attached to the experience/ memorial/play is dark in true sense).
- Die-hard Dark (perfect for the category of dark tourism; especially preferred for people who are in absolute love with this category).
- Pitch Black (suitable for the stone-hearted people, who can tolerate gruesome, macabre and ghastly situation/experience/play).
- Too Dark (these are the experiential grounds, where the situation is so hideous that it's mostly avoidable and restricted by the Government in many places).

For each category, Cooke and Dickinson (2006) identified a location that would be ideal. For instance, they encourage the Rwandan Genocide Memorials for 'Die-hard Dark Travellers' while recommending the Auschwitz concentration camp for 'Pitch Black Travellers'. The authors classified 'Too Dark Travellers' as visiting actual disaster regions that have not yet fully recovered and are not prepared to welcome any tourists.

According to Sharpley (2005), Dark Tourism could be classified in four ways as per the consumption pattern:

1. Dark Tourism as Experience (By visiting horrific places, which have been ornamented with petrifying stories for years. For examples, Cemeteries, Heritage Buildings, Memorials, Paler Fantasy, Battlefields and other War-Museums etc. Many a times 'Ghost Walks' come under this category).
2. Dark Tourism as Play (Consumption as play emphasises on the shared, community consumption of certain tourist attractions or experiences, representing "paler" sensations. A person's or a group's death serves as the catalyst for collective celebration, commemoration, or sadness. So, dark travel becomes a pilgrimage, or a journey that is followed by a sense of 'Communitas'. Many a times a Planchette with Ouija board depicts this category).
3. Dark Tourism as Integration (Dark tourism can be divided into two types: integration into the object of consumption, which involves a fascination with the larger context in which death occurs, and integration into the object of

consumption, which involves seeing sudden or violent deaths. The most sinister type of tourism is when visitors want to become one with death, either by seeing sudden or violent deaths or by travelling while expecting or knowing that they will die.)

4. Dark Tourism as Classification (In the context of dark tourism, risky travel activities can be used to gain status, but not always for death).

In this process, Sharpley (2009) also depicted Four Shades of Dark Tourism; which are comparable with the five categories of dark tourism: Black Tourism, which describes an entirely gloomy experience. Pale Tourism, which alludes to a lack of interest in passing away. Desire for Grey Tourism, which refers to fascination-driven travel to mysterious locations. The term Grey Tourism Supply describes locations created to profit on the passing of people by drawing visitors who have little genuine interest in the locations.

Some examples include Stone's (2006) proposed spectrum, which ranges from lightest to darkest sites, and Miles' (2002) distinction between dark, darker, and darkest sites. However, this strategy has come under fire for arbitrarily combining remarkably disparate visitor experiences, which some claim will lead to a growing dilution and fuzzying of the concept of dark tourism (Sharpley, 2009).

CLASSIFICATION OF DARK TOURISTS

According to Richards (2015; in Yousaf, Amin, & Santos, 2018), there are three main categories of Travellers: *Flashpackers*, *Backpackers*, and *Global Nomads*. Global nomads are inclined to seek out interactions with locals and experience daily life in the place they are visiting, sleeping in locals' houses and using local communication networks in addition to the Internet (Richards, 2015, p. 148; in Yousaf, Amin, & Santos, 2018). Lastly, according to Richards (2015), the Flashpacker is the most well-connected traveller using social media frequently and being more likely to mix and blur work and leisure. Backpackers are driven by a desire to travel independently and create new ties with other backpackers from home countries.

Black metal fans that travel to experience live music and festival atmospheres are referred to as Black packers in the study of dark tourism (Podoshen, 2013). These travellers also go to places where black metal musicians and fans have been violently attacked.

According to Krippendorf (1987; in Rucinska, 2016, p. 1459), the traveller has a variety of characteristics that make it difficult to categorize their travel behaviour. Swarbrooke (1995; in Blom, 2000) categorises tourists according to the attraction. Natural places, artefacts, amenities, and arrangements are the four attraction typologies he mentions. Adventurers are those who seek out new activities, cultures, and people, according to Burns and Holden's (1995; in Blom, 2000) classification of tourists. (Blom, 2000) continues by mentioning the following: people who crave adventure and the uncharted and terrifying.

PURPOSES OF DARK TOURISM-DEFINING THE DIMENSIONS

There are several purposes and dimensions of Dark Tourism; which have been discussed below briefly to highlight its growing popularity in public:

1. Commoditized Objective

Dark tourism, as Tarlow (2005) reminds us, has an economic component that transforms yesterday's atrocities into today's profits; as a result, souvenir shops might be found in tragic locations, replacing each person's memory with a collective memory that attracts customers.

There are three main dimensions to commodity: whether or not it is appropriate to permit any kind of money transactions, who should profit if some commercialization is permitted, and how commercialization should be accomplished without having a negative impact on the location (Seaton, 2009).

2. Remembrance of the Dark Places and the Deceased ones

Memory is not remembrance (King, 1998; in Walter, 2009). Remembrance means remembering those whose pain and passing were not yet history but that one may not have directly observed. When memory is not first-hand, it transforms into recollection, history, genealogy, or ancestry, among other likely outcomes (Walter, 2009).

3. Niche Tourism or Concept Tourism

The motivation of a *Special Interest Tourism*, *Concept Tourism* or *Niche Tourism* is to indulge in or discover a new interest in a strange or familiar place. Special interest tourism can either be the main emphasis of a vacation or just a method to fill one or two days. Dark tourism is also acknowledged as a niche type of travel that attracts travellers interested in seeing places or attractions linked to previous tragedies, atrocities, or other dark aspects of human nature (Asworth & Hartmann, 2005; Lennon & Foley, 2000; Kuznik, 2015). Military history and visiting battle sites are two examples of topics that intrigue people (Titta, 2010).

4. Edutainment

The concept of 'Edutainment' is to educate a set of discrete audience while entertaining them. It is possible to come across the dead for educational purposes with a dash of entertainment, such as through literature in the classroom or educational tourism at historical locations. In some cases, the deceased are visibly present, such as in mummy and bog body exhibits. The display is edutainment, much like all widely-attended exhibitions, museums, and historical places, as well as analogous television documentaries. The dead are utilised to instruct and amuse the masses today, just as they were in the public executions of the 18th century, along with many other remnants of the past (Walter, 2009). which Roberts (2018) and Dale and Robinson (2011) dubbed '*Dartainment*,' a term that describes both sinister attractions that make an effort to amuse (Dale &Robinson, 2011).

5.Authenticity of the Dark Theme

The authenticity of the Dark places, is often attached to the uniqueness of the places and the local tourism attached to its branding and marketing. It makes the place or the experience of it so special, that it is almost impossible to replicate the essence of it through a different emulator.

6.Interpretation of the Visitors

The visiting experience at attractions must include interpretation (Moscardo & Ballantyne, 2008, p. 237; in Sharpley& Stone, 2009). The process by which a location, an event, history, a building, a collection of artefacts, or more generally what may be referred to as heritage is granted significance and is then conveyed to the visitor by one means or another, it provided the link between an attraction and its visitors. On the one hand, effective interpretation may improve the visitor experience and satiate the need for understanding and meaning. Interpretation plays a key role in this process, acting as a filter to emotional responses to a dark site or attraction.

TRENDING ISSUES, PROBLEMS, AND CHALLENGES IN DARK TOURISM

Dark Tourism is a controversial issue in the tourism sector due to its exploitation of the suffering of others. Tour operators may use the suffering of people who have experienced tragic events as a marketing tool to attract tourists, which can be seen as insensitive and disrespectful to the victims and their families. Additionally, the influx of tourists can disrupt the lives of locals and make it difficult for them to heal from past traumas, particularly in areas that are still dealing with the aftermath of a tragedy, such as war-torn countries or communities affected by natural disasters. Furthermore, there is also the risk of commodifying death and tragedy. Some argue that turning sites of tragedy into tourist attractions can trivialize the events that occurred and reduce them to a mere spectacle. This can be seen as a form of voyeurism and can lead to a lack of respect for the victims and their families.

Dark tourism can be a way to learn about history and pay respects to victims, but it should be approached with sensitivity and respect. It is important to consider the impact of tourism on the local community and to ensure that victims and their families are not exploited for profit. There is a lot of discussion over whether it is moral to construct, advertise, or provide dark sites for tourist consumption, such as the *'Ground Zero'* viewing platform that allows voyeuristic tourists to stand next to those, who are grieving the death of loved ones (Lisle, 2004). The rights of persons, whose demise is commercialised through 'dark tourism' are a significant ethical issue that merits attention. From an ethical standpoint, it is important to consider whether trips taken in the name of social justice or global awareness are genuinely experienced as such, or whether they may instead serve to conceal the voyeuristic, recreational appeal of violence. Dark tourism can occasionally result in 'recreational grief' (West, 2004), a type of grief in which lamenting the demise or suffering of others turns into a pleasurable activity. In terms of commercial aspects, more people are eager to market or make money from 'dark' events as tourist attractions. Additionally, there is evidence that visitors are more inclined to attend dark attractions.

From literature point of view, it is quite controversial. Sharpley (2005) correctly observes that '*dark tourism literature*' remains diverse and theoretically fragile, posing more problems than its answers, despite the fact that dark tourism research stimulates significant discourses about the intersections of travelling and mortality. The dark tourism trend, according to Keil (2005: 481, in Bowman & Pezzullo, 2009), 'has so far been insufficiently characterised, and lacks a theoretical link to wider studies of violence and by-standing.' However, many of those who have condemned the phrase thus far, have made matters worse by attempting to define several 'shades' of or umbrella categories for 'darkness' without questioning the term itself (e.g. Miles, 2002; Strange & Kempa, 2003; Sharpley, 2005; Stone, 2006; Bowman &Pezzullo, 2009).

TRENDS OF DARK TOURISM IN INDIAN SUBCONTINENTS

Through the telephonic interview and online survey, it has been evident that the dark tourism is well accepted in the Indian subcontinents (80%) due to the practice of occultism and spirituality since the time immemorial. Black magic related rituals, divination, use of talisman have been always there, across all Indian subcontinents for eons. This domain of countries are well aware of the terms like 'Yoga, 'Tantra', 'Pretatma', 'PretYoni', 'Jinn','Pari', 'Huri',Sihr',etc. due to the presence of long ingrained culture of receiving death as natural course of action. The people of Indian subcontinents accept death as inevitable yet peaceful and not a dark, morbid Concept. Hence, death related discussions, after life, sorcery related to supernatural forces and divination are part of society, not only in the subaltern strata, but also in the enlightened part of society, though much subdued. It is in an underlying form with less intensities and covered under a finesse; lost its crude forms and edges. Unlike many of Western countries, the concept is not masked with darkness and negativity. That is why many a times respondents could not recognise the term 'Dark Tourism', unless mentioned about 'Ghost Walks', 'Planchette', 'Seance', 'Cemetery Walks', 'Voluntary visits to haunted places', 'Visits to controversial Monuments, Relics, and dilapidated building in organised manner'. Most of the respondents could immediately relate and expressed their intentions.

The interest level has been varied. Four types of Dark Tourism categories have been popularized in the Indian subcontinents, as per the Phase-1 and Phase-2,telephonic interview and structured survey:

1. Ghost Walks(guided tour to visit many infamous haunted places), 2.Planchette (Seance around a table to meet a departed soul by inviting or evoking), 3. Visit to Haunted Memorial(visiting one/ many notorious haunted places/ memorial/ dilapidated building/ cemetery on his/her own without any compulsion of completing an entire assisted tour like ghost walk), 4. Theme Based Horror Enactment (Horror show, light and sound to project a tragic incident).

From the chronicle of choices, respondents have shown maximum inclination towards '*haunted memorial visit without any guided tour*' because of utmost thrill factor, followed by *'Planchette'* and *'Ghost walk'*. But the situation is quite grave in reality. The respondents were questioned on their experiences with the given choices, and they blamed it on the lack of destination marketing and advertising by the organisers, local tourism institutions, and tour organisers. They also stated a lack of information on such guided tours despite queries, and government assisted tour organisers are rarely interested in conducting such tours. This is in contrast to the western culture, where government assisted tour organisers are often interested in conducting such tours.

Table 4 and Table 5 show the varied interest levels of various age group of the respondents and their actual experience in real life.

Table 4. The interest level of varied age groups

Age Group	Female (No. of Respondents)	Male (No. of Respondents)	Total	Interested in Ghost Walk	Interested in Planchette	Interested in Haunted Memorial Visit	Interested in Theme Based Horror Enactment
18-28	32	29	61	54(89%)	42 (69%)	61	48
29-39	41	26	67	62 (93%)	26 (39%)	65	50
40-50	28	27	55	40 (73%)	12 (21%)	50	23
<50	10	9	19	11 (58%)	4 (21%)	15	3
Total	111	91	202	167 (83%)	84 (42%)	191 (95%)	124 (61%)

Table 5 (a). Showing the interest versus real-life experience of the respondents

Age Group	Female (No. of Respondents)	Male (No. of Respondents)	Total	Interested in Ghost Walk	Experienced Ghost Walk
18-28	32	29	61	54	5
29-39	41	26	67	62	2
40-50	28	27	55	40	0
<50	10	9	19	11	0
Total	111	91	202	167	7

Table 5 (b). Showing the interest versus real-life experience of the respondents

Interested in Planchette	Experienced Planchette	Interested in Haunted Memorial Visit	Experienced Haunted Memorial Visit	Interested in Theme Based Horror Enactment	Experienced Theme Based Horror Enactment
42	0	61	3	48	2
26	0	65	4	50	3
12	0	50	2	23	0
4	0	15	0	3	0
84	0	191	9	124	5

EXAMPLES OF DARK TOURIST-SPOTS IN THE INDIAN SUBCONTINENTS

In the Indian subcontinents, there are several places that can attract dark tourism due to their historical significance in terms of violence, natural disasters, and political turmoil.

Recent literatures (Stone and Sharpley, 2021) have highlighted several emerging trends in dark tourism, including the rise of virtual and digital forms of dark tourism along with the on-going 'Light and Sound Effect' Programmes to give a real-life experience to the tourists. During recent pandemic virtual tours, such as videos, virtual reality, and interactive websites, have been organised in India for urban, city-centric virtual tourists, but without much robustness and vigour.

The Virtual Tours of Dark Spots

The Taj Mahal: The Taj Mahal is one of the most famous tourist attractions in India. During the pandemic, the Indian government launched a virtual tour of the Taj Mahal, allowing visitors to experience the beauty of the monument from the safety of their homes. The virtual tour of the Taj Mahal includes a 360-degree view of the monument, with narration and music to enhance the experience. The virtual tour also provides historical information and fun facts about the Taj Mahal.

The Jallianwala Bagh Massacre Site: Jallianwala Bagh is a public garden in Amritsar, Punjab, where hundreds of unarmed Indians were killed. The site has now been turned into a memorial and museum. During the pandemic, the museum launched a virtual tour, allowing visitors to explore the site and learn about the history of the massacre. The virtual tour of the site includes a walkthrough of the garden, with narration and visuals depicting the events leading up to the massacre.

The virtual tour also includes a museum with exhibits and information about the freedom struggle in India.

The Chhota Imambara: The Chhota Imambara is a Shia Muslim shrine in Lucknow, Uttar Pradesh. During the pandemic, the Uttar Pradesh government launched a virtual tour of the shrine, allowing visitors to explore the ornate architecture and learn about the history of the site.

The Cellular Jail: The Cellular Jail is a former British prison in the Andaman and Nicobar Islands, where Indian political prisoners were held during the freedom struggle. The prison has now been turned into a national memorial. During the pandemic, the government launched a virtual tour of the jail, allowing visitors to explore the cells and learn about the history of the prison. The virtual tour of the jail includes a 360-degree view of the cells, with narration and information about the prisoners who were held there. The virtual tour also includes exhibits and historical information about the freedom struggle in India.

The Dumas Beach: The Dumas Beach is a popular tourist destination in Gujarat, known for its black sand and eerie atmosphere. The beach is believed to be haunted, with many stories of paranormal activity. During the pandemic, several videos and documentaries were released, exploring the legends and myths associated with the beach. The virtual tour of the beach includes videos and documentaries about the legends and myths associated with the beach. The virtual tour also includes information about the history and culture of Gujarat.

However, here are some examples of dark tourist sites with great potential to attract tourists and worth serious contemplation during pre and post pandemic. Many of these tourist spots have 'Light and Sound Effect' to reflect the experience of the tragic incidents happened there.

Dark Tourist Spots

India

The Taj Mahal: While the Taj Mahal is primarily known as a monument of love, it is also a place of death. The Taj Mahal was built by Mughal Emperor Shah Jahan in memory of his beloved wife Mumtaz Mahal, who died during childbirth.

Chambal River: The Chambal River, which flows through Madhya Pradesh, Uttar Pradesh, and Rajasthan, is known for its association with dacoits or bandits. Many tourists visit the area to explore the history of dacoits and learn about the infamous Chambal dacoits, who were known for their Robin Hood-like actions.

Chittorgarh Fort, Rajasthan: This fort witnessed one of the most tragic events in Indian history, the jauhar of Rani Padmini and the other women of the fort to avoid being captured by the invading Sultan of Delhi, Alauddin Khilji.

Jallianwala Bagh, Amritsar: This public garden is the site of the Jallianwala Bagh massacre, which occurred on 13 April 1919 when British troops fired on a peaceful gathering of Indian protesters, killing hundreds of people. The site is now a memorial to the victims of the tragedy and attracts many visitors every year.

Dumas Beach, Gujarat: This beach is said to be haunted, and many people have reported experiencing supernatural phenomena and ghostly apparitions.

Cellular Jail, Andaman and Nicobar Islands: This colonial-era prison was used by the British to incarcerate Indian political prisoners during the Indian independence movement. The prison is known for its brutal conditions and many inmates died during their confinement. The jail is now a national memorial and is open to visitors.

Tsomgo Lake, Sikkim: This beautiful lake is located at an altitude of over 12,000 feet, and it is believed to be inhabited by spirits who can cause sickness and accidents to those who do not respect them.

Bhangarh Fort, Rajasthan: This abandoned fort is said to be cursed, and it is believed that anyone who stays there after dark will not return alive.

Kedarnath, Uttarakhand: This pilgrimage site in the Himalayas was devastated by a massive flood in 2013, which killed thousands of people and caused widespread damage. Despite the tragedy, the site remains an important destination for Hindu pilgrims and tourists.

Tsunami-Affected Areas, Tamil Nadu: The 2004 Indian Ocean earthquake and tsunami had a devastating impact on the coast of Tamil Nadu, killing thousands of people and destroying entire communities. Many of these areas have since been rebuilt, and some have become popular tourist destinations, but they also serve as a reminder of the tragedy that occurred.

Bhopal Gas Tragedy Site: Bhopal Gas Tragedy is one of the world's worst industrial disasters that occurred in 1984. Many tourists visit Bhopal to learn about the incident, visit the site of the disaster.

Bangladesh

Liberation War Museum, Dhaka: This museum is dedicated to the Bangladesh Liberation War of 1971, which led to the independence of Bangladesh from Pakistan. Visitors can see photographs, artifacts, and exhibits that tell the story of the war and the sacrifices made by the people of Bangladesh.

Rayer Bazaar Killing Field Memorial, Dhaka: This memorial commemorates the victims of a massacre that took place during the Bangladesh Liberation War. Visitors can see the mass graves and pay their respects to the victims.

Sonargaon, Narayanganj: This historic city was the capital of the Bengal Sultanate in the 14th century, and it was later ruled by the Mughal Empire. Visitors can see the ruins of the city, including the palace, the mosque, and the tombs of the rulers.

Ahsan Manzil, Dhaka: This palace was built by a wealthy merchant in the 19th century, and it was later used as the official residence of the Nawab of Dhaka. Visitors can see the palace and learn about the history of the city, its rulers and their tragic past.

Pakistan

Mohenjo-Daro, Sindh: This ancient city is believed to be one of the earliest urban settlements in the world, and it was abandoned around 1900 BCE. Visitors can see the remains of the city, including the Great Bath and the granary.

Taxila, Punjab: This archaeological site was a center of learning in ancient times, but it was also the site of battles between Alexander the Great and the local ruler. Visitors can see the remains of the city, including the Buddhist monasteries and stupas.

Khyber Pass, Khyber Pakhtunkhwa: This mountain pass has been a strategic point for military campaigns throughout history, and it has been the site of many battles. Visitors can see the forts and watchtowers that were built along the pass, as well as the markets where traders once traded goods.

Badshahi Mosque, Lahore: This mosque is one of the largest in the world, and it was built by the Mughal emperor Aurangzeb in the 17th century. It was also the site of the public flogging of political dissidents during the British colonial period.

Karachi Port Trust Cemetery, Karachi: This cemetery contains the graves of people who died during the British colonial period, including soldiers, sailors, and civilians.

Lal Shahbaz Qalandar Shrine, Sindh: This Sufi shrine is the site of a bombing that killed over 80 people in 2017. Visitors can see the shrine and pay their respects to the saint who is buried there.

Nepal

Pashupatinath Temple, Kathmandu: This Hindu temple is one of the holiest in Nepal, and it is the site of cremation ceremonies for Hindus. Visitors can see the cremation ghats and the ceremonies that take place there.

Swayambhunath Stupa, Kathmandu: This Buddhist stupa is also known as the Monkey Temple, and it is a popular tourist attraction. However, it was also the site of a massacre in 2001, when a member of the royal family opened fire on a crowd of people, killing several.

Mount Everest: This mountain is the highest in the world, and it attracts thousands of climbers and trekkers every year. However, it is also the site of many accidents and fatalities, and visitors should be aware of the risks associated with climbing and trekking in the region.

Bhaktapur Durbar Square, Bhaktapur: This historic square was damaged in the 2015 earthquake, which killed over 8,000 people in Nepal. Visitors can see the damaged buildings and learn about the history and culture of the area.

Bhutan

Bhutan is a small country with a rich cultural and natural heritage. While there are no significant dark tourism sites in Bhutan, there are some places of historical and cultural significance that visitors may find interesting:

Tashichho Dzong, Thimphu: This fortress-monastery is the seat of the Bhutanese government and the summer residence of the chief abbot. Visitors can see the intricate architecture and decoration of the building.

Wangdue Phodrang Dzong, Wangdue Phodrang: This fortress-monastery was destroyed by a fire in 2012, but it is being rebuilt with the help of local communities and international donors. Visitors can see the ongoing efforts to restore the building to its former glory.

There are several horrors theme-based Entertainment programmes, parks, tours in Indian subcontinents without much of noise creation in the marketplace and constantly failing to grab the attention of the target audience.

Ramoji Film City Horror House: Located in Hyderabad, this is a popular attraction at the Ramoji Film City. The Horror House is designed to give visitors a thrilling and spooky experience with special effects, creepy music, and live actors.

Scream Park, Mumbai: Scream Park is an indoor amusement park in Mumbai that features several horror-themed attractions, including a haunted house, a 5D theatre, and a zombie-themed laser tag arena.

Haunted Kuldhara Village Tour, Rajasthan: The abandoned village of Kuldhara in Rajasthan is said to be haunted by the ghosts of its former inhabitants. Several tour operators offer guided tours of the village at night, complete with spooky stories and legends.

Ghost Walk in Kolkata: This is a guided walking tour of the most haunted places in Kolkata, including cemeteries, haunted mansions, and old burial grounds. The tour is led by local guides who share eerie stories and legends about each location.

Fear Factor: Della Adventure, Lonavala: Della Adventure in Lonavala is an adventure park that features a horror-themed attraction called "Fear Factor." The attraction includes a haunted house, a horror maze, and a zombie apocalypse zone.

CONCLUSION

The Indian subcontinents were always in sync with the knowledge of occultism and dark theme fascination. The famous Tagore family of Bengal (the family of Nobel laureate, Bengali Polymath, social reformer, philosopher and notable 'World Poet' Rabindranath Tagore) used to practise Planchette. There are several other instances of practice of Planchette in India. One notable example is the case of the Mysore Royal Family, who were known to have practised Planchette in the late 1800s.

The Maharaja of Mysore, Chamaraja Wodeyar, and his family were said to have been fascinated by the occult and experimented with various forms of divination, including Planchette. Another example is the case of Kali Kumar, a Bengali mystic who claimed to communicate with spirits through Planchette in the early 1900s. Kali Kumar gained a reputation for his supernatural abilities and became a popular figure in Bengali society at the time. There are several other examples of individuals and groups practising Planchette throughout India during 19th and 20th centuries, including spiritualists, occultists, and members of various secret societies. Sri Aurobindo and the mother (Mirra Alfassa) were spiritual teachers and practitioners who were involved in the Indian freedom movement in the early 20th century. They also experimented with Planchette and automatic writing as a means of communicating with spiritual entities. Swami Vivekananda, a prominent Indian philosopher and spiritual leader in the late 19th and early 20th centuries, was known to have experimented with Planchette as a means of accessing spiritual knowledge and guidance.

Annie Besant was a prominent British social reformer and theosophist who lived in India during the late 19th and early 20th centuries. She was known to have practised Planchette as a means of communicating with spiritual entities and accessing esoteric knowledge. Ramakrishna Paramhansam, the great Guru (the Great Mentor of the Swami Vivekananda) was a famous Indian mystic and spiritual teacher who lived in the 19th century. He was known to have practised various forms of divination, including Planchette, as a means of accessing spiritual guidance and insights. Overall, Planchette was a popular form of divination and spiritual practice in India during the late 19th and early 20th centuries, and was embraced by many prominent figures in Indian society and spirituality.

As evident, in Indian subcontinents and in its culture, though the Dark Tourism concept has been well accepted and internalised, but in practice it is not furnished to a great extent due to lesser marketing and branding of the dark tourism industry in the contemporary society. The lack of initiatives in an organised manner has run down the worth and uniqueness of the sector, forgetting its glorious past. In recent years, people are more interested in 'grey tourism' rather than 'Opaque' type of the industry, which should have been explored more. Unfortunately, most of the governments of neighbouring countries of Indian subcontinents and India are oblivious to the need of consumerism in this particular sector and did not pay heed to the attention-grabbing activities aka promotional efforts, due to the apprehension of the associated negativity and ethical dilemma of the sector.

FUTURE RESEARCH DIRECTIONS

The chapter is envisaged mainly based on secondary information, barring a diminutive consumer survey to project the viewpoint of the people from Indian subcontinents. However, the insignificant sample size does not draw a statistically significant conclusion and could only be useful to project the trend and current status of dark tourism in developing countries.

REFERENCES

Ashworth, G., & Hartmann, R. (Eds.). (2005). *Horror and Human Tragedy Revisited: The Management of Sites of Atrocities for Tourism*. Cognizant.

Ashworth, G. J., & Isaac, R. K. (2015). Have we illuminated the dark? Shifting perspectives on dark tourism. *Tourism Recreation Research*, *40*(3), 1–17. doi:10.1080/02508281.2015.1075726

Bauman, Z. (1990). *Privatization of Ambivalence, Modernity and Ambivalence*. Polity Press.

Bauman, Z. (1992). Survival as a Social Construct. *Theory, Culture & Society*, *9*(1), 1–36. doi:10.1177/026327692009001002

Berger, P., & Luckmann, T. (1967). *The Social Construction of Reality: A Treatise in the Sociology of Knowledge*. Doubleday.

Biran, A., Poria, Y., & Oren, G. (2011). Sought experiences at (dark) heritage sites. *Annals of Tourism Research*, *38*(3), 820–841. doi:10.1016/j.annals.2010.12.001

Bird, G., Westcott, M., & Thiesen, N. (2018). Marketing Dark Heritage: Building Brands, Myth-Making And Social Marketing. In The Palgrave Handbook of Dark Tourism. Palgrave Macmillan.

Bissell, L. (2009). *Understanding Motivation and Perception at Two Dark Tourism Attractions in Winnipeg, MB* [Master's Thesis]. Retrieved from https://mspace.lib.umanitoba.ca/bitstream/handle/1993/29742/Bissell_Understanding_motivation.pdf?sequence=1

Blom, T. (2000). Morbid tourism - A Postmodern Market Niche with an Example from Althorp. *NorskGeografiskTidsskrift - Norsk Geografisk Tidsskrift, 54*(1), 29–36. doi:10.1080/002919500423564

Bowman, M. S., & Pezzullo, P. C. (2009). What's so Dark about Dark Tourism?: Death, Tours, and Performance. *Tourist Studies, 9*(3), 187–202. doi:10.1177/1468797610382699

Byran, A., & Hyde, K. F. (2013). Guest Editorial: New Perspectives On Dark Tourism. *International Journal of Culture, Tourism and Hospitality Research, 7*(3), 191–198. doi:10.1108/IJCTHR-05-2013-0032

Carrigan, A. (2014). Dark tourism and postcolonial studies: Critical intersections. *Postcolonial Studies, 17*(3), 236–250. doi:10.1080/13688790.2014.993425

Causevic, S., & Lynch, P. (2011). Phoenix Tourism: Post-Conflict Tourism Role. *Annals of Tourism Research, 38*(3), 780–800. doi:10.1016/j.annals.2010.12.004

Cohen, E. (2018). Thanatourism: A comparative approach. In The Palgrave Handbook of Dark Tourism Studies. Palgrave Macmillan. doi:10.1057/978-1-137-47566-4_6

Cohen, E., & Cohen, S. A. (2015). Beyond Eurocentrism in Tourism: A Paradigm Shift to Mobilities. *Tourism Recreation Research, 40*(2), 157–168. doi:10.1080/02508281.2015.1039331

Cooke, J., & Dickson, H. (2006). *Lonely Planet Bluelist: The Best in Travel 2007*. Lonely Planet Publications.

Corsane, G. (2005). *Heritage, Museums and Galleries: An Introductory Reader*. Routledge.

Dale, C., & Robinson, N. (2011). Dark Tourism. In P. Robinson, S. Heitmann, & P. Dieke (Eds.), *Research Themes for Tourism* (pp. 205–217). CABI. doi:10.1079/9781845936846.0205

Dann, G. (1994). Tourism: The Nostalgia Industry of the Future. In W. Theobald (Ed.), *Global Tourism: The Next Decade* (pp. 55–67). Butterworth Heinemann.

Dann, G. (1998). The Dark Side of Tourism. Centre International de Recherches et d'Etudes Touristiques.

Dann, G., & Seaton, A. (Eds.). (2001). Slavery, Contested Heritage and Thanatourism. Haworth Hospitality Press.

Dunkle, R. (2008). *Gladiators Violence and Spectacle in Ancient Rome*. Pearson Education Limited.

Foley, M., & Lennon, J. (1996). Editorial: Heart of darkness. *International Journal of Heritage Studies*, *2*(4), 195–197. doi:10.1080/13527259608722174

Foley, M., & Lennon, J. (2021). Dark Tourism: Understanding visitor motivations and site interpretation. In Tourism Management: Theory and Practice. CABI Publishing.

Fonseca, A. P., Seabra, C., & Silva, C. (2016). Dark Tourism: Concepts, Typologies And Sites. *Journal of Research Hospital*, *2*, 1–6.

Fridrich, M., Stone, P. R., & Rukesha, P. (2018). Dark Tourism, Difficult Heritage, and Memorialization: A Case of the Rwandan Genocide. In The Palgrave Handbook of Dark Tourism Studies. Palgrave Macmillan.

Harrison, D. (2003). *The Sociology of Modernization and Development*. Routledge. doi:10.4324/9780203359587

Henderson, J. (2000). War as a Tourist Attraction: The Case of Vietnam. *International Journal of Tourism Research*, *2*(4), 269–280. doi:10.1002/1522-1970(200007/08)2:4<269::AID-JTR219>3.0.CO;2-A

Kastenholz, E., Carneiro, M. J., Marques, C. P., & Oliveira, E. (2020). Dark tourism and well-being: The role of emotional experiences and cultural learning. *Journal of Destination Marketing & Management*, *17*, 100423.

Kendle, A. (2008). *Dark Tourism: A fine line between curiosity and exploitation*. Vagabondish. Retrieved from: https://www.vagabondish.com/dark-tourism-travel-tours/

Korstanje, M. E. (2016). *The Rise of Thana-capitalism and Tourism*. Routledge. doi:10.4324/9781315457482

Korstanje, M. E. (2020). The dark tourist: Consuming dark spaces in the periphery. In Tourism, Terrorism and Security. Emerald Publishing Limited.

Lennon, J. (2005). Journeys in Understanding: What Is Dark Tourism? *The Sunday Observer*. Retrieved from http://www.guardian.co.uk/travel/2005/oct/23/darktourism.observerescapesection.

Lennon, J., & Foley, M. (2002). *Dark Tourism: The Attraction of Death and Disaster*. Continuum.

Light, D. (2017). Progress in Dark Tourism and Thana-Tourism Research An Uneasy Relationship With Heritage Tourism. *Tourism Management*, *61*, 275–301. doi:10.1016/j.tourman.2017.01.011

Lisle, D. (2004). Gazing at Ground Zero: Tourism, Voyeurism and Spectacle. *Journal for Cultural Research*, *8*(1), 3–21. doi:10.1080/1479758042000797015

Merriam-Webster Inc. (n.d.). Retrieved from https://www.merriam-webster.com/dictionary/darkness

O'Neill, S. (2002). Soham Pleads with Trippers to Stay Away. *Daily Telegraph*. http:// www.telegraph.co.uk/news/main.jhtml?xml=/news/2002/08/26/nfen26.xml&sSheet=/portal/2002/08/26/ixport.html

Podoshen, J. S. (2013). Dark Tourism Motivations, Simulation, Emotional Contagion And Topographic Comparision. *Tourism Management*, *35*, 263–271. doi:10.1016/j.tourman.2012.08.002

Richards, S., Aziz, N., Bale, S., Bick, D., Das, S., Gastier-Foster, J., Grody, W. W., Hegde, M., Lyon, E., Spector, E., Voelkerding, K., & Rehm, H. L. (2015). Standards and guidelines for the interpretation of sequence variants: A joint consensus recommendation of the American College of Medical Genetics and Genomics and the Association for Molecular Pathology. *Genetics in Medicine*, *17*(5), 405–423. doi:10.1038/gim.2015.30 PMID:25741868

Roberts, C. (2018). Education the (dark) masses: dark tourism and sensemaking. In P. R. Stone, R. Hartmann, T. Seaton, R. Sharpley, & L. White (Eds.), *The Palgrave Handbook of Dark Tourism Studies* (pp. 606–637). doi:10.1057/978-1-137-47566-4_25

Rojek, C. (1993). Ways of Escape. Basingstoke: Macmillan. 1997 Indexing, Dragging and the Social Construction of Tourist Sights. In C. Rojek & J. Urry (Eds.), *Touring Cultures: Transformations of Travel and Theory* (pp. 52–74). Routledge.

Ryan, C., & Kohli, R. (2006). The Buried Village, New Zealand—An Example of Dark Tourism? *Asia Pacific Journal of Tourism Research*, *11*(3), 211–226. doi:10.1080/10941660600753240

Sather-Wagstaff, J. (2016). *Heritage that hurts: Tourists in the Memoryscapes of September 11*. Routledge. doi:10.4324/9781315427539

Seaton, A. V. (1996). Guided by the dark: From Thanatopsis to Thanatourism. *International Journal of Heritage Studies*, *2*(4), 240–242. doi:10.1080/13527259608722178

Seaton, A. V., & Lennon, J. (2004). Moral Panics, Ulterior Motives and Alterior Desires: Thanatourism in the Early 21st Century. In T. V. Singh (Ed.), *New Horizons in Tourism: Strange Experiences and Stranger Practices* (pp. 63–82). CABI. doi:10.1079/9780851998633.0063

Seaton, A. V., & Lennon, J. J. (2004). *New Horizons in Tourism: Strange Experiences And Stranger Practices*. CABI.

Seaton, T. (2009). Purposeful otherness: Approaches to the management of thanatourism. In The Darker Side of Travel Bristol: Channel View Publications.

Sharpley, R. (2020). Coronavirus and dark tourism: Managing thanatourism in the face of new risks. *Tourism Geographies*, *22*(3), 610–623.

Sharpley, R., & Stone, P. (2009). *The Darker Side of Travel: The Theory and Practice of Dark Tourism*. Channel View Publications. doi:10.21832/9781845411169

Sharpley, R., & Wright, D. (2018). Disasters and Disaster Tourism; The Role of the Media. In The Palgrave Handbook of Dark Tourism Studies. Palgrave Macmillan.

Skinner, J. (2016). Walking the Falls: Dark tourism and the significance of movement on the political tour of West Belfast. *Tourist Studies*, *16*(1), 23–39. doi:10.1177/1468797615588427

Stone, P. R. (2013). Dark tourism scholarship: A critical review. *International Journal of Culture, Tourism and Hospitality Research*, *7*(3), 316–332. doi:10.1108/IJCTHR-06-2013-0039

Stone, P. R. (2021). Dark tourism and place identity: The Role of Visitor Experience in Shaping Perceptions of Contested Places. *Tourism Geographies*, *23*(1), 52–72.

Stone, P. R., & Sharpley, R. (2021). Dark tourism: A Review of The Literature. *Journal of Hospitality and Tourism Management*, *46*, 70–83.

Strange, C., & Kempa, M. (2003). Shades of Dark Tourism: Alcatraz and Robben Island. *Annals of Tourism Research*, *30*(2), 386–403. doi:10.1016/S0160-7383(02)00102-0

Tarlow, P. (2005). Dark Tourism: The Appealing 'Dark' Side of Tourism and More. In Niche Tourism: Contemporary Issues, Trends and Cases. Elsevier.

Titta, N. (2010). *Bachelor's Thesis in Nature and Soft Adventure*. Lahti University of Applied Sciences.

Tzanelli, R. (2016). *Thanatourism and Cinematic Representations of Risk: Screening the End of Tourism*. Routledge. doi:10.4324/9781315624105

Walter, T. (1991). Modern Death: Taboo or not Taboo? *Sociology*, *25*(2), 293–310. doi:10.1177/0038038591025002009

Walter, T., Littlewood, J., & Pickering, M. (1995). Death in the News: The Public Investigation of Private Emotion. *Sociology*, *29*(4), 579–596. doi:10.1177/0038038595029004002

West, P. (2004). *Conspicuous Compassion: Why Sometimes it really is Cruel to be Kind*. CIVITAS.

White, L., & Frew, E. (Eds.). (2013). *Dark Tourism and Place Identity: Managing and Interpreting Dark Places*. Routledge. doi:10.4324/9780203134900

Wight, C. (2005). Philosophical and Methodological Praxes in Dark Tourism: Controversy, Contention and the Evolving Paradigm. *Journal of Vacation Marketing*, *12*(2), 119–129. doi:10.1177/1356766706062151

Yankholmes, A., & Hoque, N. (2020). Socio-cultural impacts of dark tourism: A Critical Review of Literature. *Journal of Heritage Tourism*, *15*(6), 601–618.

ADDITIONAL READING

Foley, M., & Lennon, J. J. (2018). Dark Tourism Scholarship: A Critical Review. In R. Sharpley & P. Stone (Eds.), *The Darker Side of Travel: The Theory and Practice of Dark Tourism* (pp. 25–37). Channel View Publications.

Haidu, A., & Törnqvist, C. (2021). Dark Tourism and social media: Authenticity, Knowledge and Affective Mediation. *Current Issues in Tourism*, *24*(8), 982–1002.

Higham, J. E. S., & Hinch, T. D. (2018). Dark Tourism and Morality. In R. Sharpley & P. Stone (Eds.), *The Darker Side of Travel: The Theory and Practice of Dark Tourism* (pp. 141–155). Channel View Publications.

Kastenholz, E., Carneiro, M. J., & Marques, C. P. (2020). Does Dark Tourism Lead to Pro-Social Behavior? A Study on The Impact of Empathy on Sustainable Behaviour. *Journal of Sustainable Tourism*, *28*(9), 1467–1486.

Lennon, J. J., & Foley, M. (2020). The "Invented" History of Dark Tourism. In N. Markwell & S. Simpson (Eds.), *Dark Tourism and Pilgrimage* (pp. 17–31). Channel View Publications.

Lepp, A., Gibson, H., & Lane, C. (2019). Dark Tourism and Ethical Considerations. *Journal of Travel Research*, *58*(3), 323–335.

Stone, P. R. (2012). Dark Tourism and Significant Other Death. *Annals of Tourism Research*, *39*(3), 1565–1587. doi:10.1016/j.annals.2012.04.007

Sutherland, K., & Pleumarom, A. (2019). The Ethics of Dark Tourism: Is It Morally Wrong to Seek Leisure from Death and Suffering? *Journal of Travel Research, 58*(3), 349–361.

Tzanelli, R. (2018). Dark Cosmopolitanism: Thanatourism and its aftermaths. In R. Sharpley & P. Stone (Eds.), *The Darker Side of Travel: The Theory and Practice of Dark Tourism* (pp. 183–195). Channel View Publications.

Wight, D., & Farrington, J. (2018). Beyond Dark Tourism: Towards a Framework for Tourism in Dark Places. In R. Sharpley & P. Stone (Eds.), *The Darker Side of Travel: The Theory and Practice of Dark Tourism* (pp. 223–235).

KEY TERMS AND DEFINITIONS

Dark Tourism: Dark Tourism, also known as Grief Tourism or Thanatourism, refers to the act of travelling to places associated with death, tragedy, or suffering. This could include visiting historical sites related to mass genocide or tragedy, such as the Auschwitz concentration camp in Poland, or more recent events, such as the 9/11 Memorial in New York City. The term "dark" does not necessarily refer to a moral judgment, but rather the morbid or macabre nature of the subject matter. Dark tourism is a growing industry, with many people seeking out experiences that offer a glimpse into the darker side of human history and tragedy.

Niche Tourism: Niche Tourism refers to a specialized or unique form of tourism that caters to the interests and preferences of a particular group of travellers. This type of tourism often focuses on a specific activity, theme, or experience, and may appeal to a smaller, more specialized market than mainstream tourism. Niche tourism can take many forms, such as Adventure Tourism, Cultural Tourism, Eco-Tourism, Culinary Tourism, Medical Tourism, Sports Tourism, Dark Tourism among many others. It often involves engaging in activities or experiences that are not typically associated with traditional mass tourism, and may require specialized knowledge, skills, or equipment. Examples of niche tourism could include wildlife safaris, trekking in remote areas, visiting historical or cultural sites, attending specialized events or festivals, or participating in extreme sports. Niche tourism can also involve specific types of accommodation, such as eco-lodges or luxury resorts, that cater to the needs and preferences of a particular group of travellers.

Occultism: Occultism refers to a broad set of beliefs and practices that deal with the mystical, supernatural, and paranormal phenomena beyond the realm of conventional understanding. It involves the study of hidden knowledge or secrets, often relating to the supernatural or mystical aspects of the universe. The term "occult" derives from the Latin word "occultus," meaning hidden or secret. Occultism encompasses a wide range of practices, including divination, ritual magic, alchemy, astrology, numerology, and various forms of spiritualism. Followers of occultism often believe in the existence of spiritual entities, such as angels, demons, and other supernatural beings, and in the power of ritualistic practices to affect the physical world. Some people practice occultism for spiritual or religious reasons, while others are drawn to it out of curiosity or a desire to explore hidden knowledge. The study and practice of occultism have a long history, dating back to ancient civilizations such as Egypt and Greece, and continue to be a subject of interest and debate in modern times.

Tourism: Tourism refers to the activity of people traveling to and staying in places outside of their usual environment for leisure, business, or other purposes. This can include activities such as visiting attractions, exploring new cultures, enjoying recreational activities, attending events or conferences, or simply relaxing and escaping from everyday life. Tourism can take place within one's own country or internationally, and it can involve a wide range of travel options, from budget-friendly to luxury accommodations and experiences. Overall, tourism is an important industry that contributes to economic growth and cultural exchange between different regions and countries.

Chapter 8
Managing Cultural Tourism:
Business Models in Grave Tourism

Francesco Carignani
https://orcid.org/0000-0002-5558-4440
University of Naples Federico II, Italy

Francesco Bifulco
https://orcid.org/0000-0002-4324-5679
University of Naples Federico II, Italy

ABSTRACT

In the field of niche tourism, some genres such as dark tourism and subgenres such as grave tourism, are attracting the attention of scholars. The contributions that use a managerial approach to analyse these genres are still limited. This work analyses some case studies of subjects who deal with the cultural enhancement of places still linked to ancient cults, in a particular on a delimited territory, such as the UNESCO historic centre of Naples, Italy. The analysis underlines their business models to show some differences in management but, above all, some similarities such as in the staff and founds. The main common point of these subjects is the involvement of local communities, especially in order to preserve the authenticity of these places still used. It is also underlined how digital can be a tool that, if used correctly, can support niche tourism in the preservation of this authenticity.

DOI: 10.4018/978-1-6684-7242-2.ch008

INTRODUCTION

In recent decades, the literature on tourism has developed extremely floridly and spans multiple fields, including management. The study of tourism management is very often focused on the management of cultural heritage, an important resource with a very strong impact on many economies. The growing interest in cultural enhancement in particular, in line with the guidelines of institutions such as UNESCO and ICOM, has led to the development of models and practices related to museums and tourism, increasingly useful for developing one's audience and attracting users, thanks also to the support of digital technology, where users become a cocreator of value but also of their own travel experience (Campos et al., 2018). The studies of recent years are therefore focusing on valorisation, also through the analysis of the business models that the various subjects of cultural tourism are developing.

For cultural-tourist enhancement, digitalization has played a central role in the last decade, in the dissemination of content, in the promotion of attraction (Bifulco, 2016), and in raising awareness of the protection of cultural heritage (Bec et al., 2019). If the impact of COVID-19 has caught many subjects in the management of cultural heritage, such as museums and archaeological areas, unprepared, the pandemic has also been a factor of great acceleration in the digitization of cultural heritage (Novakovic, 2021).

In the field of tourism and cultural heritage enhancement, in recent years, some strands of studies on niche tourism have been developing, increasingly relevant and increasingly specific to management, thanks not only to the impact of digital technologies but also to the attention in the tourism sector of factors previously little considered, such as sustainability (environmental and economic) but also social and youth policies: some examples are rural and naturalistic tourism, often fueled by youth entrepreneurship.

One of the recent lines of study is linked to so-called dark tourism. At the moment, there are several contributions in this field, but the phenomenon still has a limited number of contributions that adopt a managerial perspective, where business models, communication and the use of digital technologies still seem to be declinations related to this issue that have been little explored. Among niche tourism, dark tourism, presenting specific characteristics, becomes an interesting object of insights to understand how companies working in this sector are managed.

RQ: What are the factors that characterize tourist-cultural business linked to grave tourism?

To try to respond to this answer, it was decided to use a multiple case study with a funnel-shaped literature, first analysing the characteristics of the intangible heritage, then gradually going down in detail, with a focus on niche tourism, on a part of this, such as dark tourism and, in particular, of an even more specific sector, grave tourism, a type of tourism that is already structured in the world, just thinking of the cemetery of Père Lachaise in Paris or those of St. Louis and Lafayette in New Orleans, but which in recent years is gaining ground, involving an ever-increasing number of tourists. This analysis showed a gap in the literature regarding the managerial approach, limitedly regarding niche tourism and therefore above all regarding dark tourism and grave tourism, where this approach is extremely lacking. The chosen methodology was then specified, clarifying the choice of the subjects taken into analysis and highlighting the findings, emphasizing the salient characteristics of these subjects. In fact, the choice, within the context of grave tourism, to carry out an analysis of the chosen subjects and the relative business models highlights the way in which these places are valued, linked to a very particular centenary tradition, present in the Naples area in Italy, and therefore includes, in addition to places, the enhancement of an important intangible cultural heritage. The last part of the work highlighted what the conclusions of this analysis are and what the implications are that are suggested, also outlining what the further developments could be.

LITERATURE REVIEW

The Enhancement of Intangible Heritage

In recent years, institutions such as UNESCO first and the United Nations World Tourism Organization have given particular emphasis to intangible heritage (UNWTO, 2018), thus bringing the attention of sector scholars to this issue, generating a series of reflections on the rapid evolution of the same definition of cultural tourism, on the relationship with technologies and on how social changes affect these dynamics (Richards, 2018). As a matter of fact, the role of the cultural user has become central and technologies have often focused on capturing as much information as possible about him, with profiling strategies (Gargiulo & Carignani, 2022). In fact, normal studies on cultural heritage also focus on the distinction between what is defined as tangible heritage and what is defined as intangible heritage, as well as on environmental cultural heritage (Ahmad, 2006). However, attention has also been given to intangible heritage in management studies on cultural heritage, given the property of promoting processes linked to creativity and innovation for the enhancement of this type of heritage, with the related studies of these dynamics and processes then igniting (Cuccia, 2020). The enhancement of cultural heritage and its places therefore takes on new forms, with particular tourist dynamics, where the role of the visitor is at the centre and where real experiences are offered, linked to sensations, thanks to processes of branding places and where marketing therefore has an essential role (Campelo, 2018). That of intangible assets therefore becomes a type of tourism with a strong link with the places, becoming an existential tourism, with a strong authenticity of the original spirit, and therefore seeking a profound integration into the social fabric in which it is located, which must be valued with managerial strategies that take them into account (González, 2008). The enhancement of this type of good generates tourist satisfaction deriving from the strong authenticity of the experience, which translates as a strong indicator of tourists' intention to revisit these goods, developing a greater bond of loyalty than normal tangible goods (Park et al., 2019). If modern cultural valorisation takes sustainable development into strong consideration, different characteristics can therefore also be seen at the level of management of tangible and intangible assets, where the latter cannot be treated as the former because those forms of valorisation, but even of protection, are hardly applicable to intangible assets (Araoz, 2011). The 2003 UNESCO convention on intangible heritage also outlined the fundamental role of local communities in the dynamics of enhancing and safeguarding these assets but also underlined their rights and, in some way, their duties, stressing the active participation of citizens (Blake, 2008). The glue of the communities, however, is not identified only with goods, tangible or intangible, or with rituals and traditions, but by the productivity of the relationships that are established and where culture plays a fundamental role (Jora et al., 2019). The relationship between intangible heritage and the local social fabric therefore has extremely positive potential, creating symbiotic relationships that translate into the vision of heritage as a tourist resource, thanks to the awareness of locals but also to their empowerment by professional cultural valorisation workers (Kim et al., 2019). It has therefore been underlined how profits of an economic nature are directly linked to the benefits for the local community in an approach that therefore, directly or indirectly, involves several stakeholders (Wanda George, 2010). This involvement must not be only formal, since the involvement of multiple stakeholders, even in planning, translates into cocreation processes that are necessary for the correct protection and enhancement of tourism, especially for intangible assets (Esfehani & Albrecht, 2019). A favourable and open planning environment is in fact indicated as one of the fundamental ingredients to be able to develop good financial sustainability (Eppich & Grinda, 2019). Finally, the role of digital in the enhancement of intangible heritage should not be underestimated, especially in the most modern and recent forms, capable of being able to speak to a large and even young audience (Cunha et al., 2018). In order to involve younger users, but also, as mentioned, to create communities with stakeholders who can participate in the creation of shared value, the digital aspect is fundamental, especially if there are digital marketing strategies at the base that can be applied to the most modern technologies, such as the metaverse in the future (Carignani et al., 2023), able to exploit more and more the potential of digital ecosystems and therefore improve more and more tourist services (Yekimov, 2022).

Businesses and Niche Tourism

Niche tourism was born almost in opposition to mass tourism, taking the form of a more sustainable and less harmful type of tourism while at the same time bringing greater development opportunities for destination managers and planners (Robinson & Novelli, 2007). Over the years, it has also been divided into some macro sections that contain different types of tourism: cultural, environmental, rural, urban and others; in turn, these macro groups contain different specific typologies within them (Figure 1).

Figure 1. Niche tourism components
Source: Robinson and Novelli (2007)

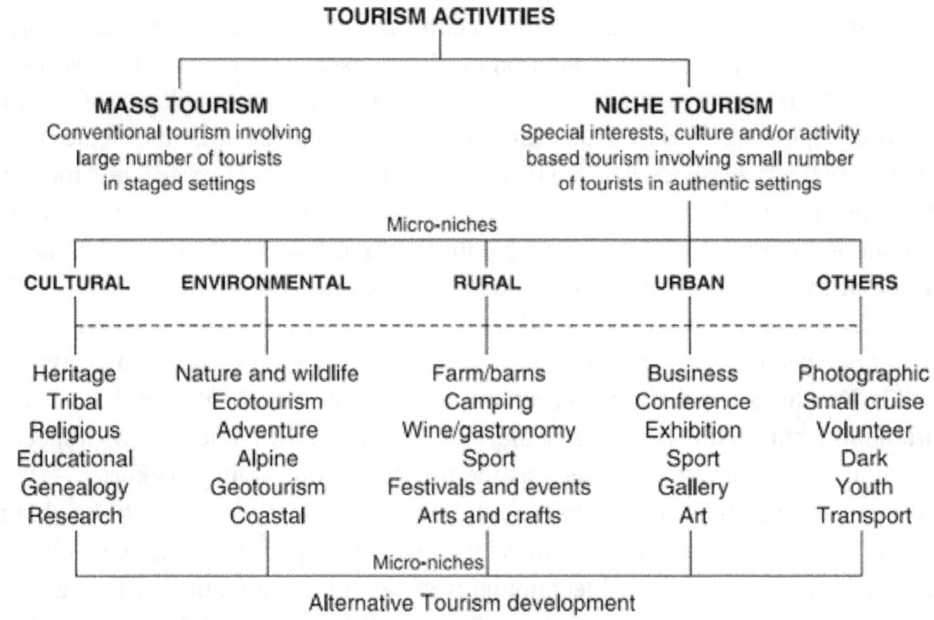

Other types of niche tourism that could enrich these micro categories are constantly developing. For example, medical tourism is growing rapidly, and studies have underlined the benefits that this type of tourism brings to local economies at an economic level (Connell, 2006). Other forms of niche tourism develop on certain themes, such as sport and well-being (Bull & Weed, 1999; Lehto et al., 2006), religious (Iliev, 2020; Preko et al., 2021) or gastronomy, and involve the cultural aspect of places (Torabi Farsani et al., 2018). It has been demonstrated that the gastronomic aspect can also be particularly relevant in the management of tourism and in the promotion of rural areas, developing strategies that can, from tourism, lead to the development of these areas considered depressed (Sidali et al., 2018).

It has also been underlined that niche tourism, in addition to requiring greater caution and attention in planning, also needs greater flexibility compared to mass tourism, which is far more standardized (Marson, 2011). In niche tourism, unlike in mass tourism, it can be much easier to develop sustainable development processes, but this requires an effort; it is not an automatic process, but it can be an important opportunity to develop more authentic forms of tourism and awareness (Novelli, 2018). At the managerial level, therefore, there is the need to develop strategies that are specific but also differentiated to be able to compete on an international market: for this reason, marketing choices must also focus on presenting the local aspect as a form of value added, such as the naturalistic or food and wine aspects of a location (Trunfio et al., 2006). These strategies must also take into account the digital aspect: the analysis of online big data therefore becomes a fundamental phase for destination management, analysing the reputation of places, especially on social networks, to increase their competitiveness and develop better strategies (Cillo et al., 2021). Destination management has underlined the need to change tourism strategies in the post-COVID-19 period, where small businesses in particular have suffered, focusing on niche tourism (Dias et al., 2022). In this sense, COVID-19 has almost underlined a point of no return towards sustainable tourism and favouring, to some extent, the development of a type of tourism model far from mass tourism (Renaud, 2020). Sustainability is therefore characterized as a key element of niche tourism, even in future perspectives that are hardly feasible today, such as space tourism (Toivonen, 2022). But this particular type of tourism cannot be promoted with normal means of promotion, but should use niche media, focused on specific audiences, with specific needs and peripheral means, such as the use of certain hashtags (Febrian, 2022). Tourism management must therefore equip itself with its own tools to be able to effectively face the challenges of niche tourism, with specific marketing strategies that make the most of digital technologies to capture the attention of increasingly independent and sophisticated tourists (Soava, 2015). The use of digital technology in the tourism sector has been extremely accelerated due to Covid-19, becoming one of the most discussed trends in the literature and developing aspects of this issue such as the impact of these technologies in the tourism sector in a sustainable way (Bijlani, 2021). The same pandemic has profoundly changed destination marketing, making the role of digital even more important, allowing us to perceive risks, user sentiment and their behaviour (Elliot & Lever, 2022).

Dark Tourism and Grave Tourism

Scholars have also demonstrated how the quality of the experience that tourists have is linked directly and indirectly to how visitors build loyalty with a place through perceived value and satisfaction: this is true for niche tourism, but especially for so-called "dark tourism" (Sharma & Nayak, 2020). So-called *thanatourism* or dark tourism, particularly grave tourism, are types of tourism that are anything but recent, developed above all in the United States, particularly in New England and New Orleans, and closely connected to the twentieth-century sentiment of nostalgia (Tarlow, 2007). The interest in grave tourism therefore has a parallel in past eras, and involves cultural and social aspects (Cohen, 2018), finding feedback and success also in narrative literature (Wiliams, 2021). This type of tourism falls within the division proposed by Robinson & Novelli (2007) on niche tourism in the "other" subcategory. In the 1990s, we started talking about dark tourism with very few contributions, and since 2014, most of the contributions have concentrated on this topic. Although the works related to this topic are still few, there have been analyses that over the years have aimed to investigate this type of tourism, linked to thanatourism, defining its characteristics and representations over the years (Seaton, 1996). In fact, there are studies that analyse this type of tourism, which can distinguish different places from each other but which cover a wide range of emotions related to the sense of dark, through numerous approaches, such as the managerial one, with the aim of understanding the motivations, experiences and visitor expectations (Stone, 2016). However, the management of these places, often linked to death and suffering, is not always easy: they require a certain sensitivity and respect. However, precisely the limited number of practices and models linked to the tourist marketing of these places, gives the idea of how it is a topic in some ways still in an embryonic stage, where they are balanced with the valorisation, often in a not simple way, the conservation and the authenticity of these spaces (Light, 2017). In most cases, in fact, cemeteries are often linked to public administrations and almost always far from the logic of cultural enhancement or from the reasoning of tourism management (Millán et al., 2019). The specific literature on what is defined as grave tourism is still extremely scarce in contributions of a managerial nature. However, attention to the topic is growing, given that from a tourist point of view, for example, the cemeteries of celebrities are in some way representing a type of modern pilgrimage (Soligo & Dickens, 2020). Even if in some cases these places are not yet perceived as attractions, they are places of popular tradition and therefore should have management more similar to that of a museum (Mionel, 2020). There are various difficulties in the tourist management of these subjects, where the support of scientific institutions is recommended, which can somehow guide the tourist development of these attractions, ensuring that the valorisation does not conflict with the sacredness of the places (Drvenkar et al., 2015). However, to analyse dark tourism at a managerial level, it is a socially complex phenomenon, and therefore, it is essential to take into account individualistic and sociocultural aspects (Biran & Hyde, 2013). Even if more than the motivations of the visitors themselves, it is the desire for a different or unusual experience that pushes people towards this kind of tourism: a relevant aspect is management, so that it can also effectively start informal learning processes (Ivanova & Light, 2018). At the same time, however, it was also discussed how the contemplation of mortality and death influence the choice of visits to some locations, especially the more authentic ones: in these cases, the management must absolutely take into account that at the basis of the choice of these visits, there is a strong feeling of spirituality, where digital devices do not necessarily have to negatively affect this spirituality, but rather, they can increase its sensitivity without altering its authenticity (Marchado Carvalho, 2023). An analysis of the motivations of the public for dark tourism is therefore useful to understand what the public seeks from the tourist experience and therefore how to improve its offerings (Min et al., 2021). In fact, some works differentiate dark tourism according to the intensity of the experience it offers, from "lightest" to "darkest" (Raine, 2013), as well as in some subcategories (Figure 2). For example, tourism linked to the subcategories of paranormal and grave tourism are two similar types but present different motivations, as well as different managerial implications, from the times of visits to the type of information that visitors expect (Millán et al., 2019).

Figure 2. Location of dark tourism
Source: Millán et al. (2019)

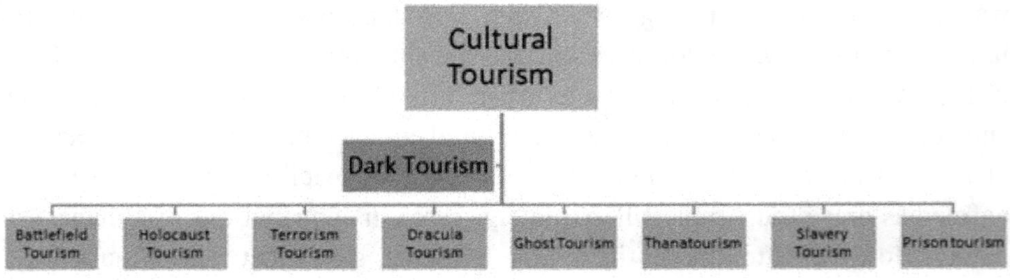

These particular characteristics of dark tourism underline how destination management is still in an incomplete stage in the analysis and management of the dynamics related to this form of tourism (Powell et al., 2018).

METHODOLOGY

It was therefore chosen to focus on the urban area of Naples in Italy, given the particular nature of the relationship between citizens and cemetery areas. Indeed, the scenario is configured in the context of a city characterized by a strong historical stratification in its historic centre, recognized by UNESCO as a World Heritage Site. The enhancement of cemeteries and churches has its roots in past eras, linked to local superstitions and cults. In fact, the popular classes of the city used to frequent these places polishing the skulls of the deceased, hoping that their souls, in gratitude, could grant them fortunes, favours, graces (cures from illnesses or good marriages) and winnings in gambling. Over time, this very particular local cult has begun to attract tourists from all over the world. In recent years, this form of strictly popular cultural enhancement has decreased, but it is always present and deeply felt, especially by the lower sections of the population, and in some cases has also been integrated by modern forms of enhancement based on managerial logic. In fact, if in Naples there are numerous cemetery areas and churches with annexed spaces for burials, there are therefore few areas that are valued with a cultural and tourist approach. This is because many of these sites are managed directly by the municipality, which struggles to maintain them and does not have the resources (physical and economic) to enhance them. It was therefore chosen to analyse the subjects who dealt with the enhancement of these places linked to this particular cult of the polishing of skulls, still so felt and so particular as to differentiate these subjects from the others normally present in grave tourism. Similar to places, it also emerges indirectly from these subjects, an enhancement of the cult and therefore of an important and secular intangible and immaterial cultural heritage, which, the action of the analysed subjects, helps to enhance and safeguard future generations. This enhancement of the cult therefore becomes an essential factor, which the analysis of the business model cannot ignore. Therefore, the chapter analyses three case studies related to grave tourism in Naples, Italy, focusing on the management and business models of these three cultural firms: La Paranza[1], Progetto Museo[2] and Respiriamo Arte[3]. A qualitative approach was adopted (Dubois & Gadde, 2002) choosing a multiple case study approach (Gummesson, 2017): three subjects were selected for analysis, using direct interviews with a manager of these realities delegated by them, as well as the use of some secondary sources such as interviews and website contents (Creswell & Creswell, 2017), to outline the recurring elements of their business models and their differences. These subjects were therefore chosen because they manage, among other sites, some of the sites linked to grave tourism in Naples: the Church of Purgatorio ad Arco (Progetto Museo), the Church of Santa Luciella ai librai (Respiriamo Arte) and the two catacombs of San Gennaro and San Gaudioso (La Paranza). It was decided to exclude the Fontanelle Cemetery, another place of worship subject to the phenomenon of cleaning skulls, because the cemetery, previously valorised, has been closed for several years due to various bureaucratic problems. We therefore chose to analyse data such as the year of foundation, the characteristics of the staff and the founders, how they interface with digital and the type of relationship they have with local communities.

MAIN EVIDENCE

The analysis shows how these tourist-cultural realities highlight different models, but in most cases, basic models, which have involved local communities, often involving groups of young people. Emphasis is therefore placed on their genesis, on the type of initial investments they have put in place, on the professionalism with which these subjects have started their activities, on the type of services they offer. There is therefore a discussion on the evolution of the models used over the years, on the growth of the staff but also on the present and future use of digital in the enhancement of these places.

As mentioned, these realities address the enhancement and management of various sites, including those linked to grave tourism: the catacombs of San Gennaro, the catacombs of San Gaudioso, the monumental complex of Santa Maria delle Anime del Purgatorio and the Church of Santa Luciella. Those of San Gennaro are a complex of Christian catacombs from the 2nd century AD, developed at the time outside the city walls and which housed the remains of the patron saint of the city. From a similar period are the catacombs of San Gaudioso. The monumental complex of Santa Maria delle Anime del Purgatorio consists of a church at the street level, a museum used in the sacristy and oratory spaces, and a hypogeum under the church, used in the 1600s to preserve the deceased of the confraternity linked to the church. Both the church and the hypogeum, where the bones of the deceased are found, are even still sacred places frequented by the faithful. The church of Santa Luciella has similar characteristics, with a space under the church used as a cemetery for the confreres of the church.

La Paranza is a cooperative founded in 2006 that manages, among other things, the catacombs of San Gennaro and the catacombs of San Gaudioso. These two sites saw 207,000 visitors in 2022. At the time of its foundation, Paranza had 8 founders, a greatly increased number up to today's 50 units, of which 40, 80%, are under 35. At the time of the foundation, the subjects were all eighteen years old, lacking a specific formal education in the field of cultural heritage or management or significant previous work experience in this sector, but they had a very strong relationship with the local community, having been born and raised in the neighbourhood hosting the two sites. To start their business, they made use of some funds linked to a private foundation, with the aim of relaunching the neighbourhood and influencing the education and creation of a professional future for those young people. Digital is widely used by La Paranza, both in the promotion on social networks and in digital marketing but also with digital systems on ticketing and on the protection of places, in particular linked to maintaining the best microclimatic conditions inside the catacombs and to the control system lighting, all completely managed digitally. The relationship with the local community, being their full members, is very close: the faithful of the area who want to pray for the souls of the deceased can enter without paying any ticket. Moreover, La Paranza is involved, with a strong sensitivity to social responsibility, in supporting various educational projects involving the young people of the neighbourhood.

Progetto Museo was founded in 1998 as an association that offers museum services of an educational nature to museums. In 2002, a virtuous path began for the enhancement of the Church of Santa Maria delle Anime del Purgatorio, the museum and the hypogeum through the institutionalization of openings and guided tours, the creation of educational paths for schools, and the training of young graduates with sporadic and planned openings when necessary. Thanks to the help of the association, the Museum Complex of Santa Maria delle Anime del Purgatorio in Arco was established in 2010, open to the public on a weekly basis and on weekends. The site welcomed approximately 21,000 visitors in 2022. Progetto Museo was founded by 4 women, and today, it has collaborations of various kinds with approximately 20 people. None of these are under 35, but it should be noted that the founders and staff are all women. The founders had already been working in cultural heritage for some time, with a mixed background, two art historians, humanities graduates and an engineering graduate. The association has always been self-financed, using private funds in 2021 for a cultural enhancement project linked to the territory. Digital is used to promote the site on social networks. Over the years, the association has launched various projects in synergy with other local cultural bodies, projects to enhance the places and above all the human resources involved in the tourism chain through a project on cultural enhancement and hospitality, called "welcoming with art". As previously illustrated, it also carried out a specific project from 2012 to 2014 involving the local community.

Respiriamo Arte is an association founded in 2013 by four Neapolitan boys. Among the sites they manage is that of the Church of Santa Luciella ai booksellers, famous because among the skulls preserved in the hypogeum under the church, there is one with deformed cartilage on the sides, so much so that it seems to have ears. The church and the hypogeum were abandoned for several years until 2019, when the association managed to reopen them and make them accessible to locals and tourists. The site in 2022 had approximately 50,000 visitors. We breathe art was founded by four young people. Today, these four have been joined by ten people with part-time contracts, seven with permanent contracts and three with occasional contracts. Among the founders and employees, ten out of fourteen are under 35 years of age, while the other four are thirty-six, while except for one man, they are all female. The background of the founders is mixed, with two of them graduates in humanities and the other two in management degrees, and all four were in their first real entrepreneurial experience. For the restoration and reopening of the church, Respiriamo Arte has resorted to private funds, focusing on the cultural project animated by young people. In the digital sphere, there is careful use of social networks as a tool for communication and promotion of the church and their activities, as well as the development of the Daria project, a virtual aid especially useful for the disabled. Respiriamo Arte has developed various strategies to involve the local community: free projects for young people and children of the neighbourhood, such as cineforums, thematic workshops, and prayer groups for the sick and for the faithful. In this regard, both the faithful and the inhabitants of the neighbourhood enter for free. Asbestos reclamation activities have also started outside the church, and the association has purchased various plants that it takes care of to beautify the piece of road where the church is located.

The picture that emerges (Table 1) highlights how the business models of the subjects analysed are characterized by some common factors. First, the initial investment is often represented by extremely scarce funds but linked to a very clear idea of valorising an asset, emphasizing its potential, without distorting the religious sense. The second factor concerns the staff of these companies: in most cases, they are people with limited or a recent education in the field of cultural-tourism enhancement, with no or little previous experience in the entrepreneurial field but extremely rooted in the territory with a significant presence of young people and women. The third factor concerns the involvement of local communities, which is pursued in an extremely active way by these companies, with the aim of appearing in the eyes of the community, not as exploiters of popular traditions, but as bearers of widespread well-being, with tangible repercussions on the territory, in terms of employment and development of neighbourhood activities: respect for religious worship is a fundamental element of this rapport with the community, given that in many cases (such as the two churches), we are not only talking about places that are still sacred but that still host religious activities. The last factor concerns the formation of these business: the internal subjects of the companies often have recourse to the help of an external expert or a private entity who believed in their project and allowed the concrete start-up from the idea to reality. The impact on the context in which they operate, economic, environmental and social, will then be highlighted, the approach towards the local community and towards the stakeholders, if and how the partnerships and the involvement of the residents have led to the value cocreation.

Table 1. The business features of the three subjects

Name	La Paranza	Progetto Museo	Respiriamo Arte
Type	Cooperative	Association	Association
Cultural sites managed	San Gennaro catacombs and San Gaudioso catacombs	Museal complex of Santa Maria delle Anime del Purgatorio ad Arco	Church of Santa Luciella ai librai
Visitors in 2022	207.000	20.910	50.000
Year of foundation	2006	1998	2013
Staff at the beginning	8	4 founders	4 founders
Current staff	50	Variable, approximately 20 in different types of collaborations	4 founders 7 permanent part-time employees 3 occasional employees
Under 35	40	None	10
Education of the founders	None	2 art historian 1 humanistic studies 1 engineer	2 managerial studies 2 humanistic studies on heritage
Previous experience of the founders	None	Yes	None
Use of private or public founds	Private	Private	Private
Use of digital	-Promotional use of social -Microclimate sensor and lighting controls -Ticketing	-Promotional use of social	-Promotional use of social -A chatbot
Relation with the local community	-Faithful of the neighbourhood do not pay -Projects with local educational realities	-Projects with other local realities -Projects to enhance human resources in the tourism chain -Involvement of the local community	-Projects for neighbourhood kids and children -Prayers groups -Neighbourhood people get in for free -Neighbourhood beautification

Source: Own elaboration

IMPLICATION

On a theoretical level, the proposed cases are framed in a particular way within grave tourism: on the one hand, they are places of a cemetery nature (completely or in part); on the other hand, they can hardly be considered as the literature frames the normal cemeteries, precisely because of the relationship with the faithful who still feel the tradition of polishing skulls. A further classification produced with respect to the grave tourism present in the literature (Millán et al., 2019) is therefore necessary, where grave tourism cannot be addressed as a tourist route framed on the paranormal but as a tourist route framed with respect to a cult still active. Within the literature, still scarce on the subject, one should therefore differentiate between the places where the cult is still active and alive and the places where it is lived in a more passive way, as we said, the normal cemeteries. On a more generic side, however, it has been shown that the use of digital in these business models linked to grave tourism seems to be completely absent in the literature, marking a strong gap to fill and something that this contribution tries to do.

At a managerial level, the cases analysed show how the subjects analysed have assumed light and flexible business models, with a small number of founders and employees. It has been highlighted how the managerial impulse behind these activities is of an entrepreneurial nature, but at the same time firmly linked to a desire to enhance not only an important cultural heritage linked to places but also a very particular intangible heritage linked to traditions and to secular cults. In this sense, the relationship with the community seems to want to underline how the managerial dynamics take these dynamics into strong consideration, thus making these companies appear in the eyes of the residents, not as lucrative activities, but as subjects implanted in the territorial fabric, which they bring strong social and economic well-being to the residents and at the same time, they also give back an image of these companies as actively involved, not only in improving the neighbourhood but also in safeguarding this cult and its traditions. The management of relations with local stakeholders, especially with local residents, therefore becomes fundamental and takes the form of direct, often informal relationships, the result of personal relationships and relationships of trust that have grown over time. However, such relationships have made residents understand the respectful intentions of companies, which otherwise could have run into NIMBY (Xie et al., 2020), gentrification (Gravari-Barbas & Guinand, 2017) and touristification processes of those places (Evans, 2002), with a loss of the sense of authenticity, which would have had extremely negative repercussions on the activity of the subjects taken into analysis. It is precisely thanks to this direct relationship with the local community, and thanks to a careful valorisation not to distort its nature, that these companies contribute to enhancing and safeguarding an important intangible heritage, such as that of the polishing of skulls. A crucial value is therefore underlined between niche tourism and the relationship with local communities, with authenticity as one of the main values of this relationship.

A further element that stands out is the composition of the founders, mostly under 35 at the time of the foundation and female. In relation to this datum, it is also interesting to note how these companies, in their initial stages, have resorted to private economic aid, highlighting a lack of vision on the part of policy makers, in not having provided economic instruments to companies that future they would have become consolidated, such as those analysed. If the digital presence is a constant element, it does not seem to be of essential importance, if not for the part of communication and promotion relating to social networks, which these realities effectively exploit to make these places known but also to inform the residents of their activities, especially the younger ones, thus integrating personal relationships with digital technology. In this sense, the technological aspect can play a fundamental role for several reasons. The first is to represent a crucial tool for connecting the communities at the base of these cults. In fact, social media seem to be extremely useful for the creation of those communities linked to cultural heritage, which share the same purposes underlying a common cultural identity (Ginzarly & Teller, 2021). The second reason is to convey a message of authenticity to visitors, who in some cases do not know what to expect: thanks to specific communication strategies, the visitor is fully informed of the type of service he is going to use (Troshin et al., 2020). In the cases analysed, visitors could think they are in the presence of sites linked to esotericism or even kitsch experiences: digital technologies and social media in particular can help prepare the visitor for the atmosphere of these visits, far from the aforementioned aspects, but full of authenticity, spirituality and authentic tradition. In general, this value of authenticity, an indispensable value also at the basis of communities, can also be supported by technologies in other forms: in fact, there are many sites in the world that suffer from overtourism, a phenomenon that undermines the experience of visiting and the authenticity of places (Skotis & Livas, 2022). It has therefore been demonstrated how digital technologies can have a positive impact on visits, if not always replacing the visit itself and a physical experience that in any case remains unique, integrating this experience with virtual methods (Sintobin, 2021), in which the visitor he does not necessarily have to physically visit the place, but he can also visit it thanks to virtual reality, holograms or multimedia tools (Frey & Frey, 2021). This approach also has a positive impact on the protection of these places, which with these technologies see the load of tourist flows decrease and therefore the physical impact on the heritage, as well as keeping track of artifacts and monuments that could disappear due to wars or natural disasters (Du & Sun, 2015; Khan et al., 2018). Similarly, time and other aspects can contribute to the disappearance of practices related to intangible heritage: also in this case technologies can play a key role in preserving them (Junping & Huixia, 2022). All these elements clearly highlight the relationship between niche tourism and the community, generating recommendations to institutions and political decision-makers who cannot ignore this link in creating effective and lasting cultural and tourism policies. In fact, if these communities passively undergo this type of tourism without being involved, there is the risk of having negative effects in the perception of the residents, taking away authenticity and generating problems and frustrations in the communities.

FINAL REMARKS AND FUTURE RESEARCH

The companies operating in the field of cultural heritage were analysed, which is a very particular typology with respect to the landscape in serious tourism, a phenomenon which is in itself little analyzed in studies in the managerial field. In fact, there are still few studies that analyse dark tourism from a managerial perspective and, specifically, of grave tourism that go beyond the analysis of the phenomenon itself, often linked to a macabre sensationalist sense and not very tied to traditions, but that analyse this type of tourism in the characters that outline its managerial aspects.

Therefore, part of the scenario present in the literature regarding niche tourism and the managerial dynamics linked to this emerging typology compared to mass tourism was traced. Then, following a funnel-shaped scheme, from niche tourism, a particularly more specific scenario emerged of dark tourism and its peculiarities, going even more specifically into grave tourism. Furthermore, it was decided to outline part of the scenario present in the literature on the enhancement of intangible tourism, specifically to enhance its particular characteristics with respect to the enhancement of the normal material cultural heritage.

The history of the city that acts as a backdrop to these places has been given prominence, precisely to contextualize the context and to better understand the type of cult behind it. This phase was essential to be able to outline the scenario in which these companies work and therefore be able, in the analysis of the business model, because the relationship with the territory and the community cannot be ignored.

The analysis therefore reveals some peculiar common features in the management of these companies, the strong presence of young people (excluding Progetto Museo) and women but also the absence of public contributions. However, the most important feature is perhaps the strong focus on the relationship with the community to convey the goal of preserving and protecting tradition and worship, thus bringing benefits to the territory (Wanda George, 2010). On the one hand, the particular nature of the cult analysed linked to these companies could represent a strong unicum compared to normal grave tourism; on the other hand, it represents an interesting case precisely because it highlights one of the essential factors of grave tourism and of tourism in a general niche, i.e., the analysis of the involvement of local communities in these business models. It would therefore be interesting to adopt this approach more and more in the future for this type of tourism, placing even more emphasis on the role of the communities to safeguard the authenticity of these places and related traditions, but also of the support that digital can give in preserving this authenticity and therefore in enhancing this type of tourism and niche tourism in general, a topic on which further research in this field could be based.

Although this study presents some limitations, such as the analysis of a limited number of cases, which represent subjects of a small size and who are found in a particular context as well as in a limited geographical area, the picture that emerges translates into some recommendations to institutions (above all cultural) and to policymakers who, in order to enhance this type of niche tourism and these intangible traditions, cannot but strongly consider the relationship between niche tourism dynamics and stakeholders, building strong relationships, above all by involving local communities, building an authentic and emotional promotion through sensitive marketing, in order to safeguard authenticity, thanks also to digital as a tool, and to involve these subjects in an optimal management and preservation of intangible cultural heritage.

REFERENCES

Ahmad, Y. (2006). The scope and definitions of heritage: From tangible to intangible. *International Journal of Heritage Studies, 12*(3), 292–300. doi:10.1080/13527250600604639

Araoz, G. F. (2011). Preserving heritage places under a new paradigm. *Journal of Cultural Heritage Management and Sustainable Development, 1*(1), 55–60. doi:10.1108/20441261111129933

Bec, A., Moyle, B., Timms, K., Schaffer, V., Skavronskaya, L., & Little, C. (2019). Management of immersive heritage tourism experiences: A conceptual model. *Tourism Management, 72*, 117–120. doi:10.1016/j.tourman.2018.10.033

Bifulco, F. (2016). *Managing cultural heritage: innovation perspectives, customer experience, resources enhancement, performance management.* McGraw-Hill Education.

Bijlani, V. A. (2021, September). *Sustainable digital transformation of heritage tourism. In 2021 IoT Vertical and Topical Summit for Tourism.* IEEE.

Biran, A., & Hyde, K. F. (2013). New perspectives on dark tourism. *International Journal of Culture, Tourism and Hospitality Research, 7*(3), 191–198. doi:10.1108/IJCTHR-05-2013-0032

Blake, J. (2008). UNESCO's 2003 Convention on Intangible Cultural Heritage: the implications of community involvement in 'safeguarding'. In Intangible heritage (pp. 59-87). Routledge.

Bull, C., & Weed, M. (1999). Niche markets and small island tourism: The development of sports tourism in Malta. *Managing Leisure, 4*(3), 142-155.

Campelo, A. (2018). Immaterial heritage and sense of place. In *Cultural Heritage* (pp. 129–138). Routledge. doi:10.4324/9781315107264-10

Campos, A. C., Mendes, J., Valle, P. O. D., & Scott, N. (2018). Co-creation of tourist experiences: A literature review. *Current Issues in Tourism, 21*(4), 369–400. doi:10.1080/13683500.2015.1081158

Carignani, F., Clemente, L., Iodice, G., & Bifulco, F. (2023). Digital Marketing in Cultural Heritage: An Approach to Metaverse. In Cultural Marketing and Metaverse for Consumer Engagement (pp. 142-163). IGI Global.

Cillo, V., Rialti, R., Del Giudice, M., & Usai, A. (2021). Niche tourism destinations' online reputation management and competitiveness in big data era: Evidence from three Italian cases. *Current Issues in Tourism, 24*(2), 177–191. doi:10.1080/13683500.2019.1608918

Cohen, E. (2018). Thanatourism: A comparative approach. The Palgrave handbook of dark tourism studies, 157-171.

Connell, J. (2006). Medical tourism: The newest of niches. *Tourism Recreation Research, 31*(1), 99–102. doi:10.1080/02508281.2006.11081252

Creswell, J. W., & Creswell, J. D. (2017). *Research design: Qualitative, quantitative, and mixed methods approaches.* Sage publications.

Cuccia, T. (2020). Intangible cultural heritage. In *Handbook of Cultural Economics* (3rd ed., pp. 294–303). Edward Elgar Publishing.

Cunha, C. R., Mendonça, V., Morais, E. P., & Carvalho, A. (2018). The role of gamification in material and immaterial cultural heritage. In *Proceedings of the 31st International Business Information Management Association Conference (IBIMA)* (pp. 6121-6129). International Business Information Management Association (IBIMA).

Dancausa Millán, M. G., Perez Naranjo, L. M., Hernandez Rojas, R. D., & Millan Vazquez de la Torre, M. G. (2019). Cemetery tourism in southern Spain: An analysis of demand. *Tourism and Hospitality Management, 25*(1), 37–52. doi:10.20867/thm.25.1.1

Dias, Á., Patuleia, M., Silva, R., Estêvão, J., & González-Rodríguez, M. R. (2022). Post-pandemic recovery strategies: Revitalizing lifestyle entrepreneurship. *Journal of Policy Research in Tourism, Leisure & Events, 14*(2), 97–114. doi:10.1080/19407963.2021.1892124

Drvenkar, N., Banožić, M., & Živić, D. (2015). Development of memorial tourism as a new concept-possibilities and restrictions. *Tourism and Hospitality Management, 21*(1), 63–77. doi:10.20867/thm.21.1.5

Du, J., & Sun, X. (2015, November). Research on Holographic and Digital Protection of Cultural Heritage Information. In *2015 International Conference on Social Science, Education Management and Sports Education* (pp. 1148-1151). Atlantis Press. 10.2991semse-15.2015.295

Dubois, A., & Gadde, L.-E. (2002). Systematic combining: An abductive approach to case research. *Journal of Business Research, 55*(7), 553–560. doi:10.1016/S0148-2963(00)00195-8

Elliot, S., & Lever, M. W. (2022). You Want to go where? Shifts in social media behaviour during the COVID-19 pandemic. *Annals of Leisure Research*, 1–15. do i:10.1080/11745398.2022.2041448

Eppich, R., & Grinda, J. L. G. (2019). Sustainable financial management of tangible cultural heritage sites. *Journal of Cultural Heritage Management and Sustainable Development, 9*(3), 282–299. doi:10.1108/JCHMSD-11-2018-0081

Esfehani, M. H., & Albrecht, J. N. (2019). Planning for intangible cultural heritage in tourism: Challenges and implications. *Journal of Hospitality & Tourism Research (Washington, D.C.), 43*(7), 980–1001. doi:10.1177/1096348019840789

Evans, G. (2002). Living in a World Heritage City: Stakeholders in the dialectic of the universal and particular. *International Journal of Heritage Studies, 8*(2), 117–135. doi:10.1080/13527250220143913

Febrian, F. (2022). Targeting Niche Tourism Using Niche Media: Expanding the Media in a Digital Era towards the Hashtag Generation. *Multidisciplinary Digital Publishing Institute Proceedings, 83*(1), 39. doi:10.3390/proceedings2022083039

Frey, B. S., & Frey, B. S. (2021). *Overcoming Overtourism*. Springer International Publishing. doi:10.1007/978-3-030-63814-6

Gargiulo, R., & Carignani, F. (2022). Novel Approaches in Profiling in Museums. In *Handbook of Research on Museum Management in the Digital Era* (pp. 228–247). IGI Global. doi:10.4018/978-1-7998-9656-2.ch012

Ginzarly, M., & Teller, J. (2021). Online communities and their contribution to local heritage knowledge. *Journal of Cultural Heritage Management and Sustainable Development, 11*(4), 361–380. doi:10.1108/JCHMSD-02-2020-0023

González, M. V. (2008). Intangible heritage tourism and identity. *Tourism Management, 29*(4), 807–810. doi:10.1016/j.tourman.2007.07.003

Gravari-Barbas, M., & Guinand, S. (Eds.). (2017). *Tourism and gentrification in contemporary metropolises: International perspectives.* Taylor & Francis. doi:10.4324/9781315629759

Gummesson, E. (2017). Case theory in business and management: Reinventing case study research. *Sage (Atlanta, Ga.).* Advance online publication. doi:10.4135/9781473920811

Iliev, D. (2020). The evolution of religious tourism: Concept, segmentation and development of new identities. *Journal of Hospitality and Tourism Management, 45*, 131–140. doi:10.1016/j.jhtm.2020.07.012

Ivanova, P., & Light, D. (2018). 'It's not that we like death or anything': Exploring the motivations and experiences of visitors to a lighter dark tourism attraction. *Journal of Heritage Tourism, 13*(4), 356–369. doi:10.1080/1743873X.2017.1371181

Jora, O. D., Iacob, M., & Apăvăloaei, M. A. (2019). Public Cultural Heritage and Private Property Rights: Building Sustainable Community Through Individuality. In *Caring and Sharing: The Cultural Heritage Environment as an Agent for Change: 2016 ALECTOR Conference, Istanbul, Turkey* (pp. 213-224). Springer International Publishing.

Junping, Q., & Huixia, Z. (2022). *Research on the protection and inheritance of intangible cultural heritage under the background of rural revitalisation.* Applied Mathematics and Nonlinear Sciences. doi:10.2478/amns.2021.2.00250

Khan, N. A., Shafi, S. M., & Ahangar, H. (2018). Digitization of cultural heritage: Global initiatives, opportunities and challenges. *Journal of Cases on Information Technology, 20*(4), 1–16. doi:10.4018/JCIT.2018100101

Kim, S., Whitford, M., & Arcodia, C. (2019). Development of intangible cultural heritage as a sustainable tourism resource: The intangible cultural heritage practitioners' perspectives. *Journal of Heritage Tourism, 14*(5-6), 422–435. doi:10.1080/1743873X.2018.1561703

Lehto, X. Y., Brown, S., Chen, Y., & Morrison, A. M. (2006). Yoga tourism as a niche within the wellness tourism market. *Tourism Recreation Research, 31*(1), 25–35. doi:10.1080/02508281.2006.11081244

Light, D. (2017). Progress in dark tourism and thanatourism research: An uneasy relationship with heritage tourism. *Tourism Management*, *61*, 275–301. doi:10.1016/j.tourman.2017.01.011

Machado Carvalho, M. A. (2023). Bone Chapels: Who might be interested in visiting and why? *Tourism Recreation Research*, 1–12. doi:10.1080/02508281.2023.2167908

Marson, D. (2011). From mass tourism to niche tourism. In *Research themes for tourism* (pp. 1–15). CABI. doi:10.1079/9781845936846.0001

Millán, G. D., Rojas, R. D. H., & García, J. S. R. (2019). Analysis of the demand of dark tourism: A case study in Córdoba (Spain). *Mediterranean Journal of Social Sciences*, *10*(1), 161–176. doi:10.2478/mjss-2019-0015

Min, J., Yang, K., & Thapa-Magar, A. (2021). Dark tourism segmentation by tourists' motivations for visiting earthquake sites in Nepal: Implications for dark tourism. *Asia Pacific Journal of Tourism Research*, *26*(8), 866–878. doi:10.1080/10941665.2021.1925315

Mionel, V. (2020). (Not so) Dark tourism: The Merry Cemetery in Săpânţa (Romania)–an expression of folk culture. *Tourism Management Perspectives*, *34*, 100656. doi:10.1016/j.tmp.2020.100656

Novakovic, J. (2021). The Role of Museums in a Digital World: Attracting Youth and Overcoming COVID19 Obstacles. *Cultural Management: Science and Education*, 59-68.

Novelli, M. (2018). Niche tourism: Past, present and future. The SAGE handbook of tourism management: Applications of theories and concepts of tourism, 1, 344-359.

Park, E., Choi, B. K., & Lee, T. J. (2019). The role and dimensions of authenticity in heritage tourism. *Tourism Management*, *74*, 99–109. doi:10.1016/j.tourman.2019.03.001

Powell, R., Kennell, J., & Barton, C. (2018). Dark Cities: A dark tourism index for Europe's tourism cities, based on the analysis of DMO websites. *International Journal of Tourism Cities*, *4*(1), 4–21. doi:10.1108/IJTC-09-2017-0046

Preko, A., Mohammed, I., Gyepi-Garbrah, T. F., & Allaberganov, A. (2021). Islamic tourism: Travel motivations, satisfaction and word of mouth, Ghana. *Journal of Islamic Marketing*, *12*(1), 124–144. doi:10.1108/JIMA-04-2019-0082

Raine, R. (2013). A dark tourist spectrum. *International Journal of Culture, Tourism and Hospitality Research*, *7*(3), 242–256. doi:10.1108/IJCTHR-05-2012-0037

Renaud, L. (2020). Reconsidering global mobility–distancing from mass cruise tourism in the aftermath of COVID-19. *Tourism Geographies*, *22*(3), 679–689. doi:10.1080/14616688.2020.1762116

Richards, G. (2018). Cultural tourism: A review of recent research and trends. *Journal of Hospitality and Tourism Management*, *36*, 12–21. doi:10.1016/j.jhtm.2018.03.005

Robinson, M., & Novelli, M. (2007). Niche tourism: an introduction. In *Niche tourism* (pp. 1–10). Routledge.

Seaton, A. V. (1996). Guided by the dark: From thanatopsis to thanatourism. *International Journal of Heritage Studies*, *2*(4), 234–244. doi:10.1080/13527259608722178

Sharma, P., & Nayak, J. K. (2020). Examining experience quality as the determinant of tourist behavior in niche tourism: An analytical approach. *Journal of Heritage Tourism*, *15*(1), 76–92. doi:10.1080/1743873X.2019.1608212

Sidali, K. L., Kastenholz, E., & Bianchi, R. (2018). Food tourism, niche markets and products in rural tourism: Combining the intimacy model and the experience economy as a rural development strategy. In *Rural Tourism* (pp. 47–65). Routledge. doi:10.4324/9781315111865-3

Sintobin, T. (2021). Traveller, tourist and the 'lost art of travelling': The debate continues. In *Routledge Handbook of the Tourist Experience* (pp. 215–234). Routledge. doi:10.4324/9781003219866-20

Skotis, A., & Livas, C. (2022). A data-driven analysis of experience in urban historic districts. *Annals of Tourism Research Empirical Insights*, *3*(2), 100052. doi:10.1016/j.annale.2022.100052

Soava, G. (2015). Development prospects of the tourism industry in the digital age. *Revista tinerilor economişti*, (25), 101-116.

Soligo, M., & Dickens, D. R. (2020). Rest in fame: Celebrity tourism in Hollywood cemeteries. *Tourism, Culture & Communication*, *20*(2-3), 141–150. doi:10.3727/109830420X15894802540214

Stone, P. R. (2006). A dark tourism spectrum: Towards a typology of death and macabre related tourist sites, attractions and exhibitions. *Tourism: An International Interdisciplinary Journal*, *54*(2), 145–160.

Tarlow, P. E. (2007). Dark tourism: The appealing 'dark'side of tourism and more. In *Niche tourism* (pp. 47–58). Routledge.

Toivonen, A. (2022). Sustainability dimensions in space tourism: The case of Finland. *Journal of Sustainable Tourism, 30*(9), 2223–2239. doi:10.1080/09669582.2020.1783276

Torabi Farsani, N., Zeinali, H., & Moaiednia, M. (2018). Food heritage and promoting herbal medicine-based niche tourism in Isfahan, Iran. *Journal of Heritage Tourism, 13*(1), 77–87. doi:10.1080/1743873X.2016.1263307

Troshin, A. S., Sokolova, A. P., Ermolaeva, E. O., Magomedov, R. M., & Fomicheva, T. L. (2020). Information technology in tourism: Effective strategies for communication with consumers. *Journal of Environmental Management & Tourism, 11*(2 (42)), 322–330. doi:10.14505//jemt.v11.2(42).10

Trunfio, M., Petruzzellis, L., & Nigro, C. (2006). Tour operators and alternative tourism in Italy: Exploiting niche markets to increase international competitiveness. *International Journal of Contemporary Hospitality Management, 18*(5), 426–438. doi:10.1108/09596110610673556

UNWTO. (2018). *Tourism and culture synergies*. UNWTO.

Wanda George, E. (2010). Intangible cultural heritage, ownership, copyrights, and tourism. *International Journal of Culture, Tourism and Hospitality Research, 4*(4), 376–388. doi:10.1108/17506181011081541

Williams, H. (2021). Communing with the Fictional Dead: Grave Tourism and the Sentimental Novel. *British Sociability in the European Enlightenment: Cultural Practices and Personal Encounters*, 41-62.

Xie, P. F., Lee, M. Y., & Wong, J. W. C. (2020). Assessing community attitudes toward industrial heritage tourism development. *Journal of Tourism and Cultural Change, 18*(3), 237–251. doi:10.1080/14766825.2019.1588899

Yekimov, S., Sobirov, B., Turdibekov, K., Aimova, M., & Goncharenko, M. (2022). Using the Digital Ecosystem in Tourism Clusters in Green Tourism. In *Ecosystems Without Borders: Opportunities and Challenges* (pp. 105–111). Springer International Publishing. doi:10.1007/978-3-031-05778-6_11

ADDITIONAL READING

Bolan, P., & Simone-Charteris, M. (2018). 'Shining a digital light on the dark': Harnessing online media to improve the dark tourism experience. The Palgrave handbook of dark tourism studies, 727-746.

Collins-Kreiner, N. (2007). Graves as attractions: Pilgrimage-tourism to Jewish holy graves in Israel. *Journal of Cultural Geography*, *24*(1), 67–89. doi:10.1080/08873630709478217

Hooper, G., & Lennon, J. J. (Eds.). (2016). *Dark tourism: Practice and interpretation*. Routledge. doi:10.4324/9781315575865

Lennon, J. J. (2017). Conclusion: Dark tourism in a digital post-truth society. *Worldwide Hospitality and Tourism Themes*, *9*(2), 240–244. doi:10.1108/WHATT-12-2016-0075

Macchia, A., Montorsi, S., Salatino, G., Albini, R., Cerilli, E., Biribicchi, C., Faella, M., Rogliani, A., de Caro, T., Lubritto, C., Vetromile, C., Di Cicco, M. R., Ambrosini, A., & Sperduti, A. (2023). Preserving Intangible Heritage through Tangible Finds: The "Skull with Ears"—Santa St. Luciella ai Librai's Church (Naples, Italy). *Heritage*, *6*(4), 3541–3566. doi:10.3390/heritage6040188

Walter, T. (2009). Dark tourism: Mediating between the dead and the living. *The darker side of travel: The theory and practice of dark tourism*, 39-55.

KEY TERMS AND DEFINITIONS

Cemetery: Place where in Western culture the body of a deceased is buried underground.

Church: Place of worship used for religious functions, often, especially in the past, they were buildings with artistic relevance. In Western culture, in the past they often housed small cemeteries and ossuaries in the underground part.

Cult of Capuzzelle: Or cult of pezzentelle souls, is an ancient Neapolitan custom which involved the polishing of the skulls in the ossuaries of cemeteries and churches in the Neapolitan subsoil. In these conditions the skull was slightly wet due to the humidity and the people believed that it sweated due to the proximity of the soul to hell in the world of the dead. They would then polish the skulls, with the hope that the related soul in the afterlife could intercede by giving them luck as a reward.

Cultural Enhancement: Process that involves practices and actions in order to promote a place, an artifact, a monument, a work, or traditions, in order to make them known to a wider audience and in a more in-depth manner.

ICOM: International Council of Museums, a worldwide network that includes museum professionals and represents an important policymaker.

Intangible Cultural Heritage: The practices, representations, expressions, knowledge, and skills, as well as the tools, artifacts, objects, and associated cultural spaces, that communities, groups and, in some cases, individuals, recognize as an integral part of their cultural heritage.

Purgatory: Western and especially Christian concept which identifies this place in the world of the dead as an intermediate waiting place between hell and heaven

UNESCO: United Nations Educational, Scientific and Cultural Organization, a body born after the Second World War in order to preserve, protect and promote cultural and artistic artifacts and prevent them from being damaged or lost, as happened in the two world wars.

World Heritage Sites: Places recognized as part of a common heritage for their cultural and historical relevance worldwide.

ENDNOTES

1 In the local dialect in Naples, it means "the group."

2 Museum project.

3 We breathe art.

Chapter 9
Antique Bookstore Marketing Strategies as Urban Cultural Landmarks:
A Case Analysis for Suzhou Antique Bookstore

Chenyang Xu
The University of Hong Kong, Hong Kong

Xuechen Gao
The University of Hong Kong, Hong Kong

Apple Hiu Ching Lam
https://orcid.org/0000-0002-2587-6979
The University of Hong Kong, Hong Kong

Dickson K. W. Chiu
https://orcid.org/0000-0002-7926-9568
The University of Hong Kong, Hong Kong

ABSTRACT

Antique bookstores record the history of urban development and are crucial for urban cultural construction. Traditional business models can no longer meet antique bookstores' survival needs in the digital era. Thus, this study investigates a traditional physical bookstore, Suzhou Antique Bookstore, as the case to re-examine its functions and values as an urban cultural landmark, proposing a new sort of sustainable niche cultural tourism. Based on interviews, the authors discuss the pros and cons of the bookstore's current position using the STP model and the 7Ps marketing mix. Despite the current strategies, there is still room for improvement in digital trends. As an urban cultural benchmark, they suggest three transformation strategies for the bookstore: product selection, experience creation, and media joint. Scant studies have proposed insights into the sustainable development of antique bookstores from the perspective of urban cultural construction, especially in Asia.

DOI: 10.4018/978-1-6684-7242-2.ch009

INTRODUCTION

Centered in the Yangtze River Delta, Suzhou is China's important historical and cultural city. According to the Special Plan for the Protection of the Historical and Cultural City of Suzhou issued by the Suzhou Natural Resources and Planning Bureau, Suzhou has a cultural heritage of thousands of years, bred countless celebrities and sages, and cultivated refined and elegant humanistic characteristics (Sky, 2020). In such an environment, Suzhou's antique book industry has always been a vital cultural label representing this ancient city's cultural heritage and the south of the lower reaches of the Yangtze river (Jiangnan area).

In 1958, Suzhou Xinhua Bookstore merged 18 famous traditional bookstores in the late Qing Dynasty, such as Literature Mountain Villa and Laiqing Pavilions, forming the current Suzhou Antique Bookstore. The Bookstore mainly acquires, collects, and sells antique books, second-hand books, and calligraphy and paintings of famous artists. Besides, it provides literature materials for high education institutions, public libraries, and academic institutions (Suzhou Antique Bookstore, 2015). Suzhou Antique Bookstore has acquired over thirty thousand volumes of old books in the past ten years, significantly contributing to the rescue of precious cultural heritage and the promotion of traditional Chinese culture. Cooperating with academic institutions and famous literati, the Bookstore has brought remarkable social benefits to promote the research and development of cultural industries in Suzhou and the Yangtze River Delta. As a result, Suzhou Antique Bookstore is a veritable cultural landmark of Suzhou.

Due to the potential cultural destruction by mass tourism and cultural preservation of the local arts and literature, this study collected the data from the interviews with the owner and staff (Jiang et al., 2019), field observation, and some online resources related to the Bookstore and associated topics. We applied the STP marketing model to re-examine the target market and positioning of Suzhou Antique Bookstore as the cultural landmark of Suzhou, developing as a sustainable niche cultural tourism with the assistance of information technologies without diminishing local culture, and used the 7Ps marketing model to analyze the advantages and disadvantages of the current marketing strategies in terms of developing urban culture. Based on the analysis results, we proposed strategies to help the Bookstore strengthen its status as an urban cultural symbol.

LITERATURE REVIEW

Development Models of the Antique Book Industry

Antique bookstores have not only economic value but also irreplaceable cultural value. Within China, experts have investigated how antique bookstores can adapt to the digital age. Antique bookstores carry a city's cultural history and play an essential role in the progress of urban civilization and the life of residents. Thus, the support of policies, the cohesion of public welfare, and the maintenance of the Bookstore's characteristics are all necessary to promote the revival of bookstores (Yang & Deng, 2018). Su et al. (2019) investigated antique bookstores in the Jiangnan area of China and suggested that since antique bookstores reflect the cultural accumulation of a city, the government, society, and consumers should work together to save these declining cultural symbols. In the context of Internet plus, Liu (2020) analyzed the antique bookstores in Nanjing and proposed a transformation model by combining online and offline channels, expanding the business scope, and creating urban cultural landmarks. Shi and Bu (2021) propose a method for antique bookstores that conforms to the digital age by integrating four levels of marketing: tools, processes, organization internals, and relationships.

For international research, Lloyd-Jones and Davies (2006) opined that antique bookstores in the British countryside should provide online sales and service models and establish a user-friendly website to ensure competitiveness. An and Seo (2013) used service quality testing techniques to analyze and study customer satisfaction in used bookstores in Korea and suggested improvement strategies for them. Other related international studies are quite limited.

The Relationships Between Bookstores, Urban Culture, and Tourism

As an urban public space, bookstores play an essential role in connecting urban culture, emotion, and social relations, which are landmark buildings that condense the memory of urban residents and are the epitome of urban culture (Wang, 2021). Zhang (2021) points out that bookstores differ from traditional libraries in that visitors' consumption experience will gradually become an emotional identity, thus helping this consumer community to establish informal social relationships and urban trust. Only recently have some libraries modeled after bookstores and cafes to attract young people's visit (Deng et al., 2019; Lo et al., 2014; Zhou et al., 2022) and emphasize communities, social connections, and social capital (Leung et al., 2022; Lei et al., 2021; Ni et al., 2023; Zhang et al., 2023).

A successful bookstore should develop from a primary book market to a landmark cultural space with multiple experiences, conveying the cultural connotation upheld by the city and becoming a cultural representative for the city's foreign exchanges (X. Li, 2021). Bookstores connect people with places, spaces, buildings, streets, and communities, linking complex relationships among people and the city's history, culture, and memory (Zhu & Sui, 2020). Culture is inscribed in the accessible physical environments and urban architectures, such as museums and themed bookstores, which make invisible culture legible and accessible (Dicks, 2004). From the perspective of specific regions, Qiu (2018) took Wuhan as an example, summed up the critical role of antique bookstores in urban cultural construction from three levels of history, emotion, and value, and concluded their irreplaceable significance in the construction of urban culture and city image. Cavaglieri and Steindel (2009) studied the antique bookstores established in the 1980s in Brazil and stated that such bookstores are essential to local society in terms of culture and information dissemination.

As antique bookstores can convey local and traditional cultures to others for cultural exchange, they can also serve as a type of cultural and heritage attraction, e.g., urban cultural landmarks, to promote culture for tourists. Nevertheless, Korstanje and Seraphin (2017) state that tourism might not pay much attention to the valuable ancient histories of cultures and traditions, and mass consumption and digital technologies might detach people from local cultures and customs. Cultures may be commoditized as a tourism product to attract tourists, thereby diminishing the authenticity of the local cultures and causing the local cultures and traditions to be faded out. Yet, Shim and Santos (2014) indicate that tourist experiences may not be affected by the degree of authenticity of the cultures prepared for them. Instead, the city's local identity may be more appealing, reflecting from the traditional cultures, customs, traditions, and the residents' daily lives (Shim & Santos, 2014). Yet, the dilemma of cultural preservation and tourism development in Kinmen (Yang & Hsing, 2001) and Qiqiao village (Zuo et al., 2023) is quite similar to our case, and feasible solutions to regional tourism may include alliance and partnership to stress the retainment of cultural uniqueness and memory and consideration of residents' lives while acquiring economic growth through tourism development.

Research Gap

According to the existing literature, although some researchers have focused solely on the antique book industry while some have focused on the relationship between ordinary bookstores and urban culture, scant studies have linked antique bookstores to urban cultural landmarks. As mass tourism can damage local cultures, recourses, and probably the destinations' authenticity, antique bookstores witnessed the transition of local histories can be repositioned as urban cultural landmarks, with the help of information technologies, as a kind of sustainable niche cultural tourism in future by retaining its cultural uniqueness while developing. Further, younger tourists are increasingly using technologies in their information search for destination planning and information during their trips (Cheng et al., 2023; Gong et al., 2017; Ni et al., 2022). This study aims to fill this research gap through the case of Suzhou Antique Bookstore by using the STP and 7Ps marketing model to analyze the current status of the Bookstore and suggest marketing strategies to enhance its position as the Suzhou cultural landmark.

STP MARKET ANALYSIS

Marketing refers to the social and managerial process by which individuals and groups create and exchange products and value with others to obtain their demands (Kotler et al., 2021). As the core of modern strategic marketing, American marketing expert Wendell Smith first proposed the STP model in 1956 and then further improved by Philip Kotler, including three dimensions of Segmentation, Targeting, and Positioning (Liu & Wu, 2013). The STP model is conducive to understanding each market segment's features and conveying the products' value to the target audience by adjusting marketing strategies (MindTools, 2022). This section uses the STP model to analyze the primary target customers of Suzhou Antique Bookstore and clear its position as Suzhou's cultural landmark.

Market Segmentation

Segmentation refers to market segmentation, as a broad market comprises subsets according to different criteria, such as customer characteristics, behaviors, and needs (Kotler et al., 2021). Based on geographical locations, the customer population of Suzhou Antique Bookstore comes from all over the country but mostly from the Yangtze River Delta, like Shanghai, Nanjing, Hangzhou, and Changzhou (Suzhou Antique Bookstore, 2015). Since ancient times, this area has cultivated many outstanding scholars and cultural celebrities. The earliest germination of capitalism in China also appeared there, which is of epic significance in Chinese cultural history (Su et al., 2019). Therefore, the Bookstore can serve as a representative landmark to arouse the cultural identity of customers who also belong to the area.

According to demographic characteristics, in addition to local consumers in Suzhou, tourists constitute a large customer group of the Bookstore. Suzhou is a famous tourist destination in China. Relying on the reputation of "Paradise is above, with Suzhou and Hangzhou below," Suzhou receives many tourists worldwide (Lv,

2016). As a cultural landmark, Suzhou Antique Bookstore is a reading place and cultural sustenance for locals, as well as a meaningful choice for tourists to learn about Suzhou and Jiangnan culture.

Regarding behavioral factors, Suzhou Antique Bookstore serves daily book-buying readers, such as students, and attracts experts and amateurs of literature and art, such as writers and painters (Jiang et al., 2019; Lo et al., 2021a; 2021b). The Suzhou Writers Association (2017) has set up the Suzhou Antique Bookstore as an activity base to jointly carry out literary activities and prosper Suzhou's cultural services. Traditional Chinese cultural products like antique books, calligraphy, and paintings sold in antique bookstores are the link to gathering celebrities.

Target Market

Targeting refers to the target market, that is, after segmenting the market, selecting the most profitable market for the organization's business as the target of marketing activities (Kotler et al., 2021). From the perspective of geographical features, the primary target market of Suzhou Antique Bookstore is the cities in the Yangtze River Delta. Due to the close geographical location, the cultural identity among the residents of Suzhou and surrounding cities is high. The Bookstore can be one of the cultural gathering places to promote the culture of Suzhou and the Jiangnan area.

Regarding demographic characteristics, the target market of Suzhou Antique Bookstore comprises two subpopulations: residents and tourists. For residents, the Bookstore serves as a daily reading and cultural market; for tourists, it can help tourists experience and understand Suzhou's cultural heritage in an immersive way, establish Suzhou's cultural image, and promote the city's popularity.

Regarding behavioral aspects, the customer base of Suzhou Antique Bookstore concentrates on traditional culture professionals and amateurs. By building the Bookstore a cultural learning and exchange base, consumers with similar hobbies and aims can establish social links (Leung et al., 2022), thereby generating emotional identification (Jacobs, 1962) and improving customer loyalty. With the word-of-mouth spread in the Literati social circle (Chan et al., 2022), the popularity of the Bookstore increases accordingly.

To sum up, the target market of Suzhou Antique Bookstore can focus on the residents of the Yangtze River Delta who love traditional culture and tourists who want to understand Suzhou culture, as well as cultural professionals and amateurs.

Market Position

Positioning refers to market position, which means through a series of marketing activities, the attributes and values of products can be perceived by consumers to find an appropriate place in the market (Kotler et al., 2021). Based on the features of target customers, to create a cultural symbol with Suzhou characteristics and give full play to the meaning and value of antique books, Suzhou Antique Bookstore needs to position itself. Like the famous Suzhou gardens, the Bookstore is a carrier of cultural development in the Suzhou and Jiangnan area with a large amount of historical, cultural, philosophical, and scientific information, reflecting the Suzhou locals' lifestyle and etiquette culture (Jiang et al., 2019). As a result, it is an ideal path for Suzhou Antique Bookstore to position itself as the city's cultural landmark and thus create a differentiated brand from other competitors. These competitors include online bookstores, such as Taobao and Amazon, and general physical bookstore chains where people can buy general reading books and literature, plus professional auction houses for cultural literature and rare books (see *Figure 1*).

Firstly, the Bookstore should become a platform to display local culture. Shaping local culture and highlighting regional differences in the new space-time scene can help the Bookstore resist the impact of popular culture and globalization (Feuchtwang, 2012). Besides, building the brand of the Bookstore from a higher level (urban culture) and strengthening the sense of presence is a way to gain the government's and society's attention (Yang & Deng, 2018). Specifically, the Bookstore can present Suzhou's conventional culture in depth through activities such as Suzhou history lectures and themed book exhibitions to build a distinctive Suzhou cultural image (Chen et al., 2018; Deng et al., 2022).

Secondly, it is wise for Suzhou Antique Bookstore to continue assuming the cultural exchange intermediary role. The Bookstore is a public cultural space for people to socialize (Leung et al., 2022). Celebrities from diverse fields, such as bibliophiles, calligraphers, painters, and sculptors, come to the Bookstore to hold book talks and lectures (Suzhou Antique Bookstore, 2015). By providing a space for knowledge dissemination and sharing (Lu et al., 2023), participants can not only contact and understand a culture but communicate and interact to realize a knowledge-based society (Zhu & Sui, 2020). In this way, the Bookstore maintains its original functions and serves as a place for cultural exchanges to explore more business channels and publicity models.

Thirdly, Suzhou Antique Bookstore should encourage the recreation of cultural values. The knowledge reproduction of bookstores can rely on book commodities for content reorganization and carrier conversion (M. Li, 2021). Through the products and activities of the Bookstore, literature fans can further appreciate the essence as recreational material. This is why Suzhou culture can continue to survive and develop in the long river of history. It is also how Suzhou Antique Bookstore builds its brand and establishes itself as a cultural benchmark.

Figure 1. Positioning map of Suzhou Antique Bookstore

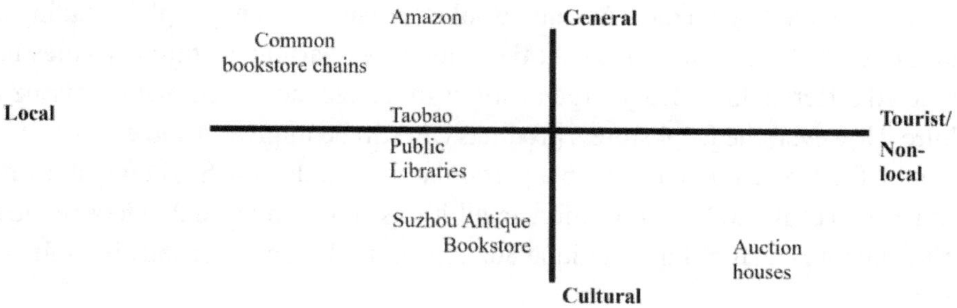

7PS MARKETING ANALYSIS

E. Jerome McCarthy first published the 4Ps marketing theory in 1960, combining four basic strategies: Product, Price, Place, and Promotion (Rowley, 2006). After that, some extended marketing mixes have been proposed by other researchers, such as the 7Ps service marketing theory comprising Personnel (i.e., people), Physical Assets (i.e., physical evidence), and procedures (i.e., process) (Goldsmith, 1999), to offer more comprehensive marketing perspectives to the researchers and practitioners. This section analyzes the marketing strategies of Suzhou Antique Bookstore through the 7Ps model, together with the merits and limitations from the perspective of the urban cultural landmark of Suzhou.

Product

Products are created by an organization to satisfy people's wishes and needs and are acquired by those willing to pay (Išoraitė, 2016). For information centers, such as libraries and bookstores, in addition to tangible products, intangible services are also 'products' that can be provided, such as training courses (Alipour-Hafezi et al., 2013). As a physical bookstore, tangible books and stationery are the primary products Suzhou Antique Bookstore owns. The Bookstore mainly sells old books, second-hand books, history and philosophy books, art books, picture albums, and the Four Treasures of the Study (Suzhou Antique Bookstore, 2015). As the primary products, besides antique books, which are of research value, the Bookstore also provides literature to academic institutions and public libraries (Suzhou Antique Bookstore, 2015). In addition, the Bookstore holds cultural activities periodically. For example, the Bookstore offers study tour programs for foreigners to experience Suzhou's traditional handicrafts (Jiang et al., 2019).

A successful bookstore usually integrates urban culture into its operations, and its unique cultural and operating characteristics make it a symbol in the city. The products provided by Suzhou Antique Bookstore should meet people's reading and creation needs (Tse et al., 2022). At the same time, various lecture activities have contacted different knowledge organizations to realize the dissemination of Suzhou culture. However, the Bookstore's products can still be improved and expanded. For example, the Bookstore can set up a particular area to display Suzhou's innovative products (Lu et al., 2023). In addition, exhibitions and promotional videos related to Suzhou history can highlight a unique Suzhou cultural promotion and dissemination center.

Price

When considering price setting, it is necessary to make consumers perceive the value of the products (Pratt, 2022). Since the price can be flexibly adjusted according to the market, which might affect consumers' willingness to purchase, it significantly impacts the organization (Išoraitė, 2016). According to the research of Suzhou Antique Bookstore, the pricing standard of books in the Bookstore is unclear. Specifically, the Bookstore sells at suggested retail prices for new books and bestsellers. However, the pricing is relatively subjective for second-hand and antique books. The price has only been penciled on the cover of the books, and most traces have been lost. Customers need to ask the staff to determine the price of these antique books. Other services, such as cultural activities, are held for free, providing more experience opportunities for potential customers of the Bookstore.

Suzhou Antique Bookstore's pricing aligns with the market environment for bestsellers and cultural products (such as the Four Treasures of the Study), which already have a mature market. However, there is still a big gap in the pricing of vintage books. Generally speaking, the length of publication and the condition of the books are the considerations of book pricing (Harwell, 2017). The bookstores should formulate a more transparent system of old book pricing to improve the purchasing experience.

Place

Place means the distribution method through which products enter the market (Pratt, 2022). Suzhou Antique Bookstore is located on Renmin Road, Gusu District, Suzhou City. As one of the streets with the longest history in Suzhou, it has witnessed changes and development in the city (Netease, 2021). Being an independent bookstore, the Bookstore sells books through direct distribution with no distributors or chain stores. The Bookstore uses its premise to provide consumers access to books and products. Customers visit the Bookstore to purchase the needed products and enjoy the services provided. Zhang Hua, the owner of Suzhou Antique Bookstore, stated that there is no online shopping channel which may make readers lack an intuitive experience of the appearance of antique books before their visits (Gao & Chiu, 2023; Jiang et al., 2019).

Appropriate venues and environments and effective linking channels can enable consumers to obtain higher satisfaction when using services (Alipour-Hafezi et al., 2013; Chan et al., 2020; Lo et al., 2014; Yip et al., 2019). Suzhou Antique Bookstore has a superior geographical location since the city's main road represents the city's history. It is convenient for local citizens and worth visiting for tourists. As for the distribution model, the current sales channels of the Bookstore tend to be traditional physical bookstores. The current way has both benefits and drawbacks. The merit is that consumers can enjoy products and services in the bookstore with decoration in Suzhou style to have a higher sense of experience. However, there has been a shift in the behaviors of consumers in the digital age, and some transactions have moved online (Chan & Chiu, 2022; Gorokhova, 2021), which may cause the physical store profit to shrink. Suzhou Antique Bookstore can explore online distribution methods to meet the needs of consumers in different regions.

Promotion

Promotion refers to behaviors that encourage purchases, build customer loyalty, and help organizations achieve higher sales (Todorova, 2015). Promotion includes advertising, public relations, and sales (Rowley, 2006). The Suzhou Antique Bookstore sets up a membership system, and the experience service of cultural products can be obtained through cumulative consumption (Jiang et al., 2019). Regarding public relations, the Bookstore has opened a WeChat public account to promote information and communicate with consumers (Jiang et al., 2019). Moreover, the bookstore invites celebrities to hold book signings and lectures, such as collector Ma Weidu and painter Yang Mingyi (Suzhou Antique Bookstore, 2015), which is also a promotional method to attract customers.

In the digital era, the speed of information release and update is greatly accelerated, and organizations can easily publish multimodal digital promotions to customers through various websites and social media (Cheng et al., 2020; Churchill & Barratt-Pugh, 2020; Lam et al., 2023; Wang et al., 2021; Yang et al., 2022). The Bookstore attracts potential customers through promotion behaviors while connecting with regular users. However, more promotion methods can allow the Bookstore to develop Suzhou's urban culture. For example, the production of Suzhou historical and cultural promotional videos with the Bookstore scenarios elements can enable consumers to understand Suzhou while knowing the bookstore and elevate the positioning of the bookstore to the city level (Zuo et al., 2023). In addition, more social media platforms (Cheung et al., 2022; Jiang et al., 2023; Lam et al., 2023), such as Weibo, can expose the Bookstore to more potential users through expanded promotion coverage.

People

People refer to the staff members as inseparable from the service quality as they may significantly influence customers' perceptions of a product's and service provider's quality (Goldsmith, 1999; Naikwadi & Chaskar, 2012). The Bookstore currently has around one staff per floor only during regular weekday business hours, possibly causing difficulty handling customers during peak hours. With a relatively indifferent service attitude, most staff members are middle-aged and even older residents. Furthermore, their key job requirement is their familiarity with ancient literature and traditional culture, significantly affecting their suitability for the positions. They can handle the cashier and book organization, yet they do not have much expertise in the Bookstore's operations and products.

Failing to offer a high-quality service can influence customers' experience and, thus, lose the opportunity to establish customer relationships through communication (Chan & Chiu, 2023; Guo et al., 2022; Ni et al., 2023; Tsang & Chiu, 2022). Eventually, the Bookstore may result in a high customer churn. Therefore, the Bookstore should hire enthusiastic staff who are familiar with the Bookstore's history, operations, and products with their histories and offer them regular training on some updates of new operations and new products (Goldsmith, 1999) to promote the traditional culture successfully and thus enhance itself as an urban cultural landmark.

Process

Process refers to the service operations that add value to the products by applying behaviors and functions, such as the procedures of buying and using products for customers (Goldsmith, 1999). The Bookstore uses manual barcode scanning for book registrations checking in and out. According to Chen (2016), the age of the books or dust stains may cause scanning barcodes to yield low recognition rates. Likewise, manual accounting documentation and old-fashioned paper-and-pencil form of book sales are unusual and highly susceptible to data collection and storage bias. Any unintentional data loss or negligence by the staff could impact the operations and financial results.

Regarding book sales, the modern technology used in the Bookstore is merely for the payment stage: WeChat and Alipay payment QR codes. To conclude, the Bookstore uses an outdated process for storing data and information highly prone to error and loss. Goldsmith (1999) stated that information technology is a key component of this factor. Therefore, to be elevated as an urban cultural landmark, the Bookstore can first increase modern payment methods, such as credit cards, to allow payment by foreign tourists interested in traditional Chinese cultures and calligraphies. Besides, it should run an electronic point-of-sale or point-of-service system to facilitate its business operations, such as reduction of accounting data collection and storage mistakes and errors. The smoother the business operations, the more efficient the service provided to the customers. It can enhance the positioning of the Bookstore as an urban cultural landmark.

Physical Evidence

Physical evidence refers to the surroundings in which the service is provided and any tangible items that make it easier to perform and exchange the service (Al Muala, 2012), resulting in a comfortable atmosphere and environment for overall service quality, including but not limited to the in-store ambiance, background music, seating comfort, the layout of the service facilities, and the appearance of the staff (Khan, 2014).

The Bookstore's premise design is in keeping with its cultural character and goods. White walls and tiles embellish the bookstore's facade, while the interior mixes modern design and arrangement with the elegant aesthetics of the old literati. Bestsellers dominate the first floor of the bookshop, organized into sections by subjects like a typical bookstore, like art, culture, history, and philosophy, allowing readers of all ages to browse and choose. This floor's books should be labeled with categories and tags. The second floor offers traditional cultural goods, including brushes, ink blots, stone tablets, topiaries, literary gems, and books on calligraphy and painting. Second-handed books, ancient books, and postcards from the last century are the main items on the third floor. They are organized based on the classification rule of the time they are published, the content, and the purpose of usage, such as the dictionary section, local history section, and the section on minority languages.

Finding a solution to the Bookstore's space design issue is necessary. Customers must retrace their steps during the site tour to access the Bookstore's numerous shelf locations. Because of the position of the staircase, some sections of the books are only reachable by actively searching. Fabio and Elena (2014) suggest the push effect, pointing out the importance of placing goods within reach of customers and how packed aisles can impact sales. To enhance the comfort and experience of customers, the Bookstore can also provide places for reading, recreation, and cultural exchange, such as book talks and sharing, seminars on traditional culture and literature, to disseminate cultural values through various channels and thus reposit the Bookstore as an urban cultural landmark.

DISCUSSIONS AND SUGGESTIONS

The STP and 7Ps marketing models contribute to the analysis of the market status of Suzhou Antique Bookstore and the merits and problems of its current marketing strategies from the perspective of urban cultural landmarks. Based on the analysis, this section first discusses the development trends of the Bookstores and their surroundings and then suggestions on the development strategies of the bookstore to strengthen the position of Suzhou's cultural landmark.

Development Trends

The COVID-19 pandemic (Huang et al., 2021; 2022; 2023) has recently changed people's reading and information habits (Cheng et al., 2022; Chiu & Ho, 2022a; Dai et al., 2023; Yi & Chiu, 2023; Yu et al., 2022), affected the operations of cultural institutions (Chin & Chiu, 2023; Chiu & Ho, 2022b; Meng et al., 2023; Xue et al., 2023; P. Y. Yu et al., 2023), and hit the retail book market worldwide (Cheung et al., 2022; Sung & Chiu, 2022). In 2021, the overall market grew by 1.65% compared to 2020 but decreased by 3.51% compared to 2019 (OpenBook, 2022). Similarly, the situation of Suzhou Antique Bookstore is pessimistic. However, under the context of accelerating the construction of socialist culture in China, the antique books circulated in Suzhou Antique Bookstore, as one of the significant carriers of the cultural industry, play a leading role in the inheritance and development of Chinese traditional culture. At present, exploring the development direction of combining Internet technologies and social networks with traditional cultural products and further enhancing the influence of Jiangnan culture is the priority for Suzhou's cultural industry (Zhang, 2022; Zuo et al., 2023). Further, due to funding available and environmental issues, sustainability is another core concern for development trends (Chung et al., 2020; Ho et al., 2023; H. H. K. Yu et al., 2023).

Select Book Lists and Products to Promote the Suzhou Culture

Suzhou Antique Bookstore should avoid converging with other ordinary bookstores and take on the role of spreading Suzhou's image. Besides an open-access catalog (Chung & Chiu, 2016; Sun et al., 2022), the Bookstore can set up a particular area to display books related to Suzhou and Jiangnan culture (Lu et al., 2023) and creative products related to the city, such as postcards (Meng et al., 2023). Visual representation of the City reflects the real urban space and imagines and reorganizes a city from a specific perspective (Zukin, 1995). Thus, residents and tourists visiting the Bookstore can take the books and creative products of Suzhou culture as souvenirs and spread the urban culture on a larger scale. Secondly, the Bookstore can consider a particular area to promote the works of local writers in the Suzhou and Jiangnan area. Based in Suzhou, such book lists promote unique humanities and create a city brand (Zuo et al., 2023). The Bookstore can also cooperate with local celebrities for book promotion and cultural activities (Lu et al., 2023) so that these celebrities can expand their influence while the Bookstore can attract more visitors. These arrangements can achieve a win-win situation.

Enhance Space Experience to Create Urban Memory of Suzhou

Since Suzhou Antique Bookstore utilizes offline distribution as currently its primary sales channel, enhancing consumers' experience and satisfaction in the Bookstore is a critical way to encourage consumption. Specifically, in the digital age, Suzhou Antique Bookstore can use its own story as a clue to shoot a promotional film on the theme of Suzhou's history and culture and play it on a loop (Zuo et al., 2023). By allocating a space that records urban development and changes in the Bookstore, it is easy to evoke the collective memory of the mnemonic community (Qiu, 2018). For Suzhou citizens, such a film can arouse a sense of cultural identity. In addition, the video makes it easier to immerse the audience in content using multiple senses than just text. Therefore, tourists can also intuitively feel the urban culture of Suzhou and thus be attracted. As a result, customer satisfaction and sales of Suzhou Antique Bookstore can be further improved. More importantly, the urban cultural landmark role of Suzhou Antique Bookstore has been realized.

Using Various Media to Expand the Cultural Influence of Suzhou

In addition to the WeChat public platform, Suzhou Antique Bookstore can use various media to promote itself and Suzhou culture. In China, Weibo has beaten many local competitors to become the dominant social media leader (Che & Ip, 2018). The Bookstore can also create its official account on popular media platforms, such as Weibo, Xiaohongshu, and Bilibili, to introduce products and services and promote Suzhou's urban culture, especially with short-form videos popular to the younger generation (Chan et al., 2020; Cheng et al., 2023; Lam et al., 2023). This enables the Bookstore to gain more audiences, thereby discovering more potential customers. At the same time, some of these social platforms (like Xiaohongshu) also have e-commerce functions, and the Bookstore can develop online sales channels to meet the demands of consumers worldwide. Besides, the Bookstore can cooperate with TV stations to conduct special reports on the Bookstore in combination with the urban culture. TV programs are a common way to obtain information for middle-aged and older adults less exposed to mobile devices. As a result, citizens can elevate the image of Suzhou Antique Bookstore to the level of Suzhou city, thereby benefiting the Bookstore to play the role of an urban cultural landmark.

CONCLUSION AND OUTLOOK

This study innovatively expounds on the marketing strategies of Suzhou Antique Bookstore from the perspective of Suzhou's urban cultural landmark and puts forward suggestions for further development. Through the STP model, this study located the target customers of the Bookstore, including residents of the Yangtze River Delta, tourists who want to know Suzhou culture, literature experts, and amateurs, and thus proposed a suitable market positioning, that is, to create a differentiated bookstore with the urban cultural imprints. Then, through the 7Ps marketing model, this study analyzed the advantages and disadvantages of the current marketing strategies of the Bookstore as the role of the urban cultural landmark and put forward three suggestions for the future: product selection, experience creation, and media joint. In this way, Suzhou Antique Bookstore can be more closely connected with the urban culture of Suzhou. To reduce the cost of some implementations and operations, the Bookstore may consider collaboration with libraries and universities and use more interns and volunteers (Guo et al., 2022; Li & Chiu, 2022; Tsang & Chiu, 2022; Yew et al., 2022).

Case studies, field research, and paper reading are all required academic research methods. The analysis should be based on the specific operating conditions of Suzhou Antique Bookstore. It is a suitable research path to understand the organization through multiple channels in the early stage, determine the research direction of the target organization, inspect the current status of the organization's operation, determine the models and methodology used to analyze the organization and form an analysis text on the target organization.

In the marketing research of information centers, many models can be used, such as STP, 7Ps, and 4C. There are both advantages and disadvantages to these marketing mixes. As for the 7Ps theory used in this paper, the advantage is that it is very representative. The seven analysis dimensions, i.e., product, price, place, promotion, people, physical evidence, and process, can comprehensively cover the service marketing dimensions of Suzhou Antique Bookstore and are intuitive and easy to understand. In addition, the 7Ps strategy helps maintain the interests of the Bookstore itself. But the shortcomings of the 7Ps theory are also evident. Compared with the 4C model starting from the perspective of consumers, the 7Ps emphasize that the organization's goods and services can always be paid for by consumers, lacking the perception of consumer dimension. In addition, the interview data collected from the Bookstore mainly concentrated on the management of how to develop the Bookstore as a tourist attraction. It might not comprehensively analyze the internal sales, market share, and competition.

Future studies can adopt more marketing mixes and matrices to analyze Suzhou Antique Bookstore, for example, the 4C theory and BCG Matrix, further improving the strategies of the Bookstore from the perspectives of consumers and market growth, not only improving the profits of the Bookstore as a profit-making organization but also strengthening the role of the representative of Suzhou urban culture. Customer focus groups and surveys can help further understand their needs and preferences. Besides, we are interested in analyzing public opinions on social media about cultural activities and organizations (Deng & Chiu, 2023; Li et al., 2023; Liu et al., 2023; Wang et al., 2022).

REFERENCES

Al Muala, A. (2012). Assessing the relationship between marketing mix and loyalty through tourists satisfaction in Jordan curative tourism. *American Academic & Scholarly Research Journal*, *4*(2), 7-23.

Alipour-Hafezi, M., Ashrafi-Rizi, H., Kazempour, Z., & Shahbazi, M. (2013). Using 4P marketing model in academic libraries: An experience. *International Journal of Information Science and Management, 11*(2), 45-58.

An, Y.-S., & Seo, K.-K. (2013). A study on service satisfaction factor analysis of an online secondhand bookstore. *Jouranl of Digital Convergence*, *11*(11), 251–256. doi:10.14400/JDPM.2013.11.11.251

Cavaglieri, M., & Steindel, G. E. (2009). Um lugar para observar, conversar, ler, comprar-livros e outros suportes de informação e lazer: Uma análise dos sebos da cidade de Florianópolis [A place to observe, talk, read, buy-books and other information and leisure media: An analysis of the tallow of the city of Florianópolis]. *The Information Society*, *19*(3), 55–64.

Chan, M. K. Y., Chiu, D. K. W., & Lam, E. T. H. (2020). Effectiveness of overnight learning commons: A comparative study. *Journal of Academic Librarianship*, *46*(7), 102253. doi:10.1016/j.acalib.2020.102253 PMID:34173399

Chan, M. M. W., & Chiu, D. K. W. (2022). Alert driven customer relationship management in online travel agencies: event-condition-actions rules and key performance indicators. In A. Naim & S. Kautish (Eds.), *Building a brand image through electronic customer relationship management* (pp. 268–303). IGI Global. doi:10.4018/978-1-6684-5386-5.ch012

Chan, T. T. W., Lam, A. H. C., & Chiu, D. K. W. (2020). From Facebook to Instagram: Exploring user engagement in an academic library. *Journal of Academic Librarianship*, *46*(6), 102229. doi:10.1016/j.acalib.2020.102229 PMID:34173399

Chan, V. H. Y., Ho, K. K. W., & Chiu, D. K. W. (2022). Mediating effects on the relationship between perceived service quality and public library app loyalty during the COVID-19 era. *Journal of Retailing and Consumer Services*, *67*, 102960. doi:10.1016/j.jretconser.2022.102960

Che, X., & Ip, B. (2018). *Social networks in China*. Chandos Publishing.

Chen, D. (2016). Intelligent application system of warehouse management based on RFID technology. *China Digital Cable TV*, *12*, 1372–1374.

Chen, Y., Chiu, D. K. W., & Ho, K. K. W. (2018). Facilitating the learning of the art of Chinese painting and calligraphy at Chao Shao-an Gallery. *Micronesian Educators*, *26*, 45–58.

Cheng, J., Yuen, A. H., & Chiu, D. K. W. (2022). (in press). Systematic review of MOOC research in mainland China. *Library Hi Tech*. Advance online publication. doi:10.1108/LHT-02-2022-0099

Cheng, W., Tian, R., & Chiu, D.K.W. (2023). Travel vlogs influencing tourist decisions: Information preferences and gender differences. *Aslib Journal of Information Management*, ahead-of-print. doi:10.1108/AJIM-05-2022-0261

Cheng, W. W. H., Lam, E. T. H., & Chiu, D. K. W. (2020). Social media as a platform in academic library marketing: A comparative study. *Journal of Academic Librarianship*, *46*(5), 102188. doi:10.1016/j.acalib.2020.102188

Cheung, L. S. N., Chiu, D. K. W., & Ho, K. K. W. (2022). A quantitative study on utilizing electronic resources to engage children's reading and learning: Parents' perspectives through the 5E instructional model. *The Electronic Library*, *40*(6), 662–679. doi:10.1108/EL-09-2021-0179

Cheung, T. Y., Ye, Z., & Chiu, D. K. W. (2021). Value chain analysis of information services for the visually impaired: A case study of contemporary technological solutions. *Library Hi Tech*, *39*(2), 625–642. doi:10.1108/LHT-08-2020-0185

Cheung, V. S. Y., Lo, J. C. Y., Chiu, D. K. W., & Ho, K. K. W. (2023). Evaluating social media's communication effectiveness on travel product promotion: Facebook for college students in Hong Kong. *Information Discovery and Delivery*, *51*(1), 66–73. doi:10.1108/IDD-10-2021-0117

Chin, G. Y. L., & Chiu, D. K. W. (2023). RFID-based Robotic Process Automation for Smart Museums with an Alert-driven Approach. In R. Tailor (Ed.), *Application and Adoption of Robotic Process Automation for Smart Cities*. IGI Global.

Chiu, D. K. W., & Ho, K. K. W. (2022a). Special selection on contemporary digital culture and reading. *Library Hi Tech, 40*(5), 1204–1209. doi:10.1108/LHT-10-2022-516

Chiu, D. K. W., & Ho, K. K. W. (2022b). Editorial: 40th anniversary: contemporary library research. *Library Hi Tech, 40*(6), 1525–1531. doi:10.1108/LHT-12-2022-517

Chung, A. C. W., & Chiu, D. K. (2016). OPAC Usability Problems of Archives: A Case Study of the Hong Kong Film Archive. [IJSSOE]. *International Journal of Systems and Service-Oriented Engineering, 6*(1), 54–70. doi:10.4018/IJSSOE.2016010104

Chung, C., Chiu, D. K. W., Ho, K. K. W., & Au, C. H. (2020). Applying social media to environmental education: Is it more impactful than traditional media? *Information Discovery and Delivery, 48*(4), 255–266. doi:10.1108/IDD-04-2020-0047

Churchill, N., & Barratt-Pugh, C. (2020). The digital entanglement of humanities, literacy, and storytelling. In K. W. Kung (Ed.), *Reconceptualizing the digital humanities in Asia* (pp. 141–154). Springer. doi:10.1007/978-981-15-4642-6_9

Dai, C., & Chiu, D. K. W. (2023). Impact of COVID-19 on reading behaviors and preferences: Investigating high school students and parents with the 5E instructional model. *Library Hi Tech*. Advance online publication. doi:10.1108/LHT-10-2022-0472

Deng, Q., Allard, B., Lo, P., Chiu, D. K., See-To, E. W., & Bao, A. Z. (2019). The role of the library café as a learning space: A comparative analysis of three universities. *Journal of Librarianship and Information Science, 51*(3), 823–842. doi:10.1177/0961000617742469

Deng, S., & Chiu, D. K. W. (2023). Analyzing Hong Kong Philharmonic Orchestra's Facebook Community Engagement with the Honeycomb Model. In M. Dennis & J. Halbert (Eds.), *Community Engagement in the Online Space* (pp. 31–47). IGI. Global. doi:10.4018/978-1-6684-5190-8.ch003

Deng, W., Chin, G. Y.-l., Chiu, D. K. W., & Ho, K. K. W. (2022). Contribution of Literature Thematic Exhibition to Cultural Education: A Case Study of Jin Yong's Gallery. *Micronesian Educators, 32*, 14–26.

Dicks, B. (2004). *Culture on display: The production of contemporary visitability*. McGraw-Hill Education.

Fabio, M., & Elena, D. (2014). *Handbook of Research on Retailer-Consumer Relationship Development*. Business Science Reference.

Feuchtwang, S. (2012). *Making place: State projects, globalisation and local responses in China*. Taylor and Francis. doi:10.4324/9781843147671

Gao, X., & Chiu, D. K. W. (2023). Integration of cultural retailing with new technologies and media: A case study of the Suzhou Antique Bookstore. In J. D. Santos & I. V. Pereira (Eds.), *Management and Marketing for Improved Retail Competitiveness and Performance*. IGI Global.

Goldsmith, R. E. (1999). The personalised marketplace: Beyond the 4Ps. *Marketing Intelligence & Planning*, *17*(4), 178–185. doi:10.1108/02634509910275917

Gong, J. Y., Schumann, F., Chiu, D. K. W., & Ho, K. K. W. (2017). Tourists' mobile information seeking behavior: An investigation on China's youth. *International Journal of Systems and Service-Oriented Engineering*, *7*(1), 58–76. doi:10.4018/IJSSOE.2017010104

Gorokhova, T. (2021). The influence of the development of digital technologies on consumer behaviour. *Ukrainian Black Sea Region Agrarian Science*, *112*(4), 45–54. doi:10.31521/2313-092X/2021-4(112)-5

Guo, Y., Lam, A. H. C., Chiu, D. K. W., & Ho, K. K. W. (2022). Perceived quality of WhatsApp reference service: A quantitative study from user perspectives. *Information Technology and Libraries*, *41*(3). Advance online publication. doi:10.6017/ital.v41i3.14325

Harwell, J. (2017). Rhyme or reason? Patterns in book pricing by format. *The Journal of Electronic Publishing : JEP*, *20*(1). Advance online publication. doi:10.3998/3336451.0020.104

Huang, P.-S., Paulino, Y. C., So, S., Chiu, D. K. W., & Ho, K. K. W. (2021). Editorial - COVID-19 Pandemic and Health Informatics (Part 1). *Library Hi Tech*, *39*(3), 693–695. doi:10.1108/LHT-09-2021-324

Huang, P.-S., Paulino, Y. C., So, S., Chiu, D. K. W., & Ho, K. K. W. (2022). Guest editorial: COVID-19 Pandemic and Health Informatics Part 2. *Library Hi Tech*, *40*(2), 281–285. doi:10.1108/LHT-04-2022-447

Huang, P.-S., Paulino, Y. C., So, S., Chiu, D. K. W., & Ho, K. K. W. (2023). Guest editorial: COVID-19 Pandemic and Health Informatics Part 2. *Library Hi Tech*, *41*(1), 1–6. doi:10.1108/LHT-02-2023-585

Išoraitė, M. (2016). Marketing mix theoretical aspects. *International Journal of Research - Granthaalayah, 4*(6), 25-37.

Jacobs, J. M. (1962). *The death and life of great American cities*. Taylor & Francis.

Jiang, T., Lo, P., Cheuk, M. K., Chiu, D. K. W., Chu, M. Y., Zhang, X., Zhou, Q., Liu, Q., Tang, J., Zhang, X., Sun, X., Ye, Z., Yang, M., & Lam, S. K. (2019). 文化新語:兩岸四地傑出圖書館、檔案館及博物館傑出工作者訪談 [New Cultural Dialog: Interviews with Outstanding Librarians, Archivists, and Curators in Greater China]. Systech Publications.

Jiang, X., Chiu, D. K. W., & Chan, C. T. (2023). Application of the AIDA model in social media promotion and community engagement for small cultural organizations: A case study of the Choi Chang Sau Qin Society. In M. Dennis & J. Halbert (Eds.), *Community Engagement in the Online Space* (pp. 48–70). IGI Global. doi:10.4018/978-1-6684-5190-8.ch004

Khan, M. T. (2014). The concept of 'marketing mix'and its elements. *International Journal of Information, Business and Management*, *6*(2), 95–107.

Korstanje, M., & Seraphin, H. (2017). Revisiting the sociology of consumption in tourism. In S. K. Dixit (Ed.), *The Routledge handbook of consumer behaviour in hospitality and tourism* (pp. 16–25). Routledge.

Kotler, P., Armstrong, G., & Opresnik, M. O. (2021). *Principles of marketing* (18th ed.). Pearson.

Lam, A. H. C., Ho, K. K. W., & Chiu, D. K. W. (2023). Instagram for student learning and library promotions? A quantitative study using the 5E Instructional Model. *Aslib Journal of Information Management*, *75*(1), 112–130. doi:10.1108/AJIM-12-2021-0389

Lei, S. Y., Chiu, D. K. W., Lung, M. M., & Chan, C. T. (2021). Exploring the aids of social media for musical instrument education. *International Journal of Music Education*, *39*(2), 187–201. doi:10.1177/0255761420986217

Leung, T. N., Luk, C. K. L., Chiu, D. K. W., & Ho, K. K. W. (2022). User perceptions, academic library usage, and social capital: A correlation analysis under COVID-19 after library renovation. *Library Hi Tech*, *40*(2), 304–322. doi:10.1108/LHT-04-2021-0122

Li, K. K., & Chiu, D. K. W. (2022). A Worldwide Quantitative Review of the iSchools' Archival Education. *Library Hi Tech*, *40*(5), 1497–1518. doi:10.1108/LHT-09-2021-0311

Li, M. (2021). Analysis on the scene construction of urban physical bookstore based on the perspective of knowledge communication. *Chinese Editors Journal,* (10), 43-46+52.

Li, S., Xie, Z., Chiu, D. K. W., & Ho, K. K. W. (2023c). Sentiment Analysis and Topic Modeling Regarding Online Classes on the Reddit Platform: Educators versus Learners. *Applied Sciences (Basel, Switzerland)*, *13*(4), 2250. doi:10.3390/app13042250

Li, X. (2021). Urban public cultural space as a dialogue: An analysis of the role of physical bookstores. *Journal of News Research*, *12*(04), 251–252.

Liu, Q., & Wu, X. (2013). Research on Chengdu vegetables marketing based on STP Model. *Asian Social Science*, *9*(4), 221–226. doi:10.5539/ass.v9n4p221

Liu, Y. (2020). The cultural persistence and future of antique bookstores under the context of "Internet plus": A case analysis of bookstores in Nanjing. *Renwen Tianxia*, (17), 56–60.

Liu, Y., Chiu, D. K. W., & Ho, K. K. W. (2023). Short-Form Videos for Public Library Marketing: Performance Analytics of Douyin in China. *Applied Sciences (Basel, Switzerland)*, *13*(6), 3386. doi:10.3390/app13063386

Lloyd-Jones, A., & Davies, A. J. (2006). An investigation of eMarketing within the second hand book trade. *BLED 2006 Proceedings*, Article 33. https://aisel.aisnet.org/bled2006/33

Lo, P., Cheuk, M. K., Lam, S. K., & Chiu, D. K. W. (2021a). 文武之道【上冊】:從城市森林中尋訪對閱讀有想法的武者與文化人 [The Tao of Arts and Warriorship, Vol. 1: Looking for martial artists and cultural people with reading ideas in a metropolis]. Systech Publications.

Lo, P., Chiu, D. K. W., & Chu, W. (2014). Modeling Your College Library after a Commercial Bookstore? The Hong Kong Design Institute Library Experience. *Community & Junior College Libraries*, *19*(3-4), 59–76. doi:10.1080/02763915.2014.915186

Lo, P., Hsu, W.-E., Wu, S. H. S., Travis, J., & Chiu, D. K. W. (2021b). *Creating a Global Cultural City via Public Participation in the Arts: Conversations with Hong Kong's Leading Arts and Cultural Administrators*. Nova Science Publishers.

Lu, S.S., Tian, R., & Chiu, D.K.W. (2023). Why do people not attend public library programs in the current digital age? *Library Hi Tech,* ahead-of-print. doi:10.1108/LHT-04-2022-0217

Lv, H. (2016). Introduce to the G20 the proverb "There is heaven above, and Suzhou and Hangzhou below". *Hangzhou Quality Life*, (04), 39–40.

Meng, Y., Chu, M. Y., & Chiu, D. K. W. (2022). The impact of COVID-19 on museums in the digital era: Practices and challenges in Hong Kong. *Library Hi Tech*, *41*(1), 130–151. doi:10.1108/LHT-05-2022-0273

MindTools. (2022). *The Segmentation, Targeting and Positioning (STP) marketing model.* https://www.mindtools.com/a5llt9t/segmentation-targeting-and-positioning-model

Naikwadi, V. A., & Chaskar, P. M. (2012). Plan: For marketing library and information services. *Indian Streams Research Journal*, *2*(10), 1–6.

Netease. (2021). *Renmin Road, the main axis of Suzhou, witnesses the development and changes of Suzhou.* https://www.163.com/dy/article/G37N4E3C0515ESFT.html

Ni, J., Chiu, D. K. W., & Ho, K. K. W. (2022). Information search behavior among Chinese self-drive tourists in the smartphone era. *Information Discovery and Delivery*, *50*(3), 285–296. doi:10.1108/IDD-05-2020-0054

Ni, Y., Lam, A. H. C., & Chiu, D. K. W. (2023). Leveraging Online Communities for Building Social Capital in University Libraries: A Case Study of Fudan University Medical Library. In Balance and Boundaries in Creating Meaningful Relationships in Online Higher Education. IGI Global.

OpenBook. (2022). *2021 China book retail market report.* http://www.cnfaxie.org/detail.html?id=26&contentId=535

Pratt, M. (2022). *What is the 4P marketing matrix?* https://www.business.org/marketing/sales/marketing-101-4p-matrix/

Qiu, M. (2018). Antique bookstores as media: A study based on Wuhan Bookstores. *Press Outpost*, (2), 15–17.

Rowley, J. E. (2018). *Information Marketing*. Routledge.

Shi, Y., & Bu, H. (2021). A preliminary study on the marketing mode of antique bookstores conforming to the trend of The Times. *Time-Honored Brand Marketing*, (8), 25–26.

Shim, C., & Santos, C. A. (2014). Tourism, place and placelessness in the phenomenological experience of shopping malls in Seoul. *Tourism Management*, *45*, 106–114. doi:10.1016/j.tourman.2014.03.001

Sky. (2020). *Special plan for the protection of the historical and cultural city of Suzhou (2035).* Suzhou Natural Resources and Planning Bureau. http://suzhou.bendibao.com/news/20201017/81860.shtm

Su, S., Zhao, Q., & Tian, H. (2019). Saving the antique book Industry in Jiangnan Area: A case analysis of bookstores in Nanjing, Suzhou, Yangzhou and Hangzhou. *Market Weekly*, (6), 143–145.

Sun, X., Chiu, D. K. W., & Chan, C. T. (2022). Recent Digitalization Development of Buddhist Libraries: A Comparative Case Study. In S. Papadakis & A. Kapaniaris (Eds.), *The Digital Folklore of Cyberculture and Digital Humanities* (pp. 251–266). IGI Global. doi:10.4018/978-1-6684-4461-0.ch014

Sung, Y. Y. C., & Chiu, D. K. W. (2022). E-book or print book: Parents' current view in Hong Kong. *Library Hi Tech*, *40*(5), 1289–1304. doi:10.1108/LHT-09-2020-0230

Suzhou Antique Bookstore. (2015). *An ancient city, a time-honored bookstore, and a cultural label of the city*. https://mp.weixin.qq.com/s/Nk1aX2bHkteR813hREm7ww

Suzhou Writers Association. (2017). Yearly Report. *Su Zhou Literary and Art Association Joint Report*, 68-71.

Todorova, G. (2015). Marketing communication mix. *Trakia Journal of Sciences*, *13*(Suppl.1), 368–374. doi:10.15547/tjs.2015.s.01.063

Tsang, A. L. Y., & Chiu, D. K. W. (2022). Effectiveness of virtual reference services in academic libraries: A qualitative study based on the 5E Learning Model. *Journal of Academic Librarianship*, *48*(4), 102533. doi:10.1016/j.acalib.2022.102533

Tse, H. L., Chiu, D. K. W., & Lam, A. H. C. (2022). From reading promotion to digital literacy: An analysis of digitalizing mobile library services with the 5E Instructional Model. In A. Almeida & S. Esteves (Eds.), *Modern Reading Practices and Collaboration Between Schools, Family, and Community* (pp. 239–256). IGI Global. doi:10.4018/978-1-7998-9750-7.ch011

Wang, J., Deng, S., Chiu, D. K. W., & Chan, C. T. (2022). Social network customer relationship management for orchestras: A case study on Hong Kong Philharmonic Orchestra. In N. B. Ammari (Ed.), *Social customer relationship management (Social-CRM) in the era of Web 4.0* (pp. 250–268). IGI Global. doi:10.4018/978-1-7998-9553-4.ch012

Wang, W., Lam, E. T. H., Chiu, D. K. W., Lung, M. M., & Ho, K. K. W. (2021). Supporting Higher Education with Social Networks: Trust and Privacy vs. Perceived Effectiveness. *Online Information Review*, *45*(1), 207–219. doi:10.1108/OIR-02-2020-0042

Wang, Y. (2021). From place to space: The development and construction of physical bookstores as urban public space: A case study of Hefei Physical Bookstores. *Journal of Changzhou Institute of Technology*, *39*(05), 82–87.

Wong, A. K.-k., & Chiu, D. K. W. (2023). Digital transformation of museum conservation practices: A value chain analysis of public museums in Hong Kong. In R. Pettinger, B. B. Gupta, A. Roja, & D. Cozmiuc (Eds.), *Handbook of Research on the Digital Transformation Digitalization Solutions for Social and Economic Needs* (pp. 226–242). IGI Global. doi:10.4018/978-1-6684-4102-2.ch010

Xue, B., Lam, A. H. C., & Chiu, D. K. W. (2023). Redesigning Library Information Literacy Education with the BOPPPS Model: A Case Study of the HKUST. In R. Taiwo, B. Idowu-Faith, & S. Ajiboye (Eds.), *Transformation of Higher Education Through Institutional Online Spaces*. IGI. Global.

Yang, M. C., & Hsing, W. C. (2001). Kinmen: Governing the culture industry city in the changing global context. *Cities (London, England)*, *18*(2), 77–85. doi:10.1016/S0264-2751(00)00059-7

Yang, W., & Deng, S. (2018). A look at the countermeasures for supporting antique bookstores from the perspective of bookstores integration. *View on Publishing*, (12), 10–13.

Yew, A., Chiu, D. K. W., Nakamura, Y., & Li, K. K. (2022). Quantitative Comparison of LIS Programs Accredited by ALA and CILIP. *Library Hi Tech*, *40*(6), 1721–1745. doi:10.1108/LHT-12-2021-0442

Yi, Y., & Chiu, D.K.W. (2023). Public information needs during the COVID-19 outbreak: A qualitative study in mainland China. *Library Hi Tech*, ahead-of-print. doi:10.1108/LHT-08-2022-0398

Yu, H. H. K., Chiu, D. K. W., & Chan, C. T. (2023). Resilience of symphony orchestras to challenges in the COVID-19 era: Analyzing the Hong Kong Philharmonic Orchestra with Porter's five force model. In W. Aloulou (Ed.), *Handbook of Research on Entrepreneurship and Organizational Resilience During Unprecedented Times* (pp. 586–601). IGI Global.

Yu, H. Y., Tsoi, Y. Y., Rhim, A. H. R., Chiu, D. K. W., & Lung, M. M. W. (2022). Changes in habits of electronic news usage on mobile devices in university students: A comparative survey. *Library Hi Tech*, *40*(5), 1322–1336. doi:10.1108/LHT-03-2021-0085

Yu, P. Y., Lam, E. T. H., & Chiu, D. K. W. (2023). Operation management of academic libraries in Hong Kong under COVID-19. *Library Hi Tech*, *41*(1), 108–129. doi:10.1108/LHT-10-2021-0342

Zhang, J. (2022, November 3). Suzhou released white paper on the development of cultural industries. *Xinhua Daily*. https://www.doi.org/10.28872/n.cnki.nxhrb.2022.006001

Zhang, J., Lam, A. H. C., & Chiu, D. K. W. (2023). Evaluating the Effectiveness of Learning Commons as Third Space with the 5E Usability Model: The Case of Hong Kong University of Science and Technology Library. In C. Kaye, J. H. Writer, & J. Batsaikhan (Eds.), *Third-Space Exploration in Education*. IGI Global.

Zhang, X. (2021). Scene integration, community activation, and "Experimental field" — Research on the value of physical bookstores as knowledge production space from the perspective of urban communication. *Dongyue Tribune*, *42*(04), 131–138.

Zhou, J., Lam, E. T. H., Au, C. H., Lo, P., & Chiu, D. K. W. (2022). Library café or elsewhere: Usage of study space by different majors under contemporary technological environment. *Library Hi Tech*, *40*(6), 1567–1581. doi:10.1108/LHT-03-2021-0103

Zhu, Y., & Sui, W. (2020). Exploration and analysis of urban physical bookstore curating and communication based on the perspective of knowledge reproduction. *Editorial Friend*, (11), 46–51.

Zukin, S. (1995). *The Cultures of the Cities*. Blackwell.

Zuo, Y., Lam, A. H. C., & Chiu, D. K. W. (2023). Digital protection of traditional villages for sustainable heritage tourism: A case study on Qiqiao Ancient Village, China. In A. Masouras, C. Papademetriou, D. Belias, & S. Anastasiadou (Eds.), *Sustainable Growth Strategies for Entrepreneurial Venture Tourism and Regional Development*. IGI Global. doi:10.4018/978-1-6684-6055-9.ch009

ADDITIONAL READING

Chan, K. F. S., & Chiu, D. K. W. (2023). Parents' view of graphic novels in Hong Kong under 21st Century mobile digital environment. In D. K. W. Chiu & K. K. W. Ho (Eds.), *Emerging Technology-Based Services and Systems in Libraries, Educational Institutions, and Non-Profit Organizations*. IGI Global.

Cho, A., Lo, P., & Chiu, D. K. W. (2017). *Inside the World's Major East Asian Collections: One Belt, One Road, and Beyond*. Chandos Publishing.

Lo, P., Cho, A., & Chiu, D. K. W. (2017). *World's Leading National, Public, Monastery and Royal Library Directors – Leadership, Management, Future of Libraries*. De Gruyter.

Lo, P., Rogers, H., & Chiu, D. K. W. (2018). *Effective School Librarianship, Two-Volume Set, Successful Professional Practices from Librarians around the World*. Apple Academic Press. doi:10.1201/b22444

Mak, M. Y. C., Poon, A. Y. M., & Chiu, D. K. W. (2022). Using Social Media as Learning Aids and Preservation: Chinese Martial Arts in Hong Kong. In S. Papadakis & A. Kapaniaris (Eds.), *The Digital Folklore of Cyberculture and Digital Humanities* (pp. 171–185). IGI Global. doi:10.4018/978-1-6684-4461-0.ch010

Wong, S. W. S., & Chiu, D. K. W. (2023). Re-examining the value of remote academic library storage in the mobile digital age: A comparative study. *Portal (Baltimore, Md.)*, *23*(1), 89–109.

KEY TERMS AND DEFINITIONS

7Ps Marketing Theory: A kind of marketing mixes comprising seven elements, i.e., product, price, place, promotion, people, process, and physical evidence, to assisting marketers and researchers to develop feasible marketing strategies. Sometimes, it may also be assisted in analyzing current marketing strategies of the organizations to uncover any possible deficiencies.

Antique Bookstore: A type of bookstores majorly acquiring, collecting, and selling antique and rare books, second-hand books, and perhaps also calligraphy and paintings of famous artists.

Cultural Landmark: A place being of cultural significance for a group of people, including monuments and ancient buildings, serving as an iconic representation of, for example, a culture, belief, and historical events.

Cultural Tourism: A sort of tourism that tourists are motivated to learn and experience the tangible and intangible cultural attractions such as arts, architecture, literature, living cultures, traditions, customs, historical heritage, and cultural heritage.

STP Model: A marketing model comprising three elements: segmentation (S), targeting (T), and positioning (P), which is frequently used for creating strategies of marketing mix and product positioning. It may also be used to analyze the current situations of the organizations and reforming marketing strategies.

Suzhou Antique Bookstore: The bookstore located on Renmin Road, Gusu District, Suzhou City, China, which is one of the main streets with the longest history in Suzhou. Formerly as 18 well-known traditional bookstores in late Qing Dynasty, the Bookstore acquires and sells vintage and rare books and other cultural artifacts for local residents, tourists and cultural professionals as well as high education institutions, public libraries, and academic institutions.

Urban Culture: The culture of a town or city that may represent the behavioral patterns of the people living in that town or city in the past and present, such as living cultures with living styles, customs, and traditions.

Chapter 10
Serial Killer Tourism:
Education and Entertainment!?

Titanilla Virág Tevely
Alexandre Lámfalussy Faculty of Economics, University of Sopron, Hungary

Árpád Ferenc Papp-Váry
https://orcid.org/0000-0002-0395-4315
University of Sopron, Hungary

ABSTRACT

Dark tourism consists of a wide range of subsections, and this chapter focuses on a less explored part of it, serial killer tourism. The demand side's fascination with death and murderers from led to diverse tourism offer types, such as museums and walking tours. This research gives an overview of these attractions and an answer to why people are visiting them. To understand the topic, three research questions were formed: What is the reason behind serial killer tourism? What type of attractions are the most attractive? and What is the attitude of the consumers. Based on the research, visitors want to be educated, to prepare to face the dark reality, but also want to be entertained. By exploring this topic, tourism professionals will get insight into the visitors' motivations, how and who to promote these attractions, and how to develop new tourism products.

DOI: 10.4018/978-1-6684-7242-2.ch010

INTRODUCTION: THE FASCINATION WITH DEATH

The duality between life and death always fascinated humans. The history of mankind always been bloody: seeing public executions and cheering on gladiator games were one of the first reasons to travel (Lennon, 2018), and our civilization cannot escape from the memories of wars, genocides, and terrorism. To be curious about this side of the past can be considered as a normal, emotionally engaging behavior, rather than pathological one (Carrabine, 2011).

This curiosity and fascination come from the need to understand the one thing nobody can be sure about: death. Visiting sites associated with atrocities, murders and wars have been considered as part of heritage tourism in a broader sense, but it is also associated with 'thanatourism' or 'dark tourism', as the motivation to see these places involve a desire to have a contact with death (Tunbridge & Ashworth, 1996; Seaton, 1996; Hartmann, 2014; Light, 2017). Carvalho (2023) identified five dark tourist categories: 'dark contemplative tourist', 'dark history tourist', 'dark passive tourist', 'dark enthusiast tourist', and 'dark occasion tourist'.

Since dark tourism consists of an enormous number of types, such as genocide tourism, holocaust tourism and graveyard tourism, this research focuses primarily on a smaller, more easily identifiable subsection of dark tourism: serial killer tourism. The psychology behind the actions of serial killers is appealing, the fear of their existence, and how other people view their acts. People find them frightening and attractive at the same time, which is one of the main reasons to visit places, museums and other attractions connected to serial killers. The morbid curiosity that induces people to seek out macabre entertainment and things related to serial killers is not gender-based, as it is a topic that interests both men and women equally as a method to learn to avoid being a victim (Harrison & Frederick, 2020).

By exploring why people are fascinated with serial killers, tourism professionals will have a better sense of the motivation behind visiting dark tourism attractions. It would help local destination management organizations, websites and travel blogs to know what to offer and how to promote them for the interested tourists. The findings of this research can be helpful for destinations on how to influence destination marketing strategies on how to develop tourism products. It also may help to develop an understanding of how the tourism industry preserves the memory of the victims.

To understand better the topic, three research questions were formed that could be answered based on observations when searching for attractions connected to serial killers, on the available literature about serial killers and dark tourism, and the empirical research.

These are the following:

- What is the reason for serial killer tourism?
- What type of attractions connected to serial killers are the most attractive for the visitors?
- What is the attitude of the consumers towards the attractions connected to serial killers?

According to our research the darker side of the destination is attractive for the tourists, but only in moderation. While they want to be educated, they also want to be entertained when they are traveling, therefore both the emotionally taxing, darkest side of dark tourism and the lighter side that offers macabre fun can be attractive for travelers.

While the morality of dark tourism, especially "serial killer tourism" is questionable, the respondents were interested in both. They want to learn about these criminals, they want to feel more prepared to face the reality of serial killers, but the offer of attractions is not what they require. They think these places are unethical and only those were attractive for the respondents that are either educational or had a distant relation to the actual crimes.

The essay presents what makes these places interesting, why people are visiting them, how ethical and moral it is to make murder into attractions and how a city can use the history of its serial killer to attract more tourists. This study seeks to find an answer to the broader question about what attracts people to consume dark tourism sites, intending to examine how consumers perceive these places.

THE CONCEPT OF DARK TOURISM

Death and Dark Tourism

Dark tourism attractions becoming more and more popular tourist attractions in Western countries, due to their fascinating stories and emotional connotations, thus more destinations are integrating their dark history into their offer and exploiting human tragedy (Wright, 2018). Dark tourism is described as visiting attractions that are connected to human suffering, death, and catastrophes (Stone, 2006). Disaster tourism is used as synonym, but it is only a part of it, defined as a place, where a severe natural destruction or human tragedy struck down, resulting in the death of many people (Tóth & Papp-Váry, 2022). It is complex and different types of attractions are visited for various reasons. Miles (2002) proposed that the shades of dark tourism (light, dark, darker) should be based on the location of the site, as they are associated in different ways with death. The shades of dark tourism, according to

Stone and Sharpley (2008), depends on the degree of exploitation from the suppliers', and the degree of interest in death from the demand's side.

Visiting places of suffering is not just a morbid curiosity; tourists may understand the meaning of life through death (Handayani, 2019). Still, it cannot be denied that there is an issue is with the way it mixes leisure with tragedy (Strange & Kempa, 2003), especially considering the 'darkest' side of the dark tourism spectrum, where the death and suffering of people are more recent and prominent. This is especially important to consider when we are talking about serial killer tourism.

As Tarlow (2005) mentions, media can be an important motivating factor. The interest in tragedy, crimes, and criminals is also not a 21st century phenomenon; news and newspapers always catered to this interest, with facts and deliberately untrue articles. The crimes of many murderers – among them Jack the Ripper – were being reported, turning the spotlight on the site of the tragedy, dramatizing the dangers, exaggerating the stories to have more media coverage (Marsh & Melville, 2019).

Most of the time, when a destination promotes their heritage, they promote a positive view, the ideal image of their culture, and they cover up the negative, darker parts of their history. Media has a huge role in spreading information and influencing people. It is not surprising that it shapes the way people gather information about places to visit as well (Lennon & Foley, 2000). Destination Organization Managements (DMOs) have a leading role in managing tourism in the area, to make sure that they present the best offer for the visitors (Varghese & Paul, 2014).

According to Powell et al (2018), the differences between how one city or the other promotes the darker side of the destination were noticeable. The results indicate that in the examined cities, DMOs include dark tourism sites as offered attractions in various extent. While most cities could use their heritage as dark tourism attraction, some cities decide not to do it. It depends on the city brand. London's brand is built on diversity and openness; it can offer various entertainment types to different people (Ladkin et al., 2008), while Venice is known for being a romantic paradise for couples.

Serial Killer Tourism: A Proposed New Category

People want to experience exciting new things. The demand for unique experiences, emotions, and the opportunity to actively participate when traveling, is a growing trend in global tourism. People are interested in the history of the place, but they are appreciating it more if the guided tour or individual sightseeing is unique in some way, rather than just hearing a monotonous voice or reading a long and boring text about it. Specialist or alternative tours, which are different from tours in "tourist hotspots" are growing in popularity and can help to disperse flows of tourists to different areas of a city, thus reducing overcrowding.

This constantly changing demand is the reason for examining attractions connected to serial killers. Even if there are common factors for the visitors when they are traveling and visiting a dark tourism attraction, such as the need to be educated, the basic motivations are diverse. Everyone has their own wants and attitudes, their unique mentality, or mental and physical restriction they must face when they plan their leisure, especially when they are visiting such sites (Gibson, 2008).

Constructing attractions connected to serial killers is not a new trend. People were always fascinated by these criminals, wanting to know more about the details of the crimes, to experience the tragedy from a safe environment. The "murder hotel" of America's first serial killer, H.H. Holmes, after the man's execution, was planned to be an entertaining "murder museum". A.M. Clark, the new owner, wanted to open an exhibition, complete with a guided tour for tourists, but before it could have been constructed, the building was destroyed in a mysterious disaster. This was not a unique situation: the tanner Ed Gein's place was crowded by curious tourists after his arrest, but before it could have been turned into an attraction, it was burned down. Even so, his car was turned into a county fair display, as "Ed Gein's Death Car" (Schechter & Everitt, 1997).

Despite this fascination with death and serial killers, academic research is limited in this scope (Hodgkinson et al., 2017) and on the tourist's opinion on dark tourism (Yu & Egger, 2022), while blog posts and news reports are readily available on the topic. They offer insight into the motivation behind visiting these places, offer recommendations on the best places, and/or condemn and question the morality of the curious nature of dark tourists (Markezic, 2022; Matravers, 2018; McFarland, 2022, Schmalbruch, 2018).

It is demonstrated in this research that there are many tours capitalizing on "dark history" to attract visitors, who are mainly young adult to middle-aged white women, who want to understand and learn from the deepest, darkest side of humanity (Knight, 2019). Some think these tours are morally questionable, they popularize serial killers and use the suffering of the victims and their families as a tool for entertainment, yet it can be argued that they are effective interpretation tools for the history. The way they represent the past depends on the tour guide, they have the responsibility on what and how they cover during their tour (Lennon, 2005). In the following part of the research, ten cities are examined, that have a reputation for being eclectic cities with dark connotations or their connection to more known serial killers.

Serial killer tourism overlaps with other categories, such as true crime tourism, ghost tourism, or even literary or film tourism (Tevely & Smith, 2022). Ghost tourism is the paranormal part of the dark tourism, which can always bring new visitors to the destination to experience something supernatural (Jászberényi, 2020) and it has an obvious connection to the victims of serial killers – sometimes even to the spirit of the murderers as well.

WHAT MAKES A SERIAL KILLER?

Jeffrey Dahmner, Jack the Ripper or Myra Hindley; just three of the countless men and women terrorizing the world's population. Each operated with distinct modus operandi, the number of their victims was different, even the era and location were diverse. Yet there was one common factor: they were serial killers. Serial killers are the main characters of many famous films, such as Se7en, Natural Born Killers, Psycho, Hannibal, American Psycho, Perfume: The Story of a Murderer, as they captivate the attention of their audience. Although these movies show the viewers the diverse motivations and types of these murderers, they are fictional, created to attract and horrify their audiences. To give a more realistic view on serial killers, the following part is focusing on the explanations, definitions, and possible reasons behind these criminals.

The main definition of serial killers comes from the National Institute of Justice, to explain the differences between serial killers, mass murderers, spree killers or contract killers.

A series of two or more murders, committed as separate events, usually but not always committed by one offender acting alone. The crimes may occur over a period of time ranging from hours to years. Quite often the motive is psychological, and the offender's behavior and the physical evidence observed at the crime scenes will reflect sadistic sexual overtones. (Schechter & Everitt, 1997, pp. 73–74)

The first special unit that dealt with serial killers and their psychology was – and still is – the Behavioral Analysis Unit, a department of the FBI, that analyzes the patterns of serial killers to help the work of criminal investigators ("A brief history", 2016.). To understand the perpetrators, their motivations and their mental state, a separate discipline deals with the minds of serial killers. In some cases, the criminals suffer some kind of mental disorder, but in some cases, they are "in their right mind". The motive for serial killing in most cases is a sexual indication, but it can be driven by anger, thrill-seeking, financial gain, and attention-seeking (Morton, 2008). According to Dr. Smith, as cited by Rosewood (2015), a lot of serial killers feel inferior that leads to anger and aggression, fantasizing and planning how and on whom to take out this feeling.

Nature versus nurture is a long-standing question regarding serial killers. Were they born with it or the external factors were the main reasons for how they turned out? Many of them were abused during their childhood, had mental disorders, and the "I can't help it" defense was not without precedent, placing the blame away from themselves (Rosewood, 2015). Female serial killers, while not common occurrences, still exist. Most commonly, they are either "Black Widows", who kill their love interests – and sometimes even their children – for money, or "Angels of Death", who work in the healthcare system, murdering the people in their care (Schechter & Everitt, 1997). But there are different types of female serial killers, murdering for sex, or an easier life.

A significant number of serial killers had an unusually cruel upbringing. Physical (non-accidental internal or external injuries), psychological (humiliation, isolation), sexual (showing, involving in sexual acts) abuse on them, on their family members; most of the time it is a cycle of abuse (Schechter & Everitt, 1997; "The Issue of Child Abuse", n.d.). Childhood abuse and the trauma itself may cause brain damage and neurobiological abnormalities (Teicher, 2000), but it is important to note, that it rarely means that they would commit violent crimes (Lilienfeld et. al., 2010). Most of the serial killers had an abusive childhood, but most of the abused children would not become serial killers.

The torture and murder of small animals can be one sign of a future serial killer. Many of the caught criminals enjoyed trapping and torturing pets, experimenting with them and murdering defenseless animals. This type of sadism is one of the three early identifying marks of psychopathic behavior; the other two are the prolonged bed-wetting and pyromania (Schechter & Everitt, 1997).

EMPIRICAL DATA ABOUT ATTRACTIONS CONNECTED TO SERIAL KILLERS

Applied Research Methodology

To understand better the topic, three research questions were formed that could be answered based on observations when searching for attractions connected to serial killers, on the available literature about serial killers and dark tourism, and the empirical research.

These are the following:

- What is the reason for serial killer tourism?
- What type of attractions connected to serial killers are the most attractive for the visitors?
- What is the attitude of the consumers towards the attractions connected to serial killers?

To answer these questions, in the first part of the paper, secondary research was used to examine the data previously collected by others, and the existing literature about dark tourism and serial killers. The literature about serial killers aided in recognizing that these violent crimes are widely spread around the world and the seven dark suppliers' framework (see Figure 1) helped identify the possible attractions connected to serial killers. Based on the Stone's (2006) dark tourism spectrum (see Figure 1), the attractions connected to serial killers, as paces that has direct relation to human death lies at the darker end of the spectrum. The different perspectives showed that while parts of dark tourism are well-researched, attractions connected to serial killers are worth further investigation, as it is a field which is lacking academic research.

Figure 1. Shades of dark tourism

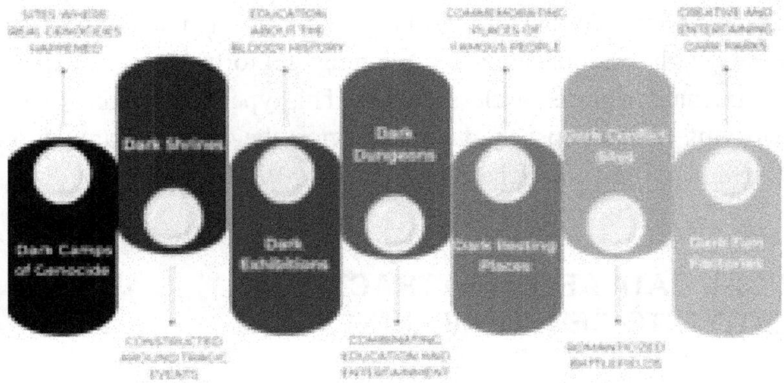

To examine the reactions and opinions of the visitors on this aspect of dark tourism, content analysis or the reviews left on walking tours was used. The choice behind these examined ten cities – Los Angeles, New Orleans, Philadelphia, Washington, D.C., London, Hannover, Venice, Vienna, Budapest and Perth – was that they offer at least two walking tours and at least one other type of attraction connected to serial killers, which was museum most of the time. Only official, at least semi-organized tourist attractions were examined, avoiding those sites that are visited by serial killer enthusiasts at the displeasure of the locals. The number of reviews was used to indicate the popularity or unpopularity of museums, walking tours and other types of attractions in the chosen cities in order to show the demand for this type of tourism.

Using content analysis on the reviews on the selected walking tours helped to evaluate the visitors' attitude. The aim was to get an idea about the consumer's motivation when visiting attractions connected to serial killers. Using online comment analysis is the most accessible way to reach dark tourists, people, who are motivated to share their experiences at the chosen sites. Both the positive and negative comments give an idea about the objective and subjective issues at the given sites, and an overview on what is attractive for them.

Comparison of Attractions Connected to Serial Killers

Serial killer tourism, like other tourism categories, has countless attraction types (Table 1).

Attractions can vary:

- Museums, e.g., Crime Through Time Museum in Littledean, where outside of other shocking objects, visitors can find tools and belongings of serial killers ("Home", 2023);
- Walking tours, e.g., Cream City Cannibal: Jeffery Dahmer Walking Tour, the tour in Milwaukee were they show the places Jeffery Dahmer hunted his victims ("Cream City Cannibal", 2023);
- Theme parks, e.g., The Amsterdam Dungeon, an experience with scenes and rides that recreates the darker side of history ("The Amsterdam Dungeon", 2023);
- Monuments, e.g., The Murder Wall, a traveling tribute to honor the victims of serial killers ("Murder Wall", 2023.);
- The childhood house of the serial killer, e.g., Ted Bundy's childhood home (Mead, 2019);
- The hunting place of the serial killer, e.g., San Francisco Bay Area, where the Zodiac Killer hunted his victims ("Zodiac Killer", 2004).

Table 1(a). Ten cities with connection to serial killers

Cities	Los Angeles	New Orleans	Philadelphia	Washington DC
Museums	The Museum of Death, LA	The Museum of Death, NO	The Eastern State Penitentiary	National Museum of Crime & Punishment
Ratings	4,0/5	3,5/5	4,5/5	4,5/5
Reviews	360	502	6437	915
Ranking	#71/784	#110/455	#5/480	Closed
Walking Tours	DTLA Murder Mystery Ghost Tour	Killers and Thrillers Tour	Cemetery & Serial Killer Tour of Philadelphia	Ghosts, Scandals and Murder
Ratings	4,5/5	5,0/5	5,0/5	5,0/5
Reviews	20	709	164	674
Walking Tours	Haunted Hollywood Walking Tour	Sinister Criminal Intentions	Dark Philly Adult Night Tour	True Crimed of Georgetown
Ratings	5,0/5	5,0/5	5,0/5	No data
Reviews	41	493	1625	
Other Products	ROSE Memorial			
Ratings	Under construction			
Reviews				
Ranking				

Source: Tripadvisor, 2023

Table 1(b). Ten cities with connection to serial killers

<table>
<tr><th>London</th><th>Hannover</th><th>Venice</th><th>Vienna</th><th>Budapest</th><th>Perth</th></tr>
<tr><td>Jack the Ripper Museum</td><td>Sprengel Museum Hannover</td><td>Mostra Serial Killer</td><td>Crime Museum Vienna</td><td>Police Museum</td><td>Fremantle Prison</td></tr>
<tr><td>4,0/5</td><td>4,0/5</td><td>3,5/5</td><td>3,0/5</td><td>4,0/5</td><td>4,5/5</td></tr>
<tr><td>1022</td><td>182</td><td>93</td><td>185</td><td>15</td><td>4155</td></tr>
<tr><td>#125/2588</td><td>#13/205</td><td>Closed</td><td>#480/987</td><td>#271/743</td><td>#2/124</td></tr>
<tr><td>The Blood and Tears Walk</td><td>Fritz Haarmann Murder Tour</td><td>Venice Ghost & Legends walking city tour</td><td>Vienna Criminal History Tour</td><td>Investigation About the First Hungarian Serial Killer</td><td>Crimes of Perth Walking Tour</td></tr>
<tr><td>5,0/5</td><td>5,0/5</td><td>4,0/5</td><td>4,9/5</td><td rowspan="2">No data</td><td>5,0/5</td></tr>
<tr><td>818</td><td>4</td><td>23</td><td>13</td><td>44</td></tr>
<tr><td>The Jack the Ripper Walking Tour</td><td>Werewolf of Hanover Crime Tour</td><td>Serial Killers of Venice Tour</td><td>Terrible Crimes in Vienna</td><td>Night-time Dark History and Vampire Tour</td><td>Torchlight Tour</td></tr>
<tr><td></td><td rowspan="2">No data</td><td rowspan="2">No data</td><td>5,0/5</td><td>5,0/5</td><td>5,0/5</td></tr>
<tr><td>1135</td><td>3</td><td>193</td><td>4</td></tr>
<tr><td>London Dungeon</td><td>The Victims of Fritz Haarmann Memorial</td><td rowspan="4"></td><td>Ungarisches Haus</td><td>Murder - The Exhibition</td><td>Museum of Perth</td></tr>
<tr><td>4,0/5</td><td rowspan="3">No data</td><td rowspan="3">No data</td><td>3,5/5</td><td>3,5/5</td></tr>
<tr><td>11476</td><td>55</td><td>16</td></tr>
<tr><td>#66/473</td><td>Closed</td><td>#100/229</td></tr>
</table>

1. Museums

The Museum of Death can be found in two places in the United States of America – in Los Angeles and in New Orleans -, both recommended as popular tourist attractions in these cities. The one in New Orleans is *the* murder museum, the original, the one that exhibits various objects from known serial killers. They include, but are not limited to:

- Letters from Jeffrey Dahmer, the Milwaukee Cannibal, who was only captured after the murder of 17 men (Philbin & Philbin, 2009);
- Paintings of the Killer Clown, John Wayne Gacy, a child rapist and a murderer of at least 33 young boys (Philbin & Philbin, 2009); and

- Photographs of the Manson family, a cult that committed nine murders, lead by Charles Manson ("Charles Manson Biography", 2021).

The objective behind the museum is to educate the visitors on the mysterious and unexplainable death ("Museum of Death", 2023.). The other one is in Los Angeles ("Museum of Death Los Angeles", 2023), which is currently under reconstruction, but the museum itself holds similar items to the one in New Orleans. The reviews on Tripadvisor ("Museum of Death Los Angeles", 2023; "Museum of Death New Orleans", 2023) are similar for both, varying from "the best experience of my life" to "waste of time". Some of the reviewers objected to the "morbidity" of the place. These places are generally not recommended for little kids as they can be disturbing and "bizarre".

The Eastern State Penitentiary in Philadelphia, USA, once a prison for known criminals – gangsters, bank robbers, and serial killers – is now one of the city's main tourist attractions. The interested tourists learn about the daily lives and backstories of the past inmates. The gothic building is the home for the Terror Behind the Walls, a haunted attraction with special effects, animatronic creatures, and performers ("Eastern State Penitentiary", 2023). According to the visitors, it is "thought-provoking" and "full of interesting architecture", but it is "overpriced" and "overrated" ("Eastern State Penitentiary Philadelphia", 2023).

While now closed, the National Museum of Crime and Punishment in Washington DC, the capitol of the States, was a privately owned specialty museum with crime scenes throughout history, objects used by criminals, such Al Capone, Bonnie and Clyde. It was an educational museum with interactive ways to learn how criminal law and its enforcement changed from witch hunting to current day organization system. Most of the visitors found this establishment "interesting place for a family with older kids", appreciating the interactive education tools and fascinating stories of criminals ("National Museum of Crime", 2023).

When talking about London, the mysterious figure of Jack the Ripper is unavoidable. The British capitol's and the visitors' fascination with this legendary figure who targeted women sex workers in 1888 is never-ending. He (presumably male) was an unidentified serial killer in Whitechapel, known for his gory murders. The Jack the Ripper Museum brings awareness to the still unsolved murders and the possible suspects. It displays the details of the police investigation, and the possible motives behind the murders. The exhibition is also designed to help the visitors gain an insight into the life of the victims – mainly the "Canonical Five" victims, Polly Nichols, Annie Chapman, Elizabeth Stride, Catherine Eddowes, and Mary Jane Kelly - and the way they lived in the Victorian-era London ("Jack the Ripper Museum", 2023). The comments highlight that visiting this museum could not only give a unique experience for those who are interested in the gory details, but for those who want to understand better the history and the living conditions of the victims. Contrary, the negative comments argue that the museum did not represent the era accurately ("Jack the Ripper Museum London", 2023).

While it is not exactly about serial killers, the Sprengel Museum Hannover in Germany, has a bronze artwork depicting the crimes of Fritz Haarmann ("Sprengel Museum", 2023). He was known as the Butcher of Hanover, or the Wolf Man. He raped, tortured and killed at least 24 boys, selling their meat to unsuspecting customers (Newton, 2006). The museum is mainly attractive for locals and visitors from Germany. It is one of the main attractions in the city, and the reactions are mainly positive, with only some mentions about the serial killer himself ("Sprengel Museum Hannover", 2023).

Venice, the City of Love may have hidden its connection to serial killers, but the Italian city is still visited by some people for its unique serial killer tours. It had a Serial Killer Exhibition, where people could learn about criminology and about famous serial killers, both from around the world and from Italy. The museum is closed for now ("Visitare Serial Killer", 2023), but while it was open, visitors thought it was "creepy but fascinating", "thought-provoking", yet "amateurish" ("Mostra Serial Killer", 2023).

The Crime Museum in Vienna, Austria, exhibits the Viennese criminal history from the Middle Ages through the Habsburg-Hungarian Monarchy to current times. It depicts how criminal justice evolved, introduces the various serial killers operating in the area, including the exhibition of the skull of Hugo Schenk (an Austrian serial killer who operated in the late 1800s). The information is mostly available in German language, but they are planning to translate them to English ("Wiener Kriminal Museum", 2023.). The visitors feel it is quite informative, but it can only interest those who wants to learn more about criminal history and/or if "you have an hour to kill" ("Crime Museum Vienna", 2023).

The Hungarian Police Museum in Budapest, Hungary exhibits the Hungarian police's work, their typical cases, some details on nation-wide known cases that are interesting because of their strange nature or the degree of interest the media gave them. It also includes information about at least five Hungarian serial killers of the past ("Rendőrmúzeum", 2023). According to the visitors, the Criminal Museum is interesting and diverse, but some thought that there are no newer cases ("Police Museum", 2023).

The Museum of Perth in Perth, Australia, while not exactly a serial killer museum, it is about the social, political, and architectural history of Perth, which means that it has temporary exhibitions about the serial killers and other crimes in its past ("Museum of Perth", 2023). It is not a widely visited attraction, but reviewers think is a great way to explore Perth and its history. The negative comments are mainly about the not clear opening hours ("Museum of Perth Perth", 2023).

2. Walking Tours

Walking tours, where tourists can visit the most famous sites by foot while hearing the myths and facts about them from a local tour guide is one of the most popular ways to get to know the darkest part of the cities. Most of the cities that has no other serial killer attraction types, have at least one true crime walking tour to present. The offer of true crime tours in these cities is seemingly endless; the following ones are just a few out of the many.

Los Angeles offers tours that go through the city's darkest places, including the Cecil Hotel – rebranded as Stay in Main -, which is one of the most notorious hotels in the world, ever since the disappearance of Elisa Lamb, and the Netflix show about the place's dark history. Among other guests, it also accommodated the Night Stalker, Richard Ramirez. He was an American burglar, rapist and serial killer who was convicted of the murder of at least 13 people, but it is suspected that there were more victims (Richard Ramirez, n.d.). The DTLA Murder Mystery Ghost Tour is mainly focused on the paranormal side of the crimes, while the Haunted Hollywood Walking Tour introduces the visitors to the celebrity gossips and tragedies, such as the Manson murders. The commenters state that participating in these tours is a way to explore Los Angeles in a unique mode and that they are "must do" experiences for true crime lovers. The latter gave an obvious positive experience for the clients, while the former had some negative comments about the high prices, and the unclear information available on the webpage. The typical visitors of these tours – based on the reviews – are young and middle-aged women, recommending the tours for couples ("DTLA", 2023; "Haunted Hollywood", 2023).

The prices and routes are very similar to each other (e.g. the French Quarters is an obligatory site to visit) in New Orleans. Some have more, some fewer places to visit, but the main deciding factor for the visitors are usually the reviews on the tour guides' attitude. Those who left 5 stars described the tours as "engaging" and the tour guides as "passionate and knowledgeable", while the less than impressed visitors felt the experience was "boring", "drawn out" and that there were no real differences between the various offers. The reviews show that the entertainment factor is necessary. Even if a city has an interesting history, and fascinating, gruesome crimes, the tour guide's style is one of the most important factors. Based on the profiles, the visitors are mainly women in their twenties-thirties, who visited with their friends and families ("New Orleans Killers", 2023; "New Orleans True", 2023).

Philadelphia offers tours to explore the dark history of the Old City, hearing about serial killers, executions, vampires and other dark tales while discovering the most popular landmarks. The Cemetery & Serial Killer Tour and the Dark Philly Adult Night Tour takes the visitors to different places, shows different sides of the city. The reviews are overwhelmingly positive (e.g. knowledgeable tour guide, creepy atmosphere, and entertaining experience), while the negative ones complained that the expectations did not meet with the reality. These tours are suggested for couples or friends, kids are advised to stay at home ("Cemetery', 2023; "Dark Philly", 2023).

The capitol of the United States of America, Washington, DC does not lack the tours connected to serial killers either. In The Ghosts, Scandals and Murder and True Crimes of Georgetown tours the businesses lead the tourists around the darkest sides of the capitol, where they can meet with ghosts, murderers and other criminals. The tours are a great way to explore the city, especially if one is interested in true crime and mysteries, and they are offered from 14 years old and above ("DC Insider", 2023; "True Crimes", 2023).

The Blood and Tears Walk and The Jack the Ripper Walking Tour in London, are both focusing on the mysterious Victorian era serial killer, but the former offers other interesting facts about other murderers and crimes as well. According to the reviewers, they are necessary to experience, as the former gives an overview on London's bloody history, while the latter demonstrates the shocking living situations of Jack the Ripper's victims and the world, they were living in. As the most popular walking tours, the visitors were more than satisfied with what they got; they used words like "bloody brilliant", "funny" and "informative" to describe their experiences. Those who were not quite impressed could only complain that there was no new information on the cases, or that they could not hear the tour guide through the noise ("The Blood and Tears", 2023; "The Jack the Ripper", 2023).

The scope of Hanover's tours – Fritz Haarmann Murder Tour Through Hanover's Old Town and Werewolf of Hanover Crime Tour – is about the life and crimes of Fritz Haarmann, the serial killer known many different monikers. It leads the visitors through the history of this murderer, covering all of his background, while introducing the 20th century Hanover. There's a lack of useful review on these tours, but the ones available are advising to seek them out and hear about the city's dark history ("Fritz Haarmann", 2023; "Werewolf of Hanover", 2023).

The Venetian serial killer tours are not as promoted as they are in other cities, as Venice rather hides his dark history, but they do exist. Venice is not just the home of Casanova, but it is also the birthplace of many murderers, unsolved mysteries and unexpected myths. The Ghost & Legends Walking Tour's focus is on the paranormal, on the ghosts of past crimes, while the Serial Killers of Venice Tour leads the participants through places where murders took place, leading to Santa Croce, where Venice's own Sweeney Todd, Baisio committed his murders. The Butcher of Santa Croce was a sausage maker in the 16th century Venice. It is not certain whether his deeds were real or just a legend, but it is said that he was making his sausages out of human meat (Thomson, 2022). The Ghost and Legends has a diverse range of reviews: some think it is a fun tour for couples who wants to explore the city at night, yet there are reviews about how "dreadful" it was ("Venice Ghost", 2023; "Tour Leader Venice", 2023).

The Vienna Criminal History Tour – only offered in German language – leads the visitors through Vienna's criminal history, explaining murder cases and their punishments, while The Terrible Crimes in Vienna is more specialized on the historical side of the crimes. According to the sites, there is no age limitation on who can participate in them, leading to the conclusion, that they are more fact-based, rather than deliberately shocking. Vienna's dark history is not as known as it is in London's case, but there are many unexplored things tourists can get information about. The few reviews on the tours indicated that they were informative and entertaining, a good was to get a deeper understanding on the city ("Vienna Criminal History", 2021; "Terrible Crimes", 2023).

Budapest, Hungary's capital city had many connections to serial killers, including, but not limited to the infamous "Blood Countess", Elizabeth Báthory, and Béla Kiss, the first *official* Hungarian serial killer. The Investigation About the First Hungarian Serial Killer is only in Hungarian, but those who participate, can follow the footsteps of Béla Kiss, whose actions were so disturbing that they pushed even the news about the World War I off of the front pages. He murdered at least 7 or 24 women in the beginning of the 20th century. Seven of his victims were discovered in seven barrels in his shed, after he was drafted. His fate is unknown: he either died in a compound or lived under pseudonym for the rest of his life (Szűcs, 2015). The Night-time Dark History and Vampire Tour introduces the myths and legends of these monsters and their connections to real-life serial killers and historical figures. The most known figure is Countess Elizabeth Bátory, a Hungarian noblewoman, who was accused of torturing and murdering 600 girls. These accusations are now considered as politically motivated, but she is still known as the noblewoman, who bathed in the blood of virgin girls (Pallardy, 2021). The tours are offered for couples as well as families. According to the visitors, the night walk was an "amazing experience", and they would recommend it to others, but some commenters lacked the available information on the tour, and that the experience was not unique enough, they did not gain any new knowledge on the cases ("Nyomozás", 2023; "Night-Time", 2023).

The Crimes of Perth Walking Tour leads the visitors around central Perth, showing the sites of murders, executions, and other crimes, while the Torchlight Tour is focusing on the Fremantle Prison, and the criminals held in captive there. The reviews for the Two Feet & a Heartbeat Tours can only be separable from their other tours, because 44 reviewers clarified in which tour they were participating in. They enjoyed the experience, but there were complaints about "false accessibility claims", and that the tour "felt like a form of punishment" ("Two Feet", 2023; "Fremantle Prison Torchlight", 2023).

3. Other Types of Attractions

As it was shown previously, there are many different types of attractions connected to serial killers. The memorials of victims are not as noticeable and notable in numbers, but they exist to show that they deserve to be remembered. That while the serial killers and their psychology is fascinating, people need to realize that they destroyed countless lives, and that the victims should be remembered instead of the murderers.

A memorial for the more than 200 black girls and women who fell victim to serial killers in the early 1980s is being constructed in South Los Angeles. It is planned to commemorate them and to honor their memories, since there is a lack of remembrance on the victims not just as a number, but as humans, who cruelly lost their lives ("The Memorial", 2021). In Hanover, a Memorial allows locals and tourists to remember on the mass grave for the 27 victims of Fritz Haarmann ("The Victims", 2023) while in London, a mural acts as a counterpoint of the fame of Jack the Ripper by commemorating the Canonical Five victims (Flood, 2020).

There are more entertainment-based attractions, such as the London Dungeon, which is a unique Dark Fun Factory attraction where history and murder meet with a Disneyland-esque entertainment ("The London Dungeon", 2023). There is a Jack the Ripper themed roller coaster in Vienna's Prater ("Jack the Ripper", 2023), and countless escape rooms and city-trail games connected to famous serial killers in various cities. The Murder Exhibition in Budapest opened its gates in 2012 and was open for a for a few months in each year until 2019. It exhibited the most famous serial killers of the world since the Middle Ages to this day. It used modern technology and special effects to make the experience both frightening, entertaining and immersive ("Murder The Exhibition", 2023).

Buildings can also be the focus of tourists. One of them is the house of the infamous Hungarian countess, Elizabeth Báthory ("Ungarisches Haus", 2023). Others can be prisons, like Perth's Word Heritage site, the Fremantle Prison ("What's On", 2023), where visitors can hear the real-life stories of the prison's inmates, or the historic Ten Bells pub ("Ten Bells", 2023) in London, where Jack the Ripper hunted his victims.

IMPLICATIONS AND RECOMMENDATIONS

This research about the niche dark tourism category, serial killer tourism, the literature and the content analysis of the visitors' reviews show that even though it is not a mainstream tourism type, it still attracts tourists. True crime enthusiasts, people who are interested in psychology, people who are fascinated by the darkest part of a place's history, paranormal believers, people who are enticed by serial killers and/or women who are trying to prepare for any eventualities are all ready to visit these attractions. Some people do it to learn about history, to have a better knowledge in how serial killers' mind works; some do it for the entertainment, to have fun; while others do because they want to experience ghosts of the killers' or their victims' haunting.

As the motivations are various, the types of attractions are different as well: museums (e.g. Museum of Crime and Punishment), tours (e.g. Jack the Ripper Tour), monuments (e.g. mural for the victims of Jack the Ripper), or escape rooms (e.g. H.H. Holmes Haunted House) that have connections to real-life serial killers. The most visited attractions are the exhibitions and the tours in those cities that have a visible connection to serial killers.

The reviews on the attractions are subjective, and consequently contradictory, which shows up in the results. There is an overwhelming positivity on the tours, while the museums are more diverse in the way visitors experience them. It is important to note that the success on the tours mainly depends on the personality of the tour guides, as they are the ones who can convince or discourage the visitors to recommend these tours. Legends, myths, facts and interesting information are nothing without somebody who can convey them in an entertaining, chilling and fascinating way.

The aim of the authors was to showcase at least one city from each continent in the research. However, the criteria for the selection was to have at least one serial killer attraction and at least two walking tours about these criminals. This lead to the conclusion that Asian and African countries cannot be introduced as these continents do not promote many attractions connected to serial killers. In opposition, the United States of America have more cities that met the criteria. This does not mean that there are no other types of attractions, or cities that has this dark connection. Many cities offer walking tours, little museums or memorials connected to serial killers, even in Asia. One place like this can be found in Bangkok. The Siriraj Medical Museum is an exhibition of forensic medicine, which includes displays of body parts of murder victims, and the mummified remains of the first known Thai serial killer, the cannibal Si Quey Sae Urng ("Siriraj Medical Museum", 2023).

As the cities based on the two criteria that were prerequisite for the selection indicate, four out of ten can be found in the United States of America. The number of serial killers – active or inactive – is the highest in the country, and their media tend to focus on these criminals, sensationalizing their crimes, naming them, drawing attention to their deeds. It is important to acknowledge that the countries that has the most famous attractions connected to serial killers are English-speaking countries that has produced the largest number of serial killers, such as the United States of America (#1) and the United Kingdom (#2) (Sheth, 2020).

FUTURE RESEARCH DIRECTIONS

Attractions connected to serial killers are arguably questionable. Most of the times serial killers are sensationalized, the entertainment and the shock factors are high while the educational part is low, especially compared to other dark tourism products. But as serial killer attractions are within the scope of dark tourism, with a better understanding of what "normal" visitors are interested in (e.g. learn about history, as the dark tourism part showed), they could improve. Of course, entertainment is important, but there should be a more balanced offer, as there are people who are interested in serial killers and their psychology but think negatively about attractions connected to them.

Qualitative research among people who visit serial killer attractions regularly, once in a while and never is proposed in order to have a more in-depth knowledge behind the specific motivations and behaviors of the visitors. While conducting quantitative research with big sample size is suggested, understand better the different attitudes in people from different backgrounds to gain a clear insight into the different segments. In this research, general conclusions about consumer behavior towards serial killer attractions can be determined, which was not done before, but a more comprehensive study could help the tourism sector more. Based on this research, it is clear, that there is a demand for these types of attractions, but it is also obvious, that many would not consider visiting places like this due to their questionable ethicality. Furthermore, a social media analysis and the examination of the local DMOs should be added to have a comprehensive overview of the topic.

CONCLUSION

Serial killers as an attraction is a niche product, but it has its own target audience. There are many people fascinated by serial killers and would want to visit sites connected to them, but since most of these places are popularizing criminals, most of them would not visit these attractions. Local DMOs and businesses promote these attractions as a unique experience but try to find the balance between arousing interest with the serial killer and commemorating their victims, because, in the end, it is about people, who lost their lives way too early. They could be a unique selling proposition for the DMOs of bigger cities to consider, but it is worth remembering that even if there are people interested in the gory part of the murders, most of the people who would be interested in visiting a place like this, would rather learn about the psychology and the history of serial killers.

The main research question was about the existence of dark tourism, including attractions connected to serial killers: What is the reason for serial killer tourism?

Dark tourism is a fascinating topic, as it involves the darkest side of humanity. Previous research has shown that the morality of dark tourism is still considered questionable, but it has been with us since ancient times, with continuous success. People have always been fascinated by death and it is not a surprise that the offer must meet the demand for this. Dark tourism, as a tourism to places where human suffering, tragedy, and death happened, includes serial killer attractions. Even if the attractions are not about specific serial killers or they are not at the sites where something tragic happened, there is a demand for attractions connected to them because serial killers interest many people.

The content analysis of the attractions in the examined cities has the answer for the second research question: What type of attractions connected to serial killers are the most attractive for the visitors?

The different places have a different connection to death and visitors have different needs. On one hand, museums are popular among the visitors, as they want to learn the backgrounds, the various crimes committed. On the other, people want to be entertained, but not by the real tragedies. When visiting a place where the tragedy happened, they are more interested in paranormal activity, while when they are interested in serial killers, they want to be educated. When they want to be entertained, they are more comfortable with imagined crimes. The Murder Exhibition in Budapest, Hungary, was a great initiation for mixing entertainment, new technologies, and education, but as many of the reviews complained about the fact that through the years it turned into a quasi-haunted house with some scary actors and more fake blood.

The third question was about consumer behavior, and their opinions about the scope of this research: What is the attitude of the consumers towards the attractions connected to serial killers?

While most of the people are concerned about the ethicality of dark tourism, they are even less sure about visiting places that exploit terrible tragedies that do not even have an educational purpose. People visit dark tourism sites to learn about history, to know about crimes against humanity and how to avoid them, but they cannot do the same at attractions connected to serial killers. Most of the times visitors are interested in serial killers because of the active serial killers in the area they live make them scared yet fascinated. This controversy leads most people to know more about these criminals, and this is what is missing from most of the attractions: a way to learn about the background of serial killers, they want to understand them and have enough information about them to be able to feel a bit safer.

This research can give local DMOs, tour agents and service providers an overview of the visitors' fascination with serial killers, which help them to know what and how to offer and promote to the potential dark tourists. It may influence the destinations' marketing strategies on how to develop dark tourism products, to be more educative and to encourage the preservation of the memory of the victims, not just the serial killers. The visitors not only expect entertainment, to be excited, horrified and titillated, but to gain quality information about these tragedies.

To conclude the research, the name serial killer tourism is proposed to describe the phenomenon. While it does not have a big market like other forms of dark tourism has, it is a specific part of dark tourism that has its unique, niche market that is big enough to have various offers in many cities. While sometimes it overlaps with ghost tourism or grave tourism, as dark tourism has many fields overlapping each other, serial killer tourism should have its own name as well.

ACKNOWLEDGMENT

We thank Prof. Dr. Melanie Kay Smith (Budapest Metropolitan University) for providing valuable insights and useful suggestions.

REFERENCES

A brief history. (2016, May 3). FBI. https://www.fbi.gov/history/brief-history

Carrabine, E. (2011). Images of Torture: Culture, Politics and Power. *Crime, Media, Culture*, *7*(1), 5–30. doi:10.1177/1741659011404418

Carvalho, M. A. M. (2023). Bone Chapels: Who Might Be Interested in Visiting and Why? *Tourism Recreation Research*, 1–12. Advance online publication. doi:10.1080/02508281.2023.2167908

Cemetery and Serial Killers Walking Tour of Philadelphia – Philadelphia Pennsylvania. (2023). Tripadvisor. https://www.tripadvisor.com/AttractionProductReview-g60795-d12859723-or10-Cemetery_and_Serial_Killers_Walking_Tour_of_Philadelphia-Philadelphia_Pennsylv.html

Charles Manson Biography. (2021, February 4). Biography. https://www.biography.com/crime-figure/charles-manson

Cream City Cannibal: Jeffery Dahmer Walking Tour. (2023). Tripadvisor. https://www.tripadvisor.com/AttractionProductReview-g60097-d23983034-Cream_City_Cannibal_Jeffery_Dahmer_Walking_Tour-Milwaukee_Wisconsin.html

Crime Museum Vienna - Vienna. (2023). Tripadvisor. https://www.tripadvisor.com/Attraction_Review-g190454-d591127-Reviews-Crime_Museum_Vienna-Vienna.html

Dark Philly Adult Night Tour - Philadelphia Pennsylvania. (2023). Tripadvisor. https://www.tripadvisor.com.sg/AttractionProductReview-g60795-d11854370-or10-Dark_Philly_Adult_Night_Tour-Philadelphia_Pennsylvania.html

DC Insider Tours – Washington DC District of Columbia. (2023). Tripadvisor. https://www.tripadvisor.com/ShowUserReviews-g28970-d3410803-r292929652-DC_Insider_Tours-Washington_DC_District_of_Columbia.html

DTLA Murder Mystery Ghost Tour – Los Angeles California. (2023). Tripadvisor. https://www.tripadvisor.com/AttractionProductReview-g32655-d16753250-DTLA_Murder_Mystery_Ghost_Tour-Los_Angeles_California.html

Eastern State Penitentiary. (2023). Eastern State Penitentiary. https://www.easternstate.org/

Eastern State Penitentiary - Philadelphia Pennsylvania. (2023). Tripadvisor. https://www.tripadvisor.com/Attraction_Review-g60795-d102763-Reviews-Eastern_State_Penitentiary-Philadelphia_Pennsylvania.html

Flood, A. (2020, February 10). *Whitechapel Mural Will Celebrate the Lives of Jack the Ripper's Victims.* The Guardian. https://www.theguardian.com/books/2020/feb/10/whitechapel-mural-jack-the-ripper-victims-the-five-hallie-rubenhold

Fremantle Prison Torchlight Tour. (2023). Ghost Tours. https://wwwghosttourbookings.com.au/ghost-tours/fremantle-prison-torchlight-tour/

Fritz Haarmann Murder Tour Through Hanover's Old Town. (2023). Viator. https://www.viator.com/tours/Hannover/Fritz-Haarmann-murder-tour-through-Hanovers-old-town/d28614-279650P1

Gibson, D. C. (2008). The Relationship Between Serial Murder and the American Tourism Industry. *Journal of Travel & Tourism Marketing, 20*(1), 45–60. doi:10.1300/J073v20n01_04

Handayani, B. (2018). Going to the Dark Sites With Intention: Construction of Niche Tourism. In M. Korstanje & B. George (Eds.), Virtual Traumascapes and Exploring the Roots of Dark Tourism (pp. 50-66). IGI Global. doi:10.4018/978-1-5225-2750-3.ch003

Harrison, M. A., & Frederick, E. J. (2020, July 1). Interested in Serial Killers? Morbid Curiosity in College Students. *Current Psychology (New Brunswick, N.J.)*. Advance online publication. doi:10.100712144-020-00896-w

Hartmann, R. (2014). Dark tourism, thanatourism and dissonance in heritage tourism management: New directions in contemporary tourism research. *Journal of Heritage Tourism*, *9*(2), 166–182. doi:10.1080/1743873X.2013.807266

Haunted Hollywood Walking Tour: True Crime, Creepy Tales – Los Angeles California. (2023). Tripadvisor. https://www.tripadvisor.com/AttractionProductReview-g32655-d12148364-Haunted_Hollywood_Walking_Tour_True_Crime_Creepy_Tales Los_Angeles_California.html

Hodgkinson, S., Prins, H., & Stuart-Bennett, J. (2017). Monsters, Madmen... and Myths: A Critical Review of the Serial Killing Literature. *Aggression and Violent Behavior*, *34*, 282–289. doi:10.1016/j.avb.2016.11.006

Home. (2023) Littledean Jail. https://www.littledeanjail.com/

Jack the Ripper. (2023). Prater. https://www.praterwien.com/nc/en/attractions/details/a/jack-the-ripper/

Jack the Ripper Museum. (2023). Jack the Ripper Museum. https://www.jacktherippermuseum.com/

Jack the Ripper Museum - London England. (2023). Tripadvisor. https://www.tripadvisor.com/Attraction_Review-g186338-d8638694-Reviews-Jack_the_Ripper_Museum-London_England.html

Jászberényi, M. (2020). *A kulturális turizmus sokszínűsége*. Akadémiai Kiadó. doi:10.1556/9789634545224

Knight, A. (2019). Why Can't We Look Away from the Worst of Humanity? *Literary Hub.* https://lithub.com/on-dark-tourism-murder-hauntings-and-the-serial-killer-capital-of-australia/

Ladkin, A., Fyall, A., Fletcher, J., & Shipway, R. (2008). London Tourism: A 'Post-Disaster' Marketing Response. *Journal of Travel & Tourism Marketing*, *23*(2–4), 95–111. doi:10.1300/J073v23n02_08

Lennon, J. (2005). Journeys Into Understanding. *The Guardian.* https://www.theguardian.com/travel/2005/oct/23/darktourism.observerescapesection

Lennon, J., & Foley, M. (2000). *Dark Tourism. The Attraction of Death and Disaster*. Cengage Learning EMEA.

Lennon, J. J. (2018). Dark Tourism Visualisation: Some Reflections on the Role of Photograpy. In P. R. Stone, R. Hartmann, T. Seaton, R. Sharpley, & L. White (Eds.), *The Palgrave Handbook of Dark Tourism Studies*. Palgrave Macmillan. doi:10.1057/978-1-137-47566-4_24

Light, D. (2017). Progress in dark tourism and thanatourism research: An uneasy relationship with heritage tourism. *Tourism Management, 61*, 275–301. doi:10.1016/j.tourman.2017.01.011

Lilienfeld, S. O., Lynn, S. J., Ruscio, J., & Beyerstein, B. L. (2010). *50 pszichológiai tévhit*. Partvonal Könyvkiadó.

Markezic, E. (2022). True Crime Tourism: The Good, the Bad, and the Bundy. *Escape*. https://www.escape.com.au/destinations/north-america/usa/true-crime-tourism-the-good-the-bad-and-the-bundy/news-story/f8fd8859f0ede988126e1f66f36feaa0

Marsh, I., & Melville, G. (2019). *Crime, Justice and the Media* (3rd ed.). Routledge., doi:10.4324/9780429432194

Matravers, D. (2018). Is It Ever Morally Acceptable to Visit a Mass Murder Site? *Open.edu*. https://www.open.edu/openlearn/history-the-arts/it-ever-morally-acceptable-visit-mass-murder-site

McFarland, F. (2022). Strange Attraction. Inside the Morbid World of 'Dark Tourists' Who Love Serial Killers Like Jeffrey Dahmer & Flock to Areas Where They Lived. *The U.S. Sun*. https://www.the-sun.com/news/6447113/dark-tourists-visit-jeffrey-dahmer-home-serial-killer/

Mead, T. (2019). *Tourists Are Flocking to Ted Bundy's Old House, and the Current Tenants Are Sick of It*. House Beautiful. https://www.housebeautiful.com/design-inspiration/real-estate/a27090576/ted-bundy-house/

Miles, W. F. (2002). Auschwitz: Museum Interpretation and Darker Tourism. *Annals of Tourism Research, 29*(4), 1175–1178. doi:10.1016/S0160-7383(02)00054-3

Morton, R. J. (2008). *Serial Murder. Multi-Disciplinary Perspectives for Investigators*. FBI Academy.

Mostra Serial Killer - Jesolo Veneto. (2023). Tripadvisor. https://www.tripadvisor.co.uk/Attraction_Review-g580246-d9823089-Reviews-Mostra_Serial_Killer-Jesolo_Veneto.html

Murder – The Exhibition – Budapest Central Hungary. (2023). Tripadvisor. https://www.tripadvisor.com/Attraction_Review-g274887-d3752902-Reviews-Murder_The_Exhibition-Budapest_Central_Hungary.html

Murder Wall. (2023). POMC. https://pomc.org/murder-wall/

Museum of Death. (2023). Museum of Death: http://www.museumofdeath.net/

Museum of Death - Los Angeles California. (2023). Tripadvisor. https://www.tripadvisor.com/Attraction_Review-g32655-d2018939-Reviews-Museum_of_Death-Los_Angeles_California.html

Museum of Death - New Orleand Louisiana. (2023). Tripadvisor. https://www.tripadvisor.com/Attraction_Review-g60864-d7951659-Reviews-Museum_of_Death-New_Orleans_Louisiana.html

Museum of Perth. (2023). Museum of Perth. https://www.museumofperth.com.au/

Museum of Perth – Perth Greater Perth Western Australia. (2023). Tripadvisor. https://www.tripadvisor.com/Attraction_Review-g255103-d8813929-Reviews-Museum_of_Perth-Perth_Greater_Perth_Western_Australia.html

National Museum of Crime & Punishment - Washington DC District of Columbia. (2023). Tripadvisor. https://www.tripadvisor.com/Attraction_Review-g28970-d941651-Reviews-National_Museum_of_Crime_Punishment-Washington_DC_District_of_Columbia.html

New Orleans Killers and Thrillers Tour – New Orleans Louisiana. (2023). Tripadvisor. https://www.tripadvisor.com/AttractionProductReview-g60864-d11469169-New_Orleans_Killers_and_Thrillers_Tour-New_Orleans_Louisiana.html

New Orleans True Murder Tour: Sinister Criminal Intentions – New Orleans Louisiana. (2023). Tripadvisor. https://www.tripadvisor.com/AttractionProductReview-g60864-d12955117-New_Orleans_True_Murder_Tour_Sinister_Criminal_Intentions-New_Orleans_Louisiana.html

Newton, M. (2006). *The Encyclopedia of Serial Killers* (2nd ed.). Infobase Publishing.

Night-Time Dark History and Vampire Walking Tour in Buda Castle District - Budapest. (2023). Tripadvisor. https://www.tripadvisor.com/AttractionProductReview-g274887-d17162384-Night_Time_Dark_History_and_Vampire_Walking_Tour_in_Buda_Castle_District-Budapest_.html

Nyomozás az első magyar sorozatgyilkos után – séta. (2023). Kertvárosi Időutazó. https://stayhappening.com/e/nyomoz%C3%A1s-az-els%C5%91-magyar-sorozatgyilkos-ut%C3%A1n-s%C3%A9ta-E2ISTS5714A

Pallardy, R. (2021, August 17). *Elizabeth Báthory.* https://www.britannica.com/biography/Elizabeth-Bathory

Philbin, T., & Philbin, M. (2009). *Killer Book of Serial Killers.* Sourcebooks, Inc.

Police Museum. (2023). Google. https://www.google.com/search?q=budapest+police+museum&biw=1462&bih=794&sxsrf=AOaemvIfi_deSrtmttn0rpuWDbwUbL7hVw%3A1634149352659&ei=6CNnYa-9J5GK9u8Ppcia2As&ved=0ahUKEwiv0bLdgMjzAhURhf0HHSWkBrsQ4dUDCA4&uact=5&oq=budapest+police+museum&gs_lcp=Cgdnd3Mtd2l6EA

Powell, R., Kennell, J., & Barton, C. (2018). Dark Cities: A Dark Tourism Index for Europe's Tourism Cities, Based on the Analysis of DMO Websites. *International Journal of Tourism Cities*, *4*(1), 4–21. doi:10.1108/IJTC-09-2017-0046

Rendőrmúzeum. (2023). http://www.rendormuzeum.com/fooldal

Richard Ramirez. (n.d.). In *Encyclopedia Britannica online.* https://www.britannica.com/biography/Richard-Ramirez

Rosewood, J. (2015). *Edmund Kemper: The True Story of the Co-ed Killer.* CreateSpace Independent Publishing Platform.

Schechter, H., & Everitt, D. (1997). *The A to Z Encyclopedia of Serial Killers.* Pocket Books.

Schmalbruch, S. (2018). US Trips for Fans of True Crime. *Insider.* https://www.insider.com/best-true-crime-vacations-2018-5

Seaton, A. V. (1996). Guided by the Dark: From Thanatopsis to Thanatourism. *International Journal of Heritage Studies*, *2*(4), 234–244. doi:10.1080/13527259608722178

Sheth, K. (2020). Countries that have produced the most serial killers. *WorldAtlas.* https://www.worldatlas.com/articles/countries-that-have-produced-the-most-serial-killers.html

Siriraj Medical Museum. (2023). Siriraj Medical Museum. https://www.si.mahidol.ac.th/sirirajmuseum/siriraj-museum-en.html

Sprengel Museum. (2023). Sprengel Museum. https://www.sprengel-museum.de/en

Sprengel Museum Hannover - Hannover Lower Saxony. (2023). Tripadvisor. https://www.tripadvisor.de/Attraction_Review-g187351-d266726-Reviews-Sprengel_Museum_Hannover-Hannover_Lower_Saxony.html

Stone, P. (2006). A Dark Tourism Spectrum: Towards a Typology of Death and Macabre Related Tourist Sites, Attractions and Exhibitions. *Tourism: An Interdisciplinary International Journal, 54*(2), 145–160. http://clok.uclan.ac.uk/27720/1/27720%20fulltext_stamped.pdf

Stone, P., & Sharpley, R. (2008). Consuming Dark Tourism: A Thanatological Perspective. *Annals of Tourism Research, 35*(2), 574–595. doi:10.1016/j.annals.2008.02.003

Strange, C., & Kempa, M. (2003). Shades of Dark Tourism: Alcatraz and Robben Island. *Annals of Tourism Research, 30*(2), 386–405. doi:10.1016/S0160-7383(02)00102-0

Szűcs, Gy. (2015). 100 éve halt meg a nőfaló cinkótai sorozatgyilkos. *Index.* https://index.hu/tudomany/tortenelem/anagyhaboru/2015/02/06/100_eve_halt_meg_a_hirhedt_cinkotai_sorozatgyilkos/

Tarlow, P. (2005). Dark Tourism: The Appealing 'Dark' Side of Tourism and More. In M. Noveilli (Ed.), *Niche Tourism: Contemporary Issues, Trends and Cases* (pp. 47–57). Elsevier. doi:10.1016/B978-0-7506-6133-1.50012-3

Teicher, M. H. (2000, October 1). *Wounds That Time Won't Heal: The Neurobiology of Child Abuse*. Cerberum. https://www.dana.org/article/wounds-that-time-wont-heal/

Ten Bells. (2023). https://www.tenbells.com/

Terrible Crimes in Vienna. Hidden Secrets of the Inner City – Vienna. (2023). Tripadvisor. https://www.tripadvisor.com/AttractionProductReview-g190454-d21193573-Terrible_crimes_in_Vienna_hidden_secrets_of_the_inner_city-Vienna.html

Tevely, T. V., & Smith, M. K. (2022). Blurring the Boundaries Between Fact and Fiction: Serial Killers in the Context of Dark Tourism. *Tourism & Heritage Journal, 4*(4), 53–75. Advance online publication. doi:10.1344/THJ.2022.4.4

The Amsterdam Dungeon (2023). https://www.thedungeons.com/amsterdam/en/

The Blood and Tears Walk – London England. (2023). Tripadvisor. https://www.tripadvisor.com/Attraction_Review-g186338-d1383076-Reviews-The_Blood_And_Tears_Walk-London_England.html

The Issue of Child Abuse. (n.d.). Childhelp. https://www.childhelp.org/child-abuse/

The Jack the Ripper Walking Tour in London – London England. (2023). Tripadvisor. https://www.tripadvisor.com/AttractionProductReview-g186338-d18935383-The_Jack_The_Ripper_Walking_Tour_in_London-London_England.html

The London Dungeon. (2023). https://www.thedungeons.com/london/

The Memorial. (2021). Rose South LA. http://rosesouthla.org/the-memorial/

The Victims of Fritz Haarmann Memorial. (2023). Atlas Obscura. https://www.atlasobscura.com/places/the-victims-of-fritz-haarmann-memorial

Thomson, R. (2022). Haunted Venice – Legends, Mysteries and Stories to Creep Yourself out about the Most Romantic Place on Earth. *Rossi Writes*. https://rossiwrites.com/italy/venice/haunted-venice-legends-mysteries-stories/

Tóth, T. Zs., & Papp-Váry, Á. F. (2022). From Nuclear Disaster to Film Tourism: The Impact of the Chernobyl Mini-Series on the Exclusion Zone. In R. Baleiro & R. Pereira (Eds.), Global Perspectives on Literary Tourism and Film-Induced Tourism (pp. 280-301). IGI Global. https://doi.org/ doi:10.4018/978-1-7998-8262-6

Tour Leader Venice. (2023). Serial Killers of Venice Tour. https://www.tourleadervenice.com/project/venice-and-the-serial-killers-tour/

True Crimes of Georgetown. (2023). DC by foot. https://freetoursbyfoot.com/true-crimes-of-georgetown/

Tunbridge, J. E., & Ashworth, G. J. (1996). *Dissonant heritage. The Management of the Past as a Resource in Conflict*. Belhaven Press.

Two Feet & a Heartbeat Tours – Perth Greater Perth Western Australia. (2023). Tripadvisor. https://www.tripadvisor.com/ShowUserReviews-g255103-d1992671-r761709116-Two_Feet_a_Heartbeat_Tours-Perth_Greater_Perth_Western_Australia.html

Ungarisches Haus. (2023). Atlas Obscura. https://www.atlasobscura.com/places/ungarisches-haus-hungarian-house

Varghese, B., & Paul, N. (2014). A Literature Review on Destination Management Organization (DMO). *ZENITH International Journal of Multidisciplinary Research, 4*(12), 82–88. http://zenithresearch.org.in/images/stories/pdf/2014/DEC/ZIJMR/9_ZIJMR_VOL4_ISSUE12_DECEMBER2014.pdf

Venice Ghost & Legends Walking City Tour – Venice Veneto. (2023). Tripadvisor. https://www.tripadvisor.com/AttractionProductReview-g187870-d13075842-Venice_Ghost_Legends_walking_city_tour-Venice_Veneto.html

Vienna Criminal History Tour in German. (2023). Get Your Guide. https://www.getyourguide.com/vienna-l7/vienna-criminal-history-tour-in-german-t51444/

Visitare Serial Killer. A Jesolo Venezia Mostra. (2023). Serial Killer. http://www.mostraserialkiller.it/serial-killer/visitare-serial-killer-a-jesolo-venezia-mostra.html

Werewolf of Hanover Crime Tour. (2023). Booking.com. https://www.booking.com/attractions/de/pr6mcl2zmkq7-werewolf-of-hanover-crime-tour.hu.html?date=2023-02-24&start_time=20%3A00&ticket_type=OFuR59frVqNc

What's On. (2023). Fremantle Prison. https://fremantleprison.com.au/

Wiener Kriminal Museum. (2023). Wiener Kriminal Museum. https://wien.kriminalmuseum.at/en/news/

Wright, D. W. M. (2018). Terror Park: A Future Theme Park in 2100. *Futures, 96*, 1–22. doi:10.1016/j.futures.2017.11.002

Yu, J., & Egger, R. (2022). Looking behind the scenes at dark tourism: A comparison between academic publications and user-generated-content using natural language processing. *Journal of Heritage Tourism, 17*(5), 548–562. doi:10.1080/1743873X.2022.2097011

Zodiac Killer. (2004). Presidio Heights. https://www.zodiackiller.com/PresidioHeights2004.html

ADDITIONAL READING

Birkbeck, C. (2014). *Media Representation of Crime and Criminal Justice*. Oxford Handbooks.

Fanning, S. E., & O'Callaghan, C. (2023). *Serial Killing on Screen*. Palgrave Macmillan. doi:10.1007/978-3-031-17812-2

Korstanje, M., & George, B. (Eds.). (2017). *Virtual Traumascapes and Exploring the Roots of Dark Tourism*. IGI Global.

Mellins, M., & Moore, S. (2021). *Critiquing Violent Crime in the Media*. Palgrave Macmillan. doi:10.1007/978-3-030-83758-7

Stone, P. (2006). A Dark Tourism Spectrum: Towards a Typology of Death and Macabre Re lated Tourist Sites, Attractions and Exhibitions. *Tourism: An Interdisciplinary International Journal*, *54*(2), 145–160.

Stone, P. R., Hartmann, R., Seaton, T., Sharpley, R., & White, L. (Eds.). (2018). *The Palgrave Handbook of Dark Tourism Studies*. Palgrave Macmillan. doi:10.1057/978-1-137-47566-4

KEY TERMS AND DEFINITIONS

Dark Fun Factory: Dark tourist attractions with the focus on interactive and entertaining performances instead of remembrance or learning.

Dark Tourism: Travels related to death, disasters, morbid events and/or human suffering, where the motivation is curiosity towards the darkest parts of human history.

Destination Management Organization: Marketing organizations that promote certain locations as attractive destinations with the intention of encouraging tourists to visit them.

Serial Killer Tourism: Visitation of places that has a connection to serial killers, like their birthplace, the crime scene, a memorial for the victims, or an exhibition that contains an object from the serial killer.

Serial Killers: Serial killers are murderers, who killed more than two people in a certain time period. Most of the time they kill in a special, unique way, and can be motivated by either sexual urges or mental health issues.

Shades of Dark Tourism: How closely related the given site is to death and suffering, what is its interpretation about the history. Places with only a loose connection (e.g., theme parks) are considered light, while sites of actual death (e.g., Auschwitz-Birkenau) are the darkest.

Walking Tours: Insightful, informative tours in an urban setting, to get to know the history and myths of the visited sites, undertaken by foot and guided by an escort.

Chapter 11
Strategy for Developing Spice Tourism:
A Study of the State of Maharashtra (India)

Harshada Rajeev Satghare
Vishwakarma University, Pune, India

ABSTRACT

Spices are always considered an integral component of the cultural heritage of the place. They provide historical, cultural, social, and geographical identities to the region. These identities are helpful in the development of tourism based on the special interest in experiences, consumption, and purchase of spices. The phenomenon can be formally named 'spice tourism'. The emerging area of spice tourism needs to be well-researched as very limited research is observed, specifically in the Indian context. Thus, identifying its dimensions, industry stakeholders, and industry framework is the need of the hour. Thus, through the case study approach, the researcher tried to design and develop destinations of Maharashtra state based on the available strengths, opportunities, and needs of potential customers. The researcher aimed to identify, develop, and promote spice tourism destinations in the state. Further, challenges of climate, spice production, global competition, and community issues were discussed for appropriate planning.

DOI: 10.4018/978-1-6684-7242-2.ch011

INTRODUCTION

Spices are not only meant to be utilized in food preparation but in many countries especially, Asian countries such as India, Sri Lanka, and Indonesia they are used as part of cultural rituals and Festivals. It is considered an element of social and cultural life (Hettiarachchi et al., 2021). Since the time of the ancient Silk Route, spices have secured their role in history and supported the development of global trade (Ariwibowo et al., 2022). In Vedas, as early as 6000 BC, shreds of evidence are available regarding the properties and use of various spices (Dhoke et al., 2020). They acted as the link between eastern and western countries. Spices are linked to the geographical characteristics of the region as they grow in very specific climates and atmospheres. Thus, spices can be considered as one of the elements providing historical, cultural, social, and geographical identities to the place (Ariwibowo et al., 2022; Jolliffe, 2014; Nair and Mohanty, 2021; Rama et al., 2021). These identities support the development of tourism in the region (Malkanthi et al., 2015; Zilihona and Mamboya, 2018).

Spices can be employed as a cultural heritage attraction, designated by United Nations Educational, Scientific, Cultural Organisation (UNESCO), for the promotion of tourism. Therefore, it can play a key role in fostering cultural heritage tourism (Ariwibowo et al., 2022; Rama et al., 2021). Being important ingredients in many regional dishes, spices are an important pillar of the food tourism industry. It is also counted as part of rural and farm tourism(Hakim andYanuwiadi, 2016). Adding to this, nowadays, specifically after the COVID-19 pandemic, spices are getting recognised widely for their medicinal benefits, resulting in providing a new identity in the health and wellness tourism sector (Shenoy, 2020; Szabo et al., 2010). The cumulative result of all this can be identified as the development of a special interest form of tourism related to spices, named 'spice tourism' (Hettiarachchi et al., 2021).

The emerging area of spice tourism needs to be explored well as very limited research is available on the topic (Hakim and Yanuwiadi, 2016; Hettiarachchi et al., 2021; Nair and Mohanty, 2021). Additionally, it is expected that after COVID-19, to avoid crowds, early travellers will seek out 'off-the-beaten-path' nature-based and outdoor destinations (Spenceley, 2021; World Travel and Tourism Council (WTTC), 2020). Thus, identifying its new dimensions, industry stakeholders, and industry framework is needed. Designing and developing destinations based on the available strengths, opportunities, and needs of potential customers will lead the path towards success in the special interest area. Further, challenges of climate, spice production, global competition and community issues should be well-addressed through appropriate planning.

There is a need of identifying more spice routes and geographical mapping of them which will facilitate research. In this vein, the present research will take the State of Maharashtra as a case study. The State is one of the leading exporters of spices from India. It ranks first globally in the export of high-quality turmeric (Ministry of Commerce and Industry, 2023). It has very well-known spice markets also. Still, the State has not identified, developed, and promoted a spice tourism destination. Thus, the chapter aimed to identify the product basket of spice tourism for Maharashtra. Categorisation of spices based on their application and usage will help in identifying the target market for the product. Further, the chapter tried to suggest a co-creative model for the development of spice tourism. The exploratory case study method will provide a guiding note to the destination planners, designers, marketers, and spice industry stakeholders on developing destinations for matching the needs and expectations of this niche market.

BACKGROUND

The word 'spice' comes from the Latin word *species*, which indicates an item of special worth more than the ordinary articles of trade(Struan, n.d.). Spice is a dried seed, fruit, root, bark, or vegetative substance primarily used for flavouring, colouring or preserving food (Lingala andDeshmukh, 2019). Directorate of Arecanut and Spices Development (DASD) (2023) defined spices as "*vegetable products or mixtures thereof, free from extraneous matter, used for flavouring, seasoning and imparting aroma in foods*". DASD is a subordinate office under the Ministry of Agriculture, Government of India. These different definitions focus on the aromatic and flavouring characteristics of spices. Though they are used in very small quantities in cooking, these characteristics of aroma and flavour make their presence notable and worthy. Globally, a large variety of spices are available which are extracted as or from the different parts of trees like the plant leaves (curry leaf), buds (clove), bark (cinnamon), rhizomes (ginger, turmeric), berries (black pepper), seeds (cumin), or even flower stigma(saffron) etc. (Nair and Mohanty, 2021; Rajanbabu and Ganesan, 2015). These spices could only grow in the tropical east covering the region from the south of China to Indonesia and southern India and Sri Lanka(Lingala and Deshmukh, 2019; Rajanbabu and Ganesan, 2015; UNESCO, n.d.).

The presence and significance of spices are visible from the time of human civilisation and the ancient Silk Route connecting eastern and western countries (UNESCO, n.d.). It was one of the crucial agricultural and export commodities since then (Ariwibowo et al., 2022; Lingala andDeshmukh, 2019). The use of spices for a variety of purposes was documented since the ancient Egyptian period (Nair andMohanty, 2021). Hettiarachchi et al (2021) mentioned 'spices' as the queen of ancient commerce due to their role in the earliest evolution of the trade. Demand for spices significantly increased in the medieval period which shaped global trade and supported the development of economies (Ariwibowo et al., 2022; Henriques, 2019; UNESCO, n.d.). These explorations, global trade and battles played role in the establishment of renowned spice trading routes, also known as the Silk Route s(Nair and Mohanty, 2021). The practice of adding spices to the food was not only the reason behind the development of the spice route but since ancient times spices were used in cultural and devotional rituals. Further, its use in medical treatments of skin ointments and anti-todes against poisons was also documented (Henriques, 2019; UNESCO, n.d.).

In modern times also, the value and number of applications of spices have increased notably. Today, spices are used as an essential component in the majority of recipes in many cultures, because of their flavour, medicinal value and rich aroma(Lingala and Deshmukh, 2019). Because of their digestive properties, they are considered essential for culinary and medicinal use (Hettiarachchi et al., 2021; Rama et al., 2021). Some of them have anti-oxidant or antibiotic or anti-microbial properties which create their strong presence in medicines for many diseases, specifically, they are used in Ayurvedic treatments. Thus, they are essential in the pharmaceutical industry. They are an important part of home remedies also(Nair and Mohanty, 2021; Shukla and Yadav, 2018). Spices are also renowned for boosting immunity and hence the world notably realized their importance during the COVID-19 pandemic. Further, few spices have preservative properties and hence, they are commonly used in foods like pickles and chutneys, etc. (Hettiarachchi et al., 2021; Shukla and Yadav, 2018). Further, it is observed that the consumption of spices is growing in countries like the United Kingdom, Europe, and North America because of the associated health benefits(Henriques, 2019).

Hettiarachchi et al., (2021) emphasised spices' use in the hospitality, cosmetics, lifestyle, and alcohol industry. All this resulted in the identification of spices as high-value and low-volume commodities of commerce in the world market and the estimated growth rate for spices demand in the world is approximately 3.19 percent (Rajanbabu and Ganesan, 2015).

As mentioned earlier, spices provide various identities like geographical, historical, cultural, social and medical to the region, hence they can be used and promoted as a tourism product. This has resulted in the introduction of the prominent phenomenon of spice tourism. Yet, the research on this emerging area of special interest is in the preliminary stages (Hettiarachchi et al., 2021; Nair and Mohanty, 2021). The researcher could not find the formal definitions of spice tourism proposed by earlier researchers.

Many of the researchers linked this phenomenon to agritourism as it includes a visit to the spice farms, experiencing the spice cultivation practices, acquiring knowledge of different spices, purchasing farm products, etc. (Barbuddhe and Singh, 2014; Chatterjee and Prasad, 2019; Hakim and Yanuwiadi, 2016; Malkanthi et al.,

2015; Nair and Mohanty, 2021; Sandaruwani and Athulagnanapala, n.d.). Spices and tourism are generally considered separate entities which are linked as 'spice garden-based agri-tourism ventures' (Hettiarachchi et al., 2021). Spice farm visit includes the interpretation of the long history of spices, the characteristics of spices, and their usage and application which is crucial for tourism programs (Hakim and Yanuwiadi, 2016).

Nair and Mohanty (2021) appreciated the role of spice tourism in sustainable rural tourism development as it could enrich the duct basket of areas hosting rural tourism. It will support the production of spices, the development of tourist accommodations in farmhouses, and the promotion of picnic areas at spice farms, spice vendor stalls, and restaurants that offer cuisine made from local spices.

Spices can play a major role in promoting cultural heritage tourism (Ariwibowo et al., 2022; Rama et al., 2021). Spice tourism strengthens the public insight and awareness of the cultural heritage of spices that has existed for generations (Rama et al., 2021). The development of the cultural heritage of spice routes can support the development of the tourism industry and the creative economy. Spice is the medium of cultural exchange (Ariwibowo et al., 2022). Further, UNESCO already identified spices as a cultural heritage attraction and working on reviving the significance of the ancient spice route. UNESCO has put forward two ways to position spices as a cultural heritage attraction. The first way is to relate spices to food and cultural attractions. The second way is to position spices within ancient trade and heritage as is done by Kerala (Nair and Mohanty, 2021).

Moreover, tourism researchers have linked spices to the different forms of tourism. Being an essential component of culinary practices of many regions, spices are linked to culinary tourism activities (Nair and Mohanty, 2021). Due to the importance of spices in healing and medicinal treatments, spices can be linked to the medical and wellness tourism sector (Shenoy, 2020; Szabo et al., 2010). Thus, if developed and marketed well for different market segments, spice tourism could become one of the most prominent special interest tourism (SIT) forms globally (Nair and Mohanty, 2021).

Being an emerging area, research on the topic of spice tourism is very limited and still in its infancy (Hakim and Yanuwiadi, 2016; Hettiarachchi et al., 2021; Nair and Mohanty, 2021).

The research conducted by Hettiarachchi et al., (2021) on perceptions of travellers towards the spice markets and their readiness to engage in spice market activities demonstrated a considerable demand for spice market establishment/s. Moreover, the researcher suggested incorporating cultural, educational, recreational, and wellness activities into spice tourism to improve the appeal of the segment.

Barbuddhe and Singh (2014) suggested that spice tourists must enjoy the fresh air and the aroma of spices in their natural forms. They should be able to purchase fresh, pure, organically grown spices and spice products. Adding to this, the researcher also urged these gardens to function as health resorts to which visitors attach the utmost importance. Further providing the experience of organic cultivation, and recycling of farm waste for raising organic spices, vermicomposting, and livestock components could enrich their experience. The inclusion of interactive and physical activities allowing visitors to enjoy simple adventurous experiences like climbing of Arecanut, plucking spices, etc., will impart a rare amusement for tourists.

Thus, it can be concluded that though the knowledge of visitors on spices is very clear, their knowledge of the concept of spice tourism is low (Hettiarachchi et al., 2021). Additionally, the geographical analysis of spice tourism research reflected that a significant number of studies are done on Sri Lanka (Hettiarachchi et al., 2021; Malkanthi et al., 2015; Sandaruwani and Athulagnanapala, n.d.). A very limited number of studies are based on India, Indonesia, China, and other regions of the world which are spice-producing and cultivating areas (Firmino, 2010; Rajanbabu and Ganesan, 2015; Shukla and Yadav, 2018). Unfortunately, no visible literature is available on the topic of interest of tourists in spice tourism so that the potential could be identified.

To sum up, there is a dire need to work on the different aspects of spice tourism including its conceptual grounds, market identification, product design and development theories, policy and planning issues, defining the role of various industry stakeholders, etc.

MAIN FOCUS OF THE CHAPTER

Spice Tourism in India

From historical times, India is regarded as the ‘land of spices’, ‘house of spices’, and ‘spice bowl of the world’ (Barbuddhe and Singh, 2014; Dhoke et al., 2020; Sontakke et al., 2018), and that role has yet remained the same (Henriques (2019), Nair and Mohanty (2021)). India has long and historical spice trading relations with the ancient civilisations of Rome and China(Lingala and Deshmukh, 2019). During the Guptas period (AD 300-550) references showed the cultivation of spices like pepper, cardamom, cloves, ginger, and turmeric(Singh et al., 2022). In 1498, the Portuguese explorer Vasco da Gama made his first sea voyage from Europe to India in the search of spices. He arrived on India’s Malabar Coast which was the popular centre of the spice trade. It was the start of direct trading between Europe and South-east Asia (Henriques, 2019).

Indian spices are used in a variety of the forms like dried seeds, leaves, flowers, barks, roots, fruits, and certain spices are ground and used in powder form(Rajanbabu and Ganesan, 2015). They are categorized into three main groups (Shukla and Yadav, 2018), namely, the basic spices (spices go into most dishes like cumin seed and coriander seed), complimentary spices (used in the combination with the basic spices and aromatics like fennel seed, carrom seed etc.) and aromatic or secondary spices (added in small quantities for aroma like cardamom, saffron, nutmeg, etc.).

India's tropical climate is varied and ideal for growing a range of different spice crops (Henriques, 2019). The country produces more than 60 spice varieties listed by International organization for Stardardization (ISO) out of the 109 global spice varieties (Lingala and Deshmukh, 2019; Rajanbabu and Ganesan, 2015; Seethalekshmy and Subramaniyan, 2021). Indian spices are popularly known for their flavour and aroma in domestic as well as international markets (Lingala and Deshmukh, 2019). With a share of 68 per cent in the export of turmeric in the world, India ranked first (Henriques, 2019). It produces many quality, rare and medicinal spices. Pepper, cardamom, ginger, turmeric, coriander, cumin, celery, fennel, fenugreek, nutmeg, spice oils and oleoresins, and mint products are the major spices shipped abroad. Indian spices reach almost more than 150 countries of the world including the United States, United Kingdom, Germany, France, Italy, Canada, Australia, the United Arab Emirates, Iran, Singapore, China, South Korea, the Middle East, Japan, and Bangladesh, among others. Value-added and processed products manufactured from spices such as spice powders, curry powder, spice mix, etc. also show a positive trend (Lingala and Deshmukh, 2019; Ministry of Commerce and Industry, 2023).

Thus, India is the world's largest producer, consumer and exporter of spices and accounts for almost half of the global spice trade. Unfortunately, the area of spice tourism didn't receive significant attention from industry practitioners and researchers. Probably, it is due to the availability of other diverse tourism products like culture, heritage, food, festival, business, medical and so on. In the Indian context, very few States like Kerala and Goa are working on such a form of tourism. Kerala has significantly focused on and developed it whereas Goa is gradually developing it but still, the tourists are not well-aware of it.

The history of spice plantations in Kerala dates back to the time of the ancient Spice Route when the Greek, Roman, Chinese, and Arab traders visited the Muziris (Kondungallur) on the Malabar Coast in search of Indian commodities. In this vein, the Spice Route Project by Kerala Tourism developed and promoted the route for tourists. UNESCO recently recognised the project and many other countries consented to the initiatives. This Project aims to educate people on the history of the region for strengthening international relations and promoting shared heritage through tourism (Nair and Mohanty, 2021).

The very scarce research literature is available on the Indian spice tourism practices focused on the spice and spice gardens of Kerala and Goa (Hettiarachchi et al., 2021; Jacob, 2008; Seethalekshmy and Subramaniyan, 2021). The research conducted by Seethalekshmy and Subramaniyan (2021) revealed that Kerala is well known for its variety and exclusive range of spices like cloves, cinnamon, cardamom and pepper among foreign tourists. Tourism stakeholders of Kerala generally conduct guided tours for tourists which take them to the spice gardens and explain the detailed aspects of the spice plant including its health benefits, growth, history, etc. These practices of spice tourism are successfully providing additional income to the farming community and supporting sustainable development.

Hettiarachchi et al. (2021) carried out comparative research on spice market of Dubai, Turkey, and Cochin (a city from Kerala). The markets sell diverse spices as well as household items. The spices from Cochin markets are considered trustworthy by travellers.

Notably, Maharashtra is one of the significant spice producers and exporter State in the country but no study or industry or government agency has explored the area of spice tourism for the State. Hence, through the present case study of Maharashtra, the researcher is suggesting a model for designing and developing the spice tourism product.

Spice Tourism in Maharashtra

Maharashtra is the third-largest State in the country and the second-largest State in the population of India(Ministry of Agriculture and Cooperation, 2012) (Figure 1). Maharashtra and Tamil Nadu have the highest numbers of foreign tourist visits, with 1.26 million and 1.23 million, respectively. Around half of the State population is dependent on agriculture for their livelihood. The average share of the agriculture and allied activities sector in the State economy is 11.9 per cent(Director of Economics and Statistics GoI, 2022).

Figure 1. Map of Maharashtra state
Source: Wikimedia Commons (2013)

Many cities of the State that were on the ancient Silk Route like Paithan (Pratishthan) (World Heritage Convention, 2010). Fonia (n.d.) in his article on the Indian link with the Silk Route mentioned the name of Shuraparaka (Nala Sopara, Mumbai, Maharashtra) as one of the trade centres.

In the present scenario, the State is one of the leading exporters of spices from India. It ranks first globally in the export of high-quality turmeric. Maharashtra has farms of chilly, turmeric, pepper, garlic, pomegranate seed, nutmeg, cardamom, ginger, fenugreek seeds, and kokum (Garcinia Indica) (Khandekar et al., 2016; Lingala and Deshmukh, 2019; Spice Board of India, n.d.). Maharashtra produces two GI-tagged turmeric varieties and one GI-tagged chilly variety. Coastal areas of Maharashtra are also known for the production of GI-tagged Kokum. Further spice production trends of the State released by the Spice Board of India estimated that approximately 1,28,475 hector land of the State will be under the production of spices (for 2021-22) including a variety of turmeric, Kokum, chilly, garlic, pepper etc. (Ministry of Commerce and Industry, 2023; Spice Board of India, 2021). Sangli, Satara, Hingoli, Nanded, and Parbhani are the major turmeric-growing districts in the State. It is one of the major crops in the Sangli district (Dhoke et al., 2020). The district significantly produces ginger also (Kadam et al., 2019). Adding to this, it has very well-known spice markets also. Thus, the State is one of the largest spice-exporting States in India (Table 1).

Table 1. Production of major spices in Maharashtra

Major Spice/State-Wise Area and Production of Spices										
Area in Hectares, Production in Tonnes										
Main Spices	2017-18		2018-19		2019-20(*)		2020-21(Final)		2021-22(Adv. Est.)	
	Area	Production	Area	Production	Area	Production	Area	Production	Area	Production
Chilly	7050	14130	5698	14030	6508	22434	5605	24484	5648	18546
Turmeric	15760	38590	17224	38310	54248	218873	59576	230741	102625	367985
Garlic	2540	14990	2659	13721	2650	13835	3779	22442	3696	22184
Tamarind	720	4660	699	4240	1106	8988	1039	9968	1245	11245
Total including others	139487	64000	244209	48000	259148	61000	309335	65000	288118	60000

Source: Spice Board of India (2021)

Still, the State has not identified, developed, and promoted a spice tourism destination. Comparatively, Kerala and the small bordering State of Goa have positioned themselves very well and their spice gardens and farms are attracting a significant number of tourists (Jacob, 2008; Seethalekshmy & Subramaniyan, 2021).

SOLUTIONS AND RECOMMENDATIONS

Conceptual Framework of Spice Tourism

After a thorough review of the literature and observation of industrial practices, the necessity of a strong conceptual framework for spice tourism is evident. To address the need, the researcher has proposed the definition of 'spice tourism' as follows:

It is a form of tourism including tourist activities that revolve around the areas exhibiting one or more spice related practices like spice cultivating, processing, trading, selling and purchasing and other supplementary activities. Additionally, these areas can also provide tourists with the opportunity of participation in the activities.

Areas associated with spice tourism activities mentioned in Figure 2.

Figure 2. Spice tourism areas and activities

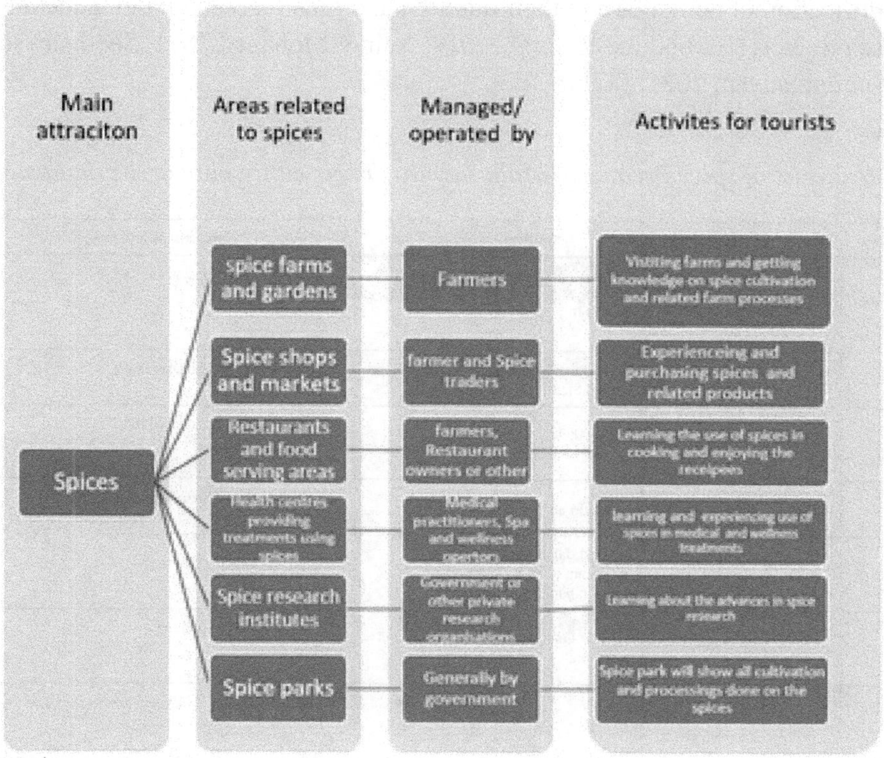

Figure 2 elaborates on the areas related to the spice attraction, the stakeholders, and activities that can be done by the tourists at the place. The first three areas can be visited by all the tourists who are interested in spices while the rest of the three areas (including health centres, research institutes and parks) may be visited by interested tourists only. Thus, with more research, more allied areas and related tourism products can be developed and promoted under this area of special interest, i.e., spice tourism. Spice parks can play a significant role in propagating knowledge about the rich biodiversity of the region, resulting in enhancing the appeal of the destination (Barbuddhe & Singh, 2014).

Spice Tourism Products of Maharashtra

Few researchers have suggested the stages of spice product design which include the identification of spice diversity, categorisation of spices based on their functions, identification of activities and their duration and then marketing and launching of these products (Barbuddhe & Singh, 2014; Nair & Mohanty, 2021; Seethalekshmy & Subramaniyan, 2021) (Table 2).

Table 2. List of spices grown in Maharashtra: Their cultivating areas and usage

SN	Spice	Cultivating Areas	Applications and Usage
1.	Turmeric	Sangali, Satara, Hingoli, Nanded, and Parbhani	- Commonly used in Indian cooking - Medicinal - Cosmetics - Preservative
2.	Ginger	Aurangabad, Sangali, more or less in the overall State	- Commonly used in Indian cooking - Medicinal
3.	Pomegranate	Pune, Ahmednagar, Pune, Sangli, Solapur, and Washim	- Commonly used in Indian cooking - Medicinal - Cosmetics
4.	Kokum	Ratnagiri and Sindhudurg	- Commonly used in Indian cooking - Medicinal
5.	Chilly	Nanded, Jalgaon, Dhule, Solapur, Kolhapur, Nagpur, Amravati, Chandrapur, and Osmanabad	- Commonly used in cooking - Medicinal
6.	Garlic	Nashik, Pune, Thane, Marathwada, and Vidarbha	- Commonly used in cooking - Medicinal - Preservative
7.	Nutmeg	Ratnagiri	- Commonly used in Indian cooking in small quantity - Medicinal - Cosmetics
8.	Cardamom	Kokan region	- Commonly used in cooking in small quantity - Medicinal
9.	Pepper	Konkan region	- Commonly used in cooking - Medicinal - Pest control
10.	Curry Leaves	Whole State	- Commonly used in cooking - Medicinal
11.	Fenugreek seeds	Central plateau and hill region of the State	Commonly used in cooking - Medicinal - Cosmetics
12.	Coriander	Whole State	- Commonly used in cooking - Medicinal
13.	Tamarind	Whole State	- Commonly used in cooking - Medicinal
14.	Onions		- Commonly used in cooking - Medicinal

Further, the State has well-known spice markets. Among them, Mumbai has famous markets like the Lalbaug spice market, Mirchi Gully, APMC, Kalbadevi and Crawford market. Further, every district has shops where tourists can purchase these spices. Adding to this, the State has a few popularly known spice–mix brands, namely, Pravin and Suhana Spices, Ravi Spices, Sururchi Spices and so on. The regional spice mix identities of the State include Goda masala, Khandeshi, and Malvani Spices. Tourists can visit the spice farms (Table 1) and purchase the earlier-mentioned products as a souvenir.

Maharashtra has many reputed medical and wellness centers which attract many domestic and international patients. Traditional wellness centres of the State utilize spices for the treatment of many diseases. Thus, the tourists experience the application of spices in medicinal treatments. Adding to this, agricultural universities as well as educational and research institutes are contributing significantly to spice-related research.

To sum up, it is evident that the State has a vast array of products to be offered to the spice tourists which, undoubtedly, create a promising spice tourism destination. Additionally, available transport infrastructure, accommodation services, availability of food variety, travel-related adequate services create a favorable atmosphere for the development of spice tourism.

SWOT analysis is essential before the design and development of any new product (the University of Minnesota, n.d.). It reflects the position of the organizations and business environment which helps decide product development strategies. Thus, Table 3 represents the SWOT analysis of Maharashtra to assess its potential for spice tourism development.

Table 3. SWOT analysis of Maharashtra for the development of spice tourism

Strengths	Weaknesses
- Rank second in foreign tourists' arrival in India - Among the top ten States of India in domestic tourists arrival - Two GI-tagged turmeric varieties - One GI tagged chilly variety - Production of GI-tagged kokum - Maharashtra has cultivation/ production of chilly, turmeric, pepper, garlic, pomegranate seed, nutmeg, cardamom, ginger, fenugreek seeds, tamarind, and kokum - The leading State in the production and export of turmeric - Well-known producers of spices mix - Significant agricultural production and development - Recommendable variety of local food and dishes - Various agricultural universities and research centres are researching spices - Adequate spice processing infrastructure - Good quality of tourism infrastructure - Adequate road and cellular network - Available trained manpower for tourism - Various government bodies are taking efforts for the development of agri and rural tourism - Well-established agri/ farm/ rural tourism centres	- The government policies or planning do not talk about the spice tourism - Tourism stakeholders do not look at spices as a tourism product - Availability of other diverse tourism products
Opportunities	**Threats**
- Increasing importance and use of spices after COVID 19 pandemic - Non-traditional spices cultivation and related research is in progress - Untapped domestic market - Developing the product under the Swadesh Darshan scheme of the Government of India	- Changing climate - Kerala has well-positioned its spice tourism business - Fierce competition from neighbouring States like Goa and Kerala

The SWOT analysis reflected that the State has tremendous potential to be developed as spice tourism which is reflected through the availability of adequate strengths with fewer numbers of weaknesses.

Table 4. Suggestive spice tourism activities and packages for the different regions of Maharashtra

SN	Region	Spices	Activities
1.	Konkan	The Region offers a true feast for spice tourists as the Region is enriched with the production of diverse and valuable spices like nutmeg, pepper, kokum, cardamom etc. Sindhudurga and Ratnagiri are prominent places for spice tourists.	- Visiting spice farms - Enjoy vegetarian and non-vegetarian dishes specifically made by using Malvani spice mix. Fishes and rice are the main food of the Region. - Drinks, namely Solkadhi and kokum syrup are made from kokum and are added delicacies. - Enjoy the different nature and culture of the Region which includes festivals, folk arts like Dashavtar, folk music, etc. - Agricultural universities of the Region like Dapoli agri-university, are doing recognisable research on spices - Due to the pleasant and pollution-free environment of the Region, many wellness centres are available in the region - Popular spice markets of Mumbai are the attraction for spice shoppers - Many tourist festivals are organised throughout the year in the Region - Ample number of homestays offer the experience of live and actual food local preparation - The Region is well-known for cashews and Alphonso mangos
2.	Pune	The Region includes Sangali and Kolhapur districts which are globally known for turmeric and chillies.	- Visiting spice farms of turmeric and chillies - Enjoying healthy as well as spicy vegetarian and non-vegetarian dishes specifically made by using Goda masala - Kolhapur is popular for its spicy non-vegetarian dishes made in red and white gravy - Spice mixes of Kolhapur are famous a - Agricultural universities and other research centres of the region are doing recognisable research on turmeric and chillies - Pune is well-known for traditional/ Ayuredic spa and wellness centres - Well-known Spice mix companies provide options for spice shoppers - Many agri-tourism farms are available in the Region
3.	Nashik	The Region is famous for chillies and pomegranate production	- Visiting spice farms of chillies, ginger, pomegranate etc. - Enjoying local cuisine - Agricultural universities and other research centres of the Region are doing recognisable research on turmeric and chillies - Many agri-tourism farms are available in the region - Vineyards of Nashik are famous for wine tourism, namely, Sula Vineyard - Spice shopping from the local market
4.	Aurangabad	The area is significant in ginger, garlic, and chilly production.	- Visiting spice farms of chillies, ginger, garlic, etc. - Enjoying local cuisine - Agricultural universities and other research centres of the Region like Parbhani agri-university, are doing recognizable research on turmeric and chillies - Many agri-tourism farms are available in the Region - Visit to the ancient caves of Ajanta and Ellora will provide a different experience of ancient trade routes and culture - Spice shopping from the local market
5.	Amravati	Famous for pomegranates and chillies.	- Visit spice farms - Visit to Lonar crater will off-the-beat experience - Enjoying local cuisine

Continued on following page

Table 4. Continued

SN	Region	Spices	Activities
6.	Nagpur	Famous for garlic and chillies.	- Visiting spice farms of chillies, ginger, garlic, etc. - Enjoying local cuisine - Non-vegetarian dishes made from Saoji spices are popular - Many agri-tourism farms are available in the Region - The Region is positioned as an eco-tourism destination that is blessed with a few significant Project tigers like Tadoba and Pench tiger reserves - Spice shopping from the local market - Haldiram is a popular snack-producing company from the Region

Marketing of Spice Tourism of Maharashtra

The target market for this niche form is European and western tourists visiting the destinations. Further school and college students can be targeted which will enhance their knowledge about spices and their importance (Barbuddhe and Singh, 2014). Spice-importing countries like the United States, United Kingdom, Germany, France, Italy, Canada, Australia, The United Arab Emirates, Iran, Singapore, China, the Middle East, Bangladesh, and other Asian countries can also be the target market (Borpuzari, 2022; Times of India, 2022). Being one of the largest producers of turmeric in the world, Maharashtra has great potential to attract the market segment which is focused on the health benefits of spices. Notably, after the COVID-19 pandemic, western countries are recommending immunity-boosting qualities of spices, specifically turmeric. Thus, it is the appropriate time to develop the product and tap the market.

Maharashtra is already a well-positioned and leading tourism State in the country (Satghare and Sawant, 2018), ranking second in foreign tourists' arrival and sixth in the ranking of domestic tourists' arrivals. This acts as a strength of the State while developing this new niche form of tourism. Creating awareness about the new product among these tourists and enabling them to visit the spice farms initially can help position the State as a spice tourism destination.

Moreover, the great potential of the domestic tourist market of India cannot be ignored. India has a population of 1210 million people (2011), making it the second most populated country in the world after China(NCERT, 2022). In 2021, 677.63 million numbers of domestic travellers were observed within the country (MoT GoI, 2021). Notably, during the COVID-19 Pandemic, the segment was less affected as compared to the foreign tourists. Further, nowadays, many travel research organisations also underlined the contribution of domestic tourism in the revival of the tourism industry after the pandemic. Moreover, domestic tourists can be considered a good market for this kind of tourism as Maharashtrian spices are commonly used by them also.

Thus, the marketing efforts should be positioning the State as an appealing spice tourism destination in the intense regional and global competition.

Role of Stakeholders in Spice Tourism of Maharashtra

Without the active involvement of all industry stakeholders the development of any kind of tourism is not possible. There is a need of involving industry stakeholders in the development of spice tourism as they will help create product awareness among tourists. Giving these spices as a souvenir will be helpful in image building of them. The stakeholders like hotels, tour operators and travel agencies can give these spices as souvenirs.

Adding to all of these, the need of giving training to the farmers to impart tourist handling skills to them is essential. They should have basic entrepreneurial skills, management skills, and communication skills to be successful in this area.

Support of regulatory policy infrastructure and tourism infrastructure will fasten the spice tourism development. Through the State having adequate tourism infrastructure, last-mile connectivity should be strengthened to connect major spice farms and to bring them on tourist maps. Therefore, a collaborative partnership between government and private players is needed. Such partnership is also expected in the area of development of world-class spice markets giving memorable spice purchase experiences to tourists. The government has to give due importance to the development of spice parks to support the development of spice tourism. Lastly, the role of the local community is pivotal in the development, therefore the community should also be educated on this special interest area (Figure 3).

Figure 3. Framework and role of industry stakeholders for the development of spice tourism

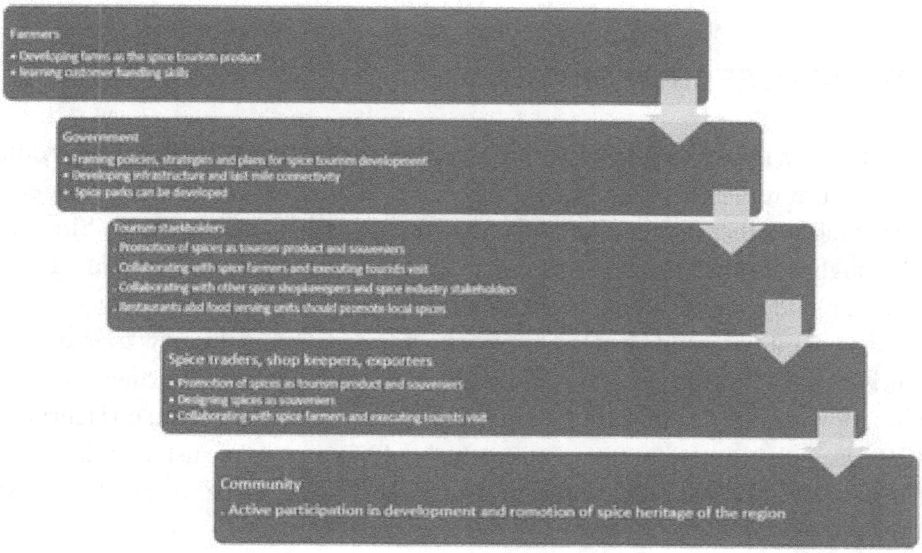

Significance of the Research

The exploratory approach of the study enables thinking beyond the conventional understanding of spice tourism.

Agriculture is the backbone of the Indian economy as a significant proportion of the population is directly or indirectly involved in agriculture. Thus, the prominent benefit of the development of spice tourism will be the generation of additional income for the agricultural society. Moreover, Maharashtra has significant count of farmer deaths in recent years (Marpakwar, 2022). The development of spice tourism in the Region will provide additional income avenues to them, supporting the financial growth of farmers. The study will focus on diverse aspects of spices like spice mixes (masala), popularity among tourists for medicinal usage of spices, and other allied attractions which will improve the involvement and engagement of correlated/ connected/ linked stakeholders in the tourism industry like spice shopkeepers, spice traders etc.

The identification of potential tourism destinations for spice tourism will add alternative destinations for Maharashtra tourism resulting in enhancing the array of tourism products. It will be of great support for strengthening the brand identity of the destinations. Moreover, it will provide exceptional experiences to tourists and encourage them to repeat travelling. Thus, the outcome of the study is practically useful to destination designers and marketers to develop destinations accordingly to attract the segment to revive the industry.

FUTURE RESEARCH DIRECTIONS

Spice tourism is an emerging niche area where very limited research has been done. Thus, future researchers could study the perspectives of farmers on the area to understand whether farmers find the initiatives appealing and beneficial. Similarly, the thoughts of other stakeholders involved in the spice trade and tourism are also worth for study and deciding the next course of action.

Sustainability should be at the forefront while developing this new tourism form as it is closely linked with the rich biodiversity of the Region. Further, the spice cultivation industry is already facing the pressure of climate change (Henriques, 2019). In such a situation, future research should focus on the sustainability of the segment and improving the contribution of spice tourism to the sustainable growth of the cultural heritage of spices and the Region.

CONCLUSION

Spice is a strong representative of the local cultural heritage which provides historical, cultural, social, and geographical identities to the Region. These identities can be helpful in the development of niche tourism providing curated experiences related to spice cultivation, processing, trading, consumption, and purchase. The evolving phenomenon is called spice tourism. The emerging area has very scares research, specifically in the Indian context. Thus, identifying its dimensions, industry stakeholders, and industry framework is the need of the hour. Thus, the researcher has taken one of the States of the country as a case study to understand the potential and suggest a way forward.

Maharashtra is a crucial tourism State and one of the top spice producers and exporters in India. The study has identified, listed and categorised the spices based on their importance. Thus, the Chapter identified the Product basket of spice tourism for Maharashtra State. SWOT analysis of the State clarified the potential of the State. Adding to this, the Chapter has provided suggestive regional packages and activities for spice tourists. Further, the Chapter has given marketing and overall development strategies for the emerging tourism form. Future research areas are guiding notes for the next spice tourism researchers.

To sum up, through the case study method, the researcher has provided the spice tourism design and development model to the tourism fraternity. Further, challenges of climate, spice production, global competition and community issues were discussed for appropriate planning. Thus, the exploratory case study method has provided a guiding note to the destination planners, designers, marketers, and spice industry stakeholders on developing destinations for matching the needs and expectations of spice tourists.

Adding to this, the researcher could explore the possibilities of developing of regional Asian spice tourism circuit connecting Sri Lanka, Nepal and Bangladesh.

ACKNOWLEDGMENT

This research received no specific grant from any funding agency in the public, commercial, or not-for-profit sectors.

REFERENCES

Ariwibowo, A., Budimen, H., & Dwinanto, A. (2022). Development of Gastronomy Tourism of Cultural Heritage Spice Route in Kota Serang, Banten. *International Forum on Spice Route 2022*. https://ocs.brin.go.id/index.php/ifsr/ifsr2022/paper/view/81

Barbuddhe, S. B., & Singh, N. P. (2014). *Agro-Eco-Tourism: A New Dimension to Agriculture*. ICAR Research.

Borpuzari, P. (2022). India's spice exports need to grow at 19.5% to meet $10 billion target by FY27. *The Economics Times*. https://economictimes.indiatimes.com/small-biz/trade/exports/insights/indias-spice-exports-need-to-grow-at-19-5-to-meet-10-billion-target-by-fy27/articleshow/92253030.cms

Chatterjee, S., & Prasad, M. V. D. (2019). The Evolution of Agri-Tourism practices in India: Some Success Stories. *Madridge Journal of Agriculture and Environmental Sciences*, *1*(1), 19–25. doi:10.18689/mjaes-1000104

Dhoke, Perke, & Karanjalkar. (2020). Economic analysis in marketing of turmeric in Sangli District. *Journal of Pharmacognosy and Phytochemistry, 9*(5), 618–620. www.phytojournal.com

Director of Economics and statistics GoI. (2022). *Economic Survey of Maharashtra 2021-22*. Author.

Directorate of Arecanut and Spices Development (DASD). (2023). *About DASD*. DASD. https://www.dasd.gov.in/index.php/content/index/home

Firmino, A. (2010). New challenges for the organic farmers in India-Tourism, spices and herbs. *Journal for Geography*. www.fao.org/nr/giahs/en/

Fonia, R. (n.d.). *India's link with silk trade roads*. UNESCO. Retrieved March 13, 2023, from https://en.unesco.org/silkroad/countries-alongside-silk-road-routes/india

Hakim, L., & Yanuwiadi, B. (2016). Identifying and Assessing Spice and Herbs Resources for Tourism Attractions: A Case Study from Banyuwangi, East Java. *Research Journal of Life Science, 3*(3), 154–159. http://rjls.ub.ac.id

Henriques, M. (2019). *How spices changed the ancient world*. BBC. https://www.bbc.com/future/bespoke/made-on-earth/the-flavours-that-shaped-the-world/

Hettiarachchi, I. C., De Silva, D. A. M., and Esham, M. (2021). International tourists' perceptions on spice markets: A guideline to introduce spice market concept into Sri Lankan context. *Journal of Agricultural Sciences - Sri Lanka, 16*(3), 410–430. doi:10.4038/jas.v16i03.9467

JacobS. (2008). Understanding Culturally Sustainable Tourism An Observed Comparison of the Models followed by Kerala and Goa. *Conference on Tourism in India - Challenges Ahead*, 75–90. https://ssrn.com/abstract=1480374

Jolliffe, L. (2014). Recognition of spices and cuisine as intangible heritage. In L. Jolliffe (Ed.), Spices and Tourism: Destinations, Attractions and Cuisines (Vol. 38, pp. 183–200). Channel View Publisher. doi:10.21832/9781845414443-013

Kadam, A. S., Suryawanshi, R. R., Dhengale, S. S., & Shinde, H. R. (2019). Economic analysis of production of ginger in Sangli district of Maharashtra. *International Journal of Chemical Studies*, *7*(6), 1456–1460. http://www.chemijournal.com

Khandekar, R. G., Salvi, B. R., Sawant, V. S., Malshe, K. V., Rathod, R. R., & Parulekar, Y. R. (2016). Recent advances in propagation of black pepper and major tree spices in Maharashtra. *Indian Journal of Arecanut, Spices and Medicinal Plants, 18*(3), 37–40.

Lingala, S., & Deshmukh, P. (2019). *A guidebook for integration of biodiversity spice sector in the Western Ghats*. https://sustainabledevelopment.in/wp-content/uploads/2020/06/1572605133IBBI-_Spice-publication_final-copy-10092019-2.pdf

Malkanthi, S. H. P., Ishana, A. S. F., Sivashankar, P., & Weeralal, J. L. K. (2015). Willingness to initiate spice-tourism in the Kolonna District Secretariat of Ratnapura district in Sri Lanka: Farmers' perspective. *Sri Lanka Journal of Food and Agriculture*, *1*(1), 35. doi:10.4038ljfa.v1i1.5

Marpakwar, P. (2022). 1,800+ deaths in 8 months_ Sharp rise in farm suicides in Maharashtra _ Mumbai News. *Times of India*. https://timesofindia.indiatimes.com/city/mumbai/1800-deaths-in-8-months-sharp-rise-in-farm-suicides-in-maharashtra/articleshow/94820572.cms

Ministry of Agriculture and Cooperation. (2012). *State Agricultural Portal Software Requirement Specifications*. http://www.msamb.com

Ministry of Commerce and Industry. (2023). 14th Edition of the World Spice Congress (WSC)to be held in Mumbai from 16-18 February 2023. *PIB GOI*. https://pib.gov.in/PressReleasePage.aspx?PRID=1890034

MoT GoI. (2021). *India tourism statistics at a glance 2021*. https://tourism.gov.in/sites/default/files/2021-09/English Tourisum 2021.pdf

Nair, B. B., & Mohanty, P. P. (2021). Positioning spice tourism as an emerging form of special interest tourism: Perspectives and strategies. *Journal of Ethnic Foods*, *8*(1), 10. Advance online publication. doi:10.118642779-021-00086-4

NCERT. (2022). Distribution, Density, Growth and Composition. In *India and Economy*. NCERT. https://ncert.nic.in/textbook/pdf/legy201.pdf

Rajanbabu, R., & Ganesan, S. (2015). Growth and Instability in Production of Indian Spices. *International Journal of Management and Social Development*, 2(9), 42–50.

Rama, S., Wulung, P., Yuliawati, A. K., Abdullah, U., & Fitriyani, E. (2021). Spice cultural heritage in geotourism trail. *Journal of Engineering Science and Technology*, 76–87.

Sandaruwani, J. A. R. C., & Athulagnanapala, W. K. (n.d.). *Impact of Tourism Operations in Spice Gardens for the Sustainable Tourism Development: A Case Study in Sri Lanka*. Academic Press.

Satghare, H., & Sawant, M. (2018). SWOT Analysis of Marketing Strategies Applied by MTDC for Promotion of Maharashtra Tourism. *Atna: Journal of Tourism Studies*, *13*(2), 79–95. doi:10.12727/ajts.20.6

Seethalekshmy, C., & Subramaniyan, M. (2021). Sustainable tourism development in spice tourism. In *Sustainable tourism development: An in-depth analysis* (Vol. 1, pp. 1–19). Multi Spectrum Publication.

Shenoy, J. (2020). Indian spices in great demand in post Covid 19 times; exports up 34%_ ASSOCHAM. *Times of India*. https://timesofindia.indiatimes.com/business/india-business/indian-spices-in-great-demand-in-post-covid-19-times-exports-up-34-assocham/articleshow/77048023.cms

Shukla, A., & Yadav, N. (2018). Role of Indian spices in Indian in history. *International Journal of Management Research and Review, 8*(11), 1–6. www.ijmrr.com

Singh, A., Hossain, F., & Burman, R. (2022). Crop husbandry in Pre-independent India. In *Indian agriculture after independence*. www.icar.org.in

Sontakke, Syed, & Sawate. (2018). Studies on extraction of essential oils from spices (Cardamom and Cinnamon). *International Journal of Chemical Studies, 6*(2), 2787–2789. https://www.ncdex.com

Spenceley, A. (2021). *The future of nature-based tourism: Impacts of COVID-19 and paths to sustainability*. Academic Press.

Spice Board of India. (2021). *Major Spice per State wise area and production of spices*. http://www.indianspices.com/sites/default/files/majorspiceStatewise2022_v2.pdf

Spice board of India. (n.d.). *India_Spice_Map*. Spice board of India. Retrieved February 20, 2023, from http://www.indianspices.com/sites/default/files/INDIA_SPICE_MAP.jpg

Struan, R. (n.d.). *The Trade in Spices*. Retrieved March 13, 2023, from https://en.unesco.org/silkroad/sites/default/files/knowledge-bank-article/the%20trade%20in%20spices.pdf

Szabo, M. R., Radu, D., Gavrilas, S., Chambre, D., & Iditoiu, C. (2010). Antioxidant and antimicrobial properties of selected spice extracts. *International Journal of Food Properties, 13*(3), 535–545. doi:10.1080/10942910802713149

Times of India. (2022). Hot trade: How Indian earns billions from its spices. *The Times of India*. https://timesofindia.indiatimes.com/business/india-business/hot-trade-how-indian-earns-billions-from-its-spices/articles how/93558184.cms

UNESCO. (n.d.). *What are the spice routes?* UNESCO. Retrieved March 14, 2023, from https://en.unesco.org/silkroad/content/what-are-spice-routes #:~:text=The%20Spice%20Routes%2C%20also%20known,across%20the %20Mediterranean%20to%20Europe

University of Minnesota. (n.d.). *4.6 SWOT Analysis – Mastering Strategic Management*. Retrieved March 19, 2023, from https://open.lib.umn.edu/strategicmanagement/chapter/4-6-swot-analysis/

Wikimedia Commons. (2013). *Map of Maharashtra*. https://commons.wikimedia.org/wiki/Category:SVG_maps_of_Maharashtra#/media/File:Maharashtra_Divisions_Eng.svg

World Heritage Convention. (2010). *Silk Road Sites in India - UNESCO World Heritage Centre*. UNESCO. https://whc.unesco.org/en/tentativelists/5492/

World Travel and Tourism Council. (2020). To Recovery and Beyond: The future of travel andtourism in the wake of Covid-19. In *World Travel and Tourism Council*. https://wttc.org/Portals/0/Documents/Reports/2020/To Recovery and Beyond-The Future of Travel Tourism in the Wake of COVID-19.pdf?ver=2021-02-25-183120-543

Zilihona, I. J., & Mamboya, S. F. (2018). Contribution of Spice Tourism to Youth Employment: A Case of Kijichi Spices Farms at Kijichi Shehia in Unguja-Zanzibar. *Rural Planning Journal, 20*(2), 81–89. https://repository.irdp.ac.tz/bitstream/handle/123456789/344/RPJ%20Vol20_Issue2_7.pdf?sequence=1&isAllowed=y

ADDITIONAL READING

Akyoo, A and Lazaro, E. (2007). The spice industry in Tanzania: General profile, supply chain structure, and food standards compliance issues. Copenhagen: Danish Institute for International Studies. DIIS Working Paper 2007: 8.

Datta, B. (2022). Positioning spices as a tourism product for foreigners: a study from the Indian perspective. *Enlightening tourism. A Pathmaking Journal, 12*(1), 272-303. doi:10.33776/et.v12i1.5453

Jolliffe, L. (2014). *Spices and Tourism: Destinations, Attractions and Cuisines*. Channel View Publications. doi:10.21832/9781845414443

Madan, M. S., & Tamil Selvan, M. (2001). Globalization and Spice Economy of India. *Indian Journal of Arecanut, Spices and Medicinal Plants, 3*(4), 262–269.

Sharangi, A. (2018). *Indian spices: The legacy, Production and processing of India's treasured export*. Springer Cham. doi:10.1007/978-3-319-75016-3

Sherman, P., & Billing, J. (1999). Darwinian gastronomy: Why we use spices. *Bioscience, 49*(6), 453–463. doi:10.2307/1313553

KEY TERMS AND DEFINITIONS

Agri and Rural Tourism: It refers to the practices of bringing tourists to farms and rural surroundings and giving them experience of agricultural and rural practices.

Food Tourism: It is counted as form of special interest tourism in which tourists travel to seek out culinary experiences.

Industry Stakeholders: Stakeholders are people or groups with an interest in the success of a business or project. Nowadays, local communities, government organisations and associations are also counted as stakeholders.

Medical and Wellness Tourism: It refers to the travelling of the person to another location specifically for medical / health related treatments or for physical/ mental rejuvenation and one's wellbeing. The form has become significantly popular after COVID-19 pandemic.

Niche Tourism/Special Interest Tourism: Tourism products/ packages/ or destinations which are designed and developed for the tourists who have very specific interest/ purpose of travel. The concept focuses more on tourists' choices and mainly evaluated through tourists' experiences.

Silk Route: Ancient trade route connecting Euro- Asia. It was generally used by traders, monks, explorers, and scholars. It played crucial role in sharing of knowledge, culture, wisdom, beliefs, inventions, etc.

Spice Tourism: It is a form of tourism including tourist activities that revolve around the areas exhibiting one or more spice related practices like spice cultivating, processing, trading, selling, and purchasing and other supplementary activities. Additionally, these areas can also provide tourists with the opportunity of participation in the activities.

SWOT Analysis: An analysis of strength, weaknesses, opportunities, and threats related to the phenomenon.

Chapter 12
Battlefield Tourism

Selçuk Yücesoy
https://orcid.org/0000-0002-1168-2508
Eskisehir Osmangazi University, Turkey

Ebru Düşmezkalender
Eskişehir Osmangazi University, Turkey

Yunus Özhasar
Eskişehir Osmangazi University, Turkey

ABSTRACT

This chapter examines the historical development, examples, motivation, and attractions of battlefield tourism while also exploring its popularity and ethical controversies. By providing visitors with the opportunity to understand the devastating effects of war and the human suffering it causes, battlefield tourism serves as an important means for individuals to connect with the past and gain a deeper understanding of history. The chapter also offers recommendations for the future of battlefield tourism.

INTRODUCTION

In the last century, in parallel with the rapid growth of the tourism sector and the number of wars and other military conflicts that took place from the beginning of the 1900s to the present, the interest in battlefield tourism has increased significantly. (Lee, 2016). World War I (WW I) is considered to be a turning point in the emergence of battlefield tourism (Dunkley, Morgan & Westwood, 2011; Pennell, 2018). Although there is evidence of organized travel to battlefields in the second half of the 1800s (Thomas Cook/Waterloo), visits to battlefields emerged on a larger

DOI: 10.4018/978-1-6684-7242-2.ch012

scale after WW I (Lloyd, 1994). From WW I to the present day, battlefield tourism products have diversified dramatically and battlefields are the largest tourist attraction destinations in the world (Smith, 1996: 248). These tourist destinations include not only battlefields and their associated memorials and cemeteries, but also numerous monuments, museums and other structures and sites commemorating wars, battles and related events or brutalities.

Battlefield tourism creates more profound experiences than traditional tourist experiences (Upton, Schänzel & Lück, 2018). These destinations remind a nation or tribe of past national suffering or national pride. Hence, battlefields are of great importance for the construction of national identity. For example, for Australians, Gallipoli is the place where soldiers died and a nation was reborn, while Turks remember Gallipoli as the place where the Turkish army defeated the Western powers (Prideaux, 2007).

Battlefields have been an important source of income for many countries today (Gibson, Yai & Pratt, 2022). Particularly in countries where major wars such as WW I and World War II (WW II) took place, there are many touristic areas where the traces of war can be seen. (Chen & Sun, 2023; Kler & Forsythe, 2022; Foreman, 2022). However, the tourism of battlefields and their presentation as a touristic product raises ethical concerns (Sharma, 2022). Therefore, the aim of this chapter is to address the fundamental relationship between tourism and war in the context of battlefields and to explore the motives of this type of tourism based on examples of tourism to battlefields.

Battlefield tourism is considered a niche tourism type for tourists with a specific interest (Dunkley, Morgan & Westwood, 2011; Simith, 1998). This type of tourism is preferred by tourists who are interested in military history and strategy, as well as historical sites, monuments, and battlefields. Battlefield have significant tourism potential in many countries, and therefore, many countries consider battlefield as an important opportunity for tourism. For example, in the UK, many battlefields have been developed as tourism destinations in commemoration of the 100th anniversary of World War I (Virgili et al., 2018).

BATTLEFIELD TOURISM, HISTORICAL DEVELOPMENT, AND IMPORTANCE

Traveling to places with cultural marks has always been attractive for people. The trips made to participate in and see these cultural events are classified as historical tourism, faith tourism, ethnic tourism, cultural heritage tourism as a sub-heading of cultural tourism. One of the most captivating forms of tourism is historical tourism, which entails a desire to gain knowledge about the significant cultural events that have shaped a society, and to witness tangible evidence of these occurrences. The primary sources for such historical exploration are often found in battlefields, as they serve as poignant reminders of the events that have shaped the course of a society's development (Doğaner, 2006, p. 3). Wars constitute a singular and paradoxical phenomenon, characterized by irrationality and brutality. Although war inflicts irreparable harm and creates tragic consequences, it is often deemed a necessary sacrifice of a minority for the betterment and security of the majority. Yet, despite the deleterious impacts on the society and the depletion of national resources, war persists. Paradoxically, the desire to visit the sites where violent conflicts occurred creates a conundrum for societies. The urge to commemorate the fallen, to honor the memory of those who sacrificed their lives, to reflect on the triumphs of their forebears, to experience a sense of pride or sadness, all motivate people to visit the battlefields. Regardless of the various reasons behind such a pilgrimage, the trend of visiting war zones has emerged as a burgeoning sub-sector of the tourism industry (Prideaux, 2007, p. 18).

War zones have been historically neglected as touristic destinations, as tourism is often associated with entertainment and leisure rather than sites of conflict and graves. However, with the passage of time since the wars, the tourist behavior has undergone a transformation. The fusion of the desire to learn about the past and to exhibit loyalty, with the act of travel, has propelled the popularity of battlefield tourism to greater heights. Tourism is increasingly regarded as a powerful force for commemorating wars and rekindling familial and national memories (Winter, 2012, p. 261). Within the framework of touristic experiences delineated by Lennon and Foley (2000), battlefield tourism has emerged as a critical form of touristic activity that has expanded to include novel experiences by visiting antiquated battlefields and museums.

Battlefield tourism is a specific form of war tourism, encompassed within the larger domain of dark tourism or thana tourism (Dunkley, Morgan, & Westwood, 2011, p. 860). It is seen as a means to emerge from the shadows of war, to confront the harsh realities and the complexities of history (Lee, 2016, p. 697). Battlefield tourism can be defined as a type of travel that entails visiting former battlefields. This mode of tourism encompasses not only visits to battlefields, but also war museums, monuments, military cemeteries, hospitals, former housing areas, and participation in commemorative ceremonies (Fathi, 2021; Ryan, 2007). In addition, engaging in war reenactments and experiencing the realities of war are also considered integral to the purview of battlefield tourism (Dunkley, Morgan & Westwood, 2011, p. 860). The extant literature on war tourism underscores the importance of the operational aspects involved in developing, managing, and interpreting battlefields, historical monuments, and other war-related attractions (Bigley et al., 2010, p. 375).

Battlefield tourism, as a practice of visiting sites of historical military battles, is not a recent phenomenon. In fact, historical evidence suggests that such visits have occurred for over a thousand years, with battlefields, military graves, monuments, and other war-related sites serving as focal points for pilgrimages, commemorative events, and educational visits. However, with the emergence of modern tourism, battlefield tourism has undergone significant expansion and popularization, emerging as a niche market that attracts a diverse range of visitors (Baldwin & Sharpley, 2009, pp. 186-187). This growth can be attributed to tourists' inherent desire to explore new destinations and gain novel experiences, as well as the continued fascination with warfare and military history. Consequently, battlefield tourism has become a prominent subfield within the larger context of heritage and cultural tourism, with various destinations offering tours, exhibitions, and educational programs to cater to this specific interest group.

The origins of battlefield tourism can be traced back to ancient times, as demonstrated by Alexander the Great's detour from his Asian invasion to visit ancient Troy and the Tomb of Achilles in 334 BC. However, the battle monuments at Waterloo, completed in 1826, are considered the most prominent and frequented battlefield sites (Baldwin & Sharpley, 2009, p. 186). Waterloo introduced a new type of battlefield tourism that revolutionized travel itineraries (Seaton, 1999, p. 139). Henry Gaze, one of the oldest travel agencies, organized the first group of tourists to visit the Brussels and Waterloo battlefields in 1854 (François, 2013). In 1856, Thomas Cook arranged the first trip to Waterloo (Miles, 2012, p. 4). Similarly, Thomas Cook organized tours to the South African Battlefields of the Boer War before the conflict ended in 1902 (Lloyd, 2014). Nonetheless, the WW I is particularly viewed as a pivotal moment in the emergence of battlefield tourism (Baldwin & Sharpley, 2009, p. 186). Baldwin and Sharpley (2009) assert that the growing desire of those who lost loved ones during the WW I to visit their graves or places of death was a crucial event in the development of battlefield tourism. In 1919, 60,000 people visited the Western Front on tours organized to the Western Front battlefields, and this trend continued in the subsequent years until the WW II (Seaton, 2000).

The disruption of tourism during the WW II caused a significant decline in visits to battlefield sites. However, since the 1960s, there has been a resurgence of interest in battlefield tourism. This can be seen in the increased publication of books on the subject, as well as the growing number of organized tours to battlefields (Miles, 2012, pp. 4-5). The 1970s marked a turning point in the growth of battlefield tourism, with the emergence of guided tours and the availability of expert information through the internet, which have since created a thriving industry (Seaton, 2000, p. 64). The majority of visitors to battlefield sites are motivated by a desire to commemorate their deceased relatives. This is exemplified by the participation of Turks, Australians, and

New Zealanders in commemorative ceremonies at Gallipoli (Baldwin & Sharpley, 2009, p. 187). The emergence of school tours in the 1990s has also contributed to the increasing interest in battlefield tourism. Overall, the commemorative aspect of battlefield tourism continues to be a significant driving force behind its popularity.

The recognition of the immense significance of the toil and sacrifice endured by families, ancestors, and nations during the great wars has led to an increasing desire among people to visit sites and gain a deeper understanding of these events. The reasons for visiting these battlefields are diverse and include special interest, excitement, risk-seeking, legitimization, verification, validation, authenticity, curiosity, fear, national pride, visualization of the event, sense of identity, personal aspirations (van der Merwe, 2014, p. 123). In addition to offering invaluable data sources for those interested in history and researching their origins and culture, battlefields also serve as important places for individuals who wish to confront the past and pay their respects to their ancestors. These visitors seek to experience the past and the future, friendship and enmity in the same environment, and view battlefield tourism as a means of combining pain, sadness, anger, and friendship. Tourism related to battlefields, which is especially concentrated in the places where WW I and II took place, has the potential to bring together the peoples who once fought each other in a friendly manner. Today, discussions surrounding these areas are focused not on being right or wrong, but on the senselessness of war. As such, it should be noted that battlefield tourism has the power to promote peace (Doğaner, 2006, p. 2). Tragic events and wars in the past have the capacity to change the roles of individuals in their own lives and reshape the heritage of the cultures and countries to which they belong (Leopold, 2007). In this regard, battlefield tourism has the potential to generate nationalistic feelings among visitors from countries involved in the war (Prideaux, 2007).

MOTIVATION AND EXPERIENCE OF BATTLEFIELD TOURISM

Motivation is a fundamental concept that pertains to the underlying motives that activate and guide human behavior (Beh & Bruyere, 2007; Boo & Jones, 2009). These motives can stem from either intrinsic factors or extrinsic factors. Within the domain of tourism, travelers' decisions to visit a particular destination are typically driven by a combination of intrinsic and extrinsic motivators. Specifically, tourists' intrinsic desires, such as self-actualization, rest, leisure, or social interaction, can influence their travel decisions. Additionally, a set of destination characteristics that elicit extrinsic and cognitive factors, such as scenery, climate, hospitality, or facilities, can further increase their desire to visit a particular place (Chen & Tsai, 2019; Yoon & Uysal, 2005). Hence, Yoon and Uysal (2005) have defined these motivators as push and pull motivations, respectively, where push motivations pertain to emotional and intrinsic desires, while pull motivations are linked to extrinsic and cognitive factors.

Throughout history, wars have occurred in various locations, resulting in military martyrdoms, monuments, and memorials that hold significant meaning for societies and impact tourists' travel motivations. Visiting battlefields is a topic that is frequently discussed in tourism literature under different categories, including special interest tourism, heritage tourism, death tourism, or dark tourism (Akbulut & Ekin, 2022). Tourists may be motivated to visit such sites for a range of reasons, including a fascination with historical events, morbid curiosity, an interest in death, a desire to maintain national or cultural identity, a sense of guilt, or superstition (Clarke & Eastgate, 2011; Dunkley, Morgan, & Westwood, 2011; Foote, 2003; Stone, 2011; Panakera, 2010). Furthermore, although visiting battlefields is typically viewed as a secular activity, it can also be influenced by religious sensitivities. In fact, war cemeteries are now considered as sacred sites and are visited for spiritual reasons, and to honor the memory of those who lost their lives in the war. Additionally, mystical narratives surrounding battlefields add a supernatural dimension to the motivation to visit these sites. For instance, the Angels of Mons narrative, which is believed to have saved the British forces from destruction during the WW I, is one of the significant examples of such mystical narratives (Winter, 2011).

Tourists visiting battlefields often have distinct emotional and cognitive responses towards the conflicts and individuals involved. Numerous studies in the literature have explored the different motivations behind battlefield visits. For instance, Çakar (2019) examined the motivations of tourists visiting the historical Gallipoli peninsula and identified factors such as a desire to learn, historical awareness, participation in commemoration ceremonies, national and cultural identity, family ties, curiosity, patriotism, a desire to understand different cultures, media, and reference groups as significant factors that motivate tourists to visit these sites. On the other hand, Foley and Lennon (1997) assert that battlefield tourism is driven by three primary purposes: education, remembrance, and entertainment.

According to Frost (2007), the interest of tourists in visiting battlefields is rooted in their desire to understand the historical significance and relevance of past wars in contemporary societies (p. 187). Therefore, the experience of visiting these sites enables tourists to gain insights into the ways in which historical events have shaped present-day societies. However, heritage sites related to wars present unique challenges that are not found in other traditional tourist attractions. For instance, visitors to war-related sites may have different perceptions and interpretations of the events, which can be challenging to manage (Henderson, 2000).

In this context Poria, Biran, and Reichel (2009) argue that visitors seek different experiences and meanings from heritage sites. Thus, interpreters need to carefully approach sensitive topics, such as death and war, while maintaining the desired balance between education and entertainment (Upton, Schänzel & Lück 2018; Ngo & Bui, 2019). Additionally, it is essential to consider that multiple sides may have an interest in visiting a battlefield, and the provision of guidance and interpretation services will be critical in ensuring that visitors receive accurate and impartial information. Failure to do so may have a negative impact on the tourism experience, as war zones can evoke stronger emotions than other dark tourism sites (Miles, 2014).

Therefore, Leopold (2007) suggests that the presentation of exhibits and displays should be objective and neutral, providing a comprehensive view of the war, including its consequences and legacy. Finally, it is argued that visiting battlefields provides visitors with valuable skills, knowledge, and a deeper appreciation and understanding of attitudes, values, and ideas related to past and contemporary conflicts (Packer, Ballantyne, & Uzzell, 2019).

SIGNIFICANT SITES OF HISTORICAL BATTLES: RECOGNIZED BATTLEFIELD DESTINATIONS

In recent years, battlefield tourism has emerged as a popular trend in the tourism industry (Fathi, 2021; Dunkley, Morgan, & Westwood, 2011). Many tourists prefer to visit historical battlefields worldwide to gain a deeper understanding of the historical and cultural significance of wars (Miles, 2012). Battlefield tourism has become an important market for the tourism industry in various countries, with tourists visiting these sites to acquire historical and cultural knowledge, including an understanding of the devastating effects of war and the promotion of peace (Zhang, 2010, p. 401).

Advancements in technology have also transformed battlefield tourism. Digital guides, mobile applications, and virtual reality technologies provide visitors with more interactive experiences at battlefield sites. Furthermore, tourists seeking to gain a more comprehensive understanding of the historical and cultural significance of war can participate in specialized tours of battlefield sites (Nguyen et al., 2019; Jung, 2020).

Battlefield tourism, which falls under the umbrella of dark tourism, encompasses visits to destinations that have witnessed wars and battles (Dunkley, Morgan, & Westwood, 2011, p. 860; Miles, 2014, p. 137; Chang, 2017, p. 1). Tourists engaging in battlefield tourism are typically motivated by historical and cultural interests (Miles, 2012). These destinations are particularly popular among tourists seeking to explore and gain a deeper understanding of the locations where significant historical events occurred (Tarlow, 2007, p. 51). D-Day Normandy, Gettysburg Battlefield, Stalingrad (now Volgograd), and Gallipoli are examples of unique battlefield sites that offer tours, exhibitions, and museums highlighting the experiences of heroes and life during wartime. The prominence of battlefield tourism has increased considerably and has been the focus of substantial research within the tourism literature (Light, 2017, p. 276). As Smith (1998) and Stone and Sharpley (2008) suggest, battlefields constitute one of the most significant categories of tourist attractions globally. The following section provides an overview of some of the world's most famous battlefield sites.

Verdun: Verdun is a small city situated on the banks of the Meuse River in northeastern France. In 1914, the Germans advanced towards Verdun from the Alsace and Lorraine regions, driving them 15 kilometers south of the city and establishing positions there. The Battle of Verdun, which lasted from February to December 1916, is regarded as one of the most brutal and destructive battles in history (Axelrod, 2016, p. vii). It claimed the lives of more than 300,000 individuals and resulted in over 400,000 wounded, according to some estimates. The 26 million shells used during the conflict have become symbolic of the battle itself (Jankowski, 2014; Virgili et al., 2018; Prost & Krumeich, 2015). Among the French, it was considered the most violent battle of WW I and the one that saw the largest number of soldiers engaged. As a result, the Battle of Verdun achieved legendary status among the populace (Prost & Krumeich, 2015). Furthermore, the large quantity of ammunition used at Verdun has made it a unique battlefield for WW I relics (Barcellini, 2009).

Waterloo: The Battle of Waterloo was fought on June 18, 1815, between Napoleon Bonaparte and a multinational allied army composed of soldiers from six different nationalities (British, Belgians, Hanoverians, Brunswickers, Dutch, and Prussians) led by British General Wellington and Prussian Blucher. The battle took place near a deserted hamlet approximately 15 kilometers from Brussels and lasted for eight hours. The outcome of the battle was the complete disintegration of the French army and the arrest of Napoleon (Black, 2010). At least 54,000 soldiers were either killed or wounded during this battle (Heinzen, 2014). The Battle of Waterloo brought an end to 23 years of military activity in Europe. Notably, the Waterloo battlefield is one of the few battlefields in Europe to gain permanent tourism status. The battle had a significant impact on travel from Britain to Europe after 1815 (Seaton, 1999). In fact, Waterloo was one of the oldest destinations in Europe included in Thomas Cook's itinerary (Brendon, 1991). The significance of Waterloo as a destination worth visiting was highlighted in the first English-language guidebook to Belgium and subsequent guidebooks. It was also the most frequently visited battlefield in Europe for a century until new battlefields emerged in the WW I (Seaton, 1999).

Gallipoli: In 1901, Australia established a federation of nations by adopting a constitution, marking the emergence of a new nation-state. One of the first acts of this new state was the expedition to Gallipoli in the Ottoman Empire, where Australian soldiers, alongside their New Zealand counterparts, formed the ANZACs (Lockstone et al., 2013, p. 301). The ANZACs landed on the shores of Anzac Cove on April 25, 1915, and withdrew on December 20 of the same year, with 8,709

Australian soldiers having lost their lives during the battle. Notably, most of these soldiers hailed from Australian towns, villages, and hamlets. The battle involved a million soldiers from both sides, with nearly half of them having been killed or wounded by the end of the war (Slade, 2003). Specifically, the number of soldiers who died or were wounded in the war is as follows: Turks, 250,000; British, 120,000; French, 47,000; Australians, 27,700; and New Zealanders, 7,500 (Fewster, Başarin, & Başarin, 2003).

Gallipoli stands out among other battlefields in that a vast majority of visitors come from overseas to commemorate ANZAC Day, the day of the landings, on April 25th, each year (Slade, 2003). First observed in Australia in 1916, ANZAC Day gained nationwide recognition when each state in Australia declared April 25 as a public holiday in 1923, setting it apart from other countries' days of national importance. Commemorative ceremonies were first held in Gallipoli in 1925, although they did not gain popularity until the 1980s (Wahlert, 2011). On the 90th anniversary of ANZAC Day in 2005, roughly 20,000 Australians participated in the ceremony, while on the 100th anniversary of the war, a commemorative ceremony was held in Gallipoli, drawing a crowd of 50,000 attendees.

Gettysburg: Gettysburg National Military Park, the largest and most significant battle of the American Civil War, represents a pivotal moment in American history (Sears, 2003). On July 1, 1863, the Union and Confederate armies engaged in a fierce battle in the town of Gettysburg, resulting in over 50,000 casualties (Cobb, 1995). Today, the park remains a crucial historical site that annually attracts 1.2 million visitors, underscoring its continued relevance and importance to American identity (National Park Service, 2023). Indeed, Gettysburg's enduring significance lies in its capacity to remind visitors of the brutal realities of war while simultaneously promoting a unifying national identity. Moreover, Gettysburg serves as a potent reminder of the cost of fratricidal conflict that must never be forgotten. Scholars have characterized Gettysburg as "the most important event in American history" (Anderson, 2006), "the foundational place of America" (Hall, 1994, p. 7), and "the symbolic heart of America" (Desjardin, 2003, p. 7), underscoring its significance as a central and defining element of the nation's collective memory.

Normandy: The Battle of Normandy, also known as D-Day, remains a significant event in global WW II history almost 80 years after its occurrence. On June 6, 1944, the Allies, including the USA, UK, Canada, Belgium, Norway, Poland, and Greece, executed the largest amphibious landing in history in Normandy to initiate a war against the Nazi army occupying French territory. The original code name for the operation was Operation Neptune, but "D-Day," a military term for the beginning of any pre-planned combat operation, became the more widely used name for this event. The Allies had planned for this assault since the latter half of 1943, and American, British, and Canadian troops landed in Normandy in 1944 and battled for 80 days (Rose, 2019). The Canadian military, with a total population of only 11 million, witnessed 1 million Canadian volunteers for military service, resulting in 6,000 Canadian casualties. For many Canadians, D-Day serves as a poignant symbol of their nation's cultural memory of WW II, as the capture of Juno Beach is considered a remarkable achievement for Canada's modest population (Keegan, 2004). Consequently, Canadian visitors regularly explore the Normandy beaches, towns, and surrounding areas to appreciate their national heritage.

Iwo Jima: Iwo Jima, a Japanese island in the Pacific, was the site of a significant battle during WW II, lasting from February 19 to March 26, 1945. The primary objective of the United States was to utilize the island as a base for an eventual invasion of Japan and to gain control of its airspace (Russ,2001). The American offensive began with a massive aerial bombardment of the island, followed by the landing of ground troops. Despite encountering limited resistance on the coastal regions, the American soldiers faced fierce opposition from the approximately 23,000 Japanese soldiers who had entrenched themselves in a network of tunnels and caves in the island's interior (Hammel, 2006). The ensuing battles were characterized by extreme violence and bloodshed, with the Americans ultimately prevailing due to their superior military capabilities and control of the skies above Iwo Jima.

The conflict at Iwo Jima is notable for various reasons. The intense and protracted nature of the fighting resulted in the loss of many lives, with 6,281 American soldiers killed and 19,217 wounded, while less than 1,000 of the Japanese soldiers were captured or wounded (Edwards & Winkler, 1997, p. 301; Carlson, 2021). Additionally, the Japanese soldiers' adherence to their code of bushido, which required them to fight to the death in defense of their homeland, resulted in the vast majority of them perishing in the battle (Davis, 2011, p. 20). Another notable aspect of the conflict was the famous photograph of six American soldiers raising the American flag atop Mount Suribachi, which has become an iconic image in American popular culture and war memorials (Mortensen, 2013). Furthermore, the story of two Japanese soldiers who remained hidden in the island's caves and tunnels for 60 years after the battle, believing the war was still ongoing, is an intriguing and unusual historical footnote (Keenan, 2013, p. 3).

In summary, the battle of Iwo Jima was a significant engagement of WW II, with the United States emerging victorious despite facing determined resistance from entrenched Japanese forces. The conflict is remembered for its brutality, the adherence of Japanese soldiers to their code of bushido, and the iconic image of the American flag being raised on Mount Suribachi.

ATTRACTIONS OF BATTLEFIELD SITES

The significance of battlefields as tourist destinations lies in their ability to provide a firsthand experience of the impact of historical events and wars. Despite the tragic and brutal events that occurred at these sites, visiting them can be seen as a form of human rights and respect. Scholars argue that it is crucial to preserve and promote the cultural heritage and historical values of these locations (Venter, 2011; Miles, 2012; Akbulut & Ekin, 2018). By doing so, we can understand and commemorate significant historical events and wars and preserve these values for future generations.

For history buffs or tourists interested in learning more about the history of war, battlefields are an attractive option. Visiting these sites allows for a deeper understanding of the historical artifacts, structures, and strategies associated with battlefields. In particular, visitors can learn about the tactics and strategies employed during battles, providing insight into the historical context and significance of the event (Driessen, Grever, & Reijnders, 2022). Overall, the preservation and promotion of battlefields as tourist destinations offer a unique opportunity to learn and reflect on our shared history, providing valuable insight into the impact of past events on the world we live in today.

Individual stories and the human aspect of battles are crucial in shaping visitors' experiences at battlefields. In fact, Gatewood and Cameron (2004) found that visitors to the Gettysburg battlefield were more strongly influenced by stories of physical hardship and drama than by technical descriptions of the battles. Personal narratives and accounts of individual experiences provide a deeper understanding of the human impact of these events, helping visitors to connect emotionally with the history of the site. By learning about the experiences of soldiers and civilians caught up in these events, visitors can gain a greater appreciation of the sacrifices and struggles faced by those involved in the conflict.

Furthermore, these stories help visitors to understand the complex and often personal motivations that drive individuals to fight in wars. This understanding can lead to greater empathy and compassion, helping visitors to view historical events from a more nuanced perspective. Overall, the inclusion of individual stories and the human aspect of battles is crucial to the interpretation of battlefields, offering a more meaningful and impactful experience for visitors.

Battlefield tourism is often considered a type of dark tourism, but some argue that it doesn't fit the definition of thana-tourism (Baldwin & Sharpley, 2009; Dunkley, Morgan, & Westwood, 2011; Ryan, 2007). Australian visitors to WWI sites have reported spiritual encounters, contradicting the idea that they're not seeking a symbolic connection with death (Scates, 2002; Slade, 2003). Slade (2003) argued that Australian and New Zealand tourists to Gallipoli should not be included in Seaton's (1996) definition of thanaturism because they are seeking to connect with their national identity rather than an experience or a connection to death. However, Scates (2002), who did not specifically examine thanaturism, found that Australian visitors to WW I sites (Gallipoli and Flanders) spoke of spiritual encounters on battlefields. These experiences included feeling the chill in the air, connecting with God's cry when it rains, and seeing ghostly figures in various landscapes. This clearly contradicts Slade's (2003) theory that Australian and New Zealand tourists do not seek a symbolic connection with death and the dead. Miles (2014), who examines tourist experiences at UK battlefields, suggests that tourist experiences are more everyday experiences, which is likely to be due to the commercialization of historic battlefields.

In terms of supply origin, war zones are on the darker end of Stone's (2006) spectrum. Destinations that have witnessed both World Wars have a shorter time scale from war to present day, and these battlefields may be on the darker end of Stone's (2006) spectrum than other battlefields. However, although the chronological proximity of the battles to the present day places them at the darker end of the scale, this does not mean that they are darker than some battlefields. For example, the Culloden battlefield is at the same point on the spectrum as the WW I and WW II fronts. This depends on the location of the battlefields. Some of the open-ended features of Culloden that Stone mentions in his spectrum can trigger darker experiences. In addition, the decisions taken by battlefield destination administrations should not be ignored when evaluating the darkness levels of battlefields. Their decisions, methods and media choices can determine the position of visitors on the dark tourism scale and how the message is received by the visitor (Miles, 2014).

War memorials, as tangible manifestations of the allure of battlefields, are constructed to commemorate the individuals who fought and lost their lives as a result of war. Taking on various forms, these memorials serve to perpetually remind society of those who perished in conflict. According to Raynor (2004) there are more than 500 battlefields in England, and according to Foard and Partida (2005) there are more than 350 battlefields in Scotland (p. 7). But very few of them attract the attention of visitors. Among the war museums in the United Kingdom, the Imperial War Museums stands out as the most remarkable one, which researches conflicts from WW I to the present day. There are also interesting museums such as the Museo Diffuso della Resistenza, della Deportazione, della Guerra, dei Diritti e della Liberta located in Italy, which focus on conflicts in more recent history. In addition, the Spasa Tunnel in Sarajevo, established as a shelter from Serbian sniper and mortar fire in 1993, is an intriguing museum that focuses on conflicts of more recent times. The Islamic Revolution and Holy Defense Museum in Tehran, Iran, which focuses on the Iran-Iraq War (1980-1988), is another extraordinary example of a war museum today, owing to the extensive use of new technologies. Lastly, the Afghanistan Memory and Dialogue Center in Kabul is the most current example of a war museum, aiming to promote peace and reconciliation in a country currently affected by internal violent conflict (Carbone, 2022, p. 572).

NATIONAL IDENTITY AND BATTLEFIELD TOURISM

The term "nation" encompasses more than a collection of people within the borders of a particular country. Nationalism theorists argue that the term can include political, social, cultural, historical, economic, economic, linguistic and religious factors (Stolcke, 1997; Parekh, 1995; Rivera, 2008). For instance, when a person is said to belong to a nation, it is generally understood that that person's personal values are linked to that country.

National identity is a complex phenomenon that has long been a subject of academic debate. The nature and significance of national identity have evolved over time and are shaped by various historical, political, and social factors (Nairn, 1997). The emergence of globalization, consumer culture, and social media, along with the fragmentation of social structures, has reinvigorated discussions about national identity in contemporary society (Spillman, 1997; Mitchell, 2001; Packer, Ballantyne, & Uzzell, 2019; Zhang et al., 2019). These issues have prompted scholars to examine the nature of social bonds and individual belonging in the context of national identity, and have highlighted the importance of understanding the role of national identity in shaping social relations and cultural practices.

The tourism sector has the potential to serve as an effective tool for promoting a country's national identity and cultural values on a global scale (Hall, 2002; White & Frew, 2013; Farmaki & Antoniou, 2017). Battlefields, as an integral part of a country's historical and cultural heritage, offer an exceptional opportunity for visitors to explore and gain a deeper understanding of the past, as well as to preserve this knowledge for future generations. However, the utilization of these sites for tourism purposes also requires a high degree of sensitivity and caution. It is crucial to approach the protection of natural and cultural resources in these areas with care in order to strike a balance between the promotion of national identity and the preservation of cultural values. Achieving this balance is essential to fully exploit the touristic potential of these sites while also safeguarding their national and cultural significance.

In their study, Jamal and Lelo (2011) explored the connection between identity and visits to destinations associated with death or suffering. They suggested that such visits can provide an opportunity for individuals to reflect on and reconstruct their own identities, whether they be personal, familial, national or international. Although death may be a central element in shaping certain personal identities, the relevance of personal identities to the experience of death may be less significant.

It is worth noting, however, that the relationship between identity and tourism is not exclusive to dark tourism, as the role of identity has been highlighted in numerous other forms of tourism, particularly in the context of heritage tourism (González, 2008; Olsen, 2003).

According to Gieling and Ong (2016), national identity refers to an individual's cognitive assessment of other group classifications. The unique nature of battlefield tourism can play a pivotal role in supporting a tourist's process of national identity formation. When tourists visit a battlefield, their sense of national identity is often strengthened. The emotional experience of visiting a battlefield can evoke a sense of unity and shared experience, thereby fostering a greater connection to one's national identity.

According to Dunkley, Morgan, and Westwood (2011), battlefield tours can be a highly meaningful experience for visitors. These tours provide a unique opportunity for tourists to understand and connect with their national identity, as well as to commemorate historical events that may be difficult to fully grasp without the context of the battlefield itself. For those seeking deep, emotional and spiritual experiences in their tourism, war heritage sites and commemorative events are often the primary motivators for visiting these destinations. When visitors perceive a particular war heritage site as part of their own personal story, they are more likely to be motivated by the emotional experience than by educational or recreational interests (Poria, Biran, & Reichel, 2006). Additionally, there are casual tourists who may not feel as strong of a connection to battlefield heritage and are primarily motivated by educational or leisure interests (Winter, 2011). However, it can be challenging to differentiate whether visitors are primarily seeking an emotional experience or visiting for educational purposes.

CONCLUSION

Wars have been a pervasive aspect of human history since the dawn of civilization and have had significant impacts on the course of humanity. WW I and WW II are two of the most notable conflicts, which left indelible marks on history. Today, many people visit former battlefields as tourists, with WW I and WW II battlefields being particularly popular destinations. Nevertheless, warfare remains a pressing issue in the 21st century (Piekarz, 2007).

Throughout history, battlefields have captured people's attention and elicited a range of responses. After WW I and II, many countries built cemeteries and monuments to commemorate fallen soldiers, and surviving veterans visited former battlegrounds to pay respects to fallen comrades (Caddick, 2005). The development of battlefield tourism has underscored the importance of proper management of these sites. Destination management organizations should respect the solemnity of these sites, help tourists understand historical events, and ensure that tragic memories of war are honored. Moreover, battlefield tourism can have significant economic benefits for war-torn regions, contributing to regional development (Foulk, 2016; Ryan, 2007; Taru, Chingombe & Mukwada, 2014).

Notably, battlefield tourism has unique characteristics that distinguish it from other forms of dark tourism (Miles, 2014; Mileva, 2018). While any tragic event or death can be a subject of dark tourism, not all such events are suitable for battlefield tourism. Battlefield tourism is concerned with the entire topography of war, including the tragic events and deaths that occurred there. Accordingly, battlefields occupy a distinct position among types of dark tourism. This section aims to contribute to the literature by highlighting the importance of battlefields as a unique form of dark tourism.

In recent years, the interest in battlefield tourism has been on the rise, despite the ethical and moral concerns raised by researchers (Hertzog, 2012; Stone, 2011). However, it cannot be denied that battlefields tourism has the potential to promote cultural exchange, educational experiences, and economic growth in areas that have been impacted by conflicts. Therefore, it is crucial for travel agencies that specialize in the tourism industry to take the lead in organizing trips to battlefields, ensuring the safety and security of the visitors, and providing them with a deep and meaningful understanding of the social, cultural, and political contexts of the conflict. By offering professional and educational tours, tourists can gain a better understanding of the impact of war on the local communities and the importance of peace and reconciliation efforts. As Atay (2008) suggests, organizing professional trips to conflict zones will not only increase the satisfaction of visiting tourists but also provide an opportunity for conscious tourism to be carried out in the war zones. Additionally, tourism organizations in destinations can directly or indirectly affect visitors' experiences by providing and supporting physical and informational access to commemorations related to historical battlefields and other attractions (Winter, 2009, p. 620).

Battlefield tourism has become a popular form of tourism for those interested in the history of warfare. Visiting historical battlefields offers a range of tourist activities, including visits to monuments, museums, and other facilities that provide insights into the historical and cultural significance of these sites. However, battlefield tourism is also associated with traumatic events, and therefore, can have both positive and negative effects. Nevertheless, battlefield tourism has the potential to benefit historical preservation, tourism development, and local economic growth. Tourists have the opportunity to learn historical facts while also observing the cultures and lifestyles of local communities. Yet, given that battlefield tourism is conducted in places where the scars of war and conflict are still felt, it can have emotional and psychological effects. Consequently, the future of battlefield tourism should be approached as a sensitive issue, in order to ensure that this form of tourism is sustainable.

REFERENCES

Akbulut, O., & Ekin, Y. (2018). Kültürel miras turizmi olarak savaş alanları turizmi: Türkiye'de yer alan savaş anıtlarının coğrafi bilgi sistemleri analizi. *Hitit Üniversitesi Sosyal Bilimler Enstitüsü Dergisi, 11*(1), 395–420. doi:10.17218/hititsosbil.397914

Akbulut, O., & Ekin, Y. (2022). Yaşayan tarih: Savaş alanları turizmi ve yeniden canlandırma toplulukları örneği. *Kent Akademisi Dergisi, 15*(2), 439–451. doi:10.35674/kent.988625

Anderson, J. T. (2006). Snow day: Memories of growing up in the Bronx. *Gettysburg Review, 19*(3), 257–311.

Atay, L. (2008). Alan kılavuzluğunun Çanakkale İli'ndeki uygulaması ve ilgili turizm mevzuatı açısından değerlendirilmesi. *Anatolia: Turizm Araştırmaları Dergisi, 19*(2), 169–176.

Axelrod, A. (2016). *The battle of Verdun*. Rowman & Littlefield.

Baldwin, F., & Sharpley, R. (2009). Battlefield tourism: bringing organised violence back to life. In The darker side of travel: the theory and practice of dark tourism (pp. 186-206). Channel View.

Barcellini, S. (2009). *La Meuse face au défi du centenaire de la grande guerre 2014-2018. Propositions pour une refondation de la politique mémorielle*. Rapport de la Mission Histoire.

Bigley, J. D., Lee, C., Chon, J., & Yoon, Y. (2010). Motivations for war-related tourism:A case of DMZ visitors in Korea. *Tourism Geographies, 12*(2), 371–394. doi:10.1080/14616688.2010.494687

Black, J. (2010). *The Battle of Waterloo*. Random House.

Blom, T. (2000). Morbid tourism-a postmodern market niche with an example from Althorp. *Norsk Geografisk Tidsskrift, 54*(1), 29–36. doi:10.1080/002919500423564

Brendon, P. (1991). *Thomas Cook: 150 years of popular tourism*. Secker (Martin) & Warburg Ltd.

Caddick-Adams, P. (2005). II. footprints in the mud: The British Army's approach to the battlefield tour experience. *Defence Studies*, *5*(1), 15–26. doi:10.1080/14702430500096368

Çakar, K. (2019). Transnational tourism experiences at Gallipoli. *Tourism Management*, *74*, 411–412.

Carbone, F. (2022). "Don't look back in anger". War museums' role in the post conflict tourism-peace nexus. *Journal of Sustainable Tourism*, *30*(2-3), 565–583. doi:10.1080/09669582.2021.1901909

Carlson, E. D. (2021). War of great distances: Allied strategy in the Pacific War, 1941–1945. In The Routledge history of the Second World War (pp. 101-116). Routledge. doi:10.4324/9780429455353

Chang, L. H. (2017). Tourists' perception of dark tourism and its impact on their emotional experience and geopolitical knowledge: A comparative study of local and non-local tourists. *Journal of Tourism Research & Hospitality*, *6*(3), 1–5. doi:10.4172/2324-8807.1000169

Chen, H. X., & Sun, Y. (2023). Applying war heritage in the national World War II history course for college students in China: An exploration of digitization strategies. *Sustainability (Basel)*, *15*(3), 2417–2439. doi:10.3390u15032417

Clarke, P., & Eastgate, A. (2011). Cultural capital, life course perspectives and Western Front battlefield tours. *Journal of Tourism and Cultural Change*, *9*(1), 31–44. doi:10.1080/14766825.2010.527346

Cobb, H. H. (1995). *American battlefields: A complete guide to the historic conflicts in words, maps, and photos*. Macmillan.

Davis, P. (2011). Total war. *The Metropolitan Review*, (Spring), 18–25.

Desjardin, T. A. (2003). *These honored dead: How the story of Gettysburg shaped American memory*. Da Capo Press.

Doğaner, S. (2006). Savaş ve turizm: Troya ve Gelibolu savaş alanları. *Türk Coğrafya Dergisi*, (46), 1–21.

Driessen, S., Grever, M., & Reijnders, S. (2022). Lessons of war. The significance of battlefield tours for the Dutch military. *Critical Military Studies*, *8*(2), 214–232. doi:10.1080/23337486.2019.1651044

Dunkley, R., Morgan, N., & Westwood, S. (2011). Visiting the trenches: Exploring meanings and motivations in battlefield tourism. *Tourism Management*, *32*(4), 860–868. doi:10.1016/j.tourman.2010.07.011

Edwards, J. L., & Winkler, C. K. (1997). Representative form and the visual ideograph: The Iwo Jima image in editorial cartoons. *The Quarterly Journal of Speech*, *83*(3), 289–310. doi:10.1080/00335639709384187

Farmaki, A., & Antoniou, K. (2017). Politicising dark tourism sites: Evidence from Cyprus. *Worldwide Hospitality and Tourism Themes*, *9*(2), 175–186. doi:10.1108/WHATT-08-2016-0041

Fathi, R. (2021). *Centenary (Battlefield Tourism). In 1914-1918-online: International Encyclopedia of the First World War*. Freie Universitat Berlin.

Fewster, K., Başarin, V., & Başarin, H. H. (2003). Gallipoli: The Turkish story. Allen & Unwin.

Foard, G., & Partida, T. (2005). *Scotland's historic fields of conflict: An assessment for historic Scotland* by *the battlefields trust.* Retrieved: https://www.battlefieldstrust.com/media/660.pdf

Foley, M., & Lennon, J. J. (1997). Dark tourism: an ethical dilemma. In Hospitality, tourism and leisure management: Issues in strategy and culture (pp. 153-164). Cassell.

Foote, K. E. (2003). *Shadowed ground: America's landscapes of violence and tragedy*. University of Texas Press. doi:10.7560/705258

Foreman, M. (2022). *From the battlefield to the big screen: Audie murphy, laurence olivier, vivien leigh and dirk bogarde in WW2*. Frontline Books.

Foulk, D. (2016). The impact of the "economy of history": The example of battlefield tourism in France. *Mondes du Tourisme*, (12), 41–57. doi:10.4000/tourisme.1338

François, P. (2013). 'The best way to see Waterloo is with your eyes shut': British'histourism', authenticity and commercialisation in the mid-nineteenth century. *Anthropological Journal on European Cultures*, *22*(1), 26–41. doi:10.3167/ajec.2013.220103

Frost, W. (2007). Refighting the Eureka Stockade: Managing a dissonant battlefield. In C. Ryan (Ed.), *Battlefield tourism: History, place and interpretation* (pp. 187–194). Elsevier. doi:10.1016/B978-0-08-045362-0.50022-1

Gatewood, J. B., & Cameron, C. M. (2004). Battlefield pilgrims at Gettysburg National Military Park. *Ethnology*, *43*(3), 193–216. doi:10.2307/3774062

Gibson, D., Yai, E., & Pratt, S. (2022). Journeying into the past to discover the potential for WWII dark tourism in the Solomon Islands. *Current Issues in Tourism*, *25*(14), 2285–2302. doi:10.1080/13683500.2021.1957787

Gieling, J., & Ong, C. E. (2016). Warfare tourism experiences and national identity: The case of Airborne Museum 'Hartenstein' in Oosterbeek, the Netherlands. *Tourism Management*, *57*, 45–55. doi:10.1016/j.tourman.2016.05.017

González, M. V. (2008). Intangible heritage tourism and identity. *Tourism Management*, *29*(4), 807–810. doi:10.1016/j.tourman.2007.07.003

Hall, D. (1994). Civil War reenactors and the postmodern sense of history. *The Journal of American Culture*, *17*(3), 7–11. doi:10.1111/j.1542-734X.1994.00007.x

Hall, D. (2002). Brand development, tourism and national identity: The re-imaging of former Yugoslavia. *Journal of Brand Management*, *9*(4), 323–334. doi:10.1057/palgrave.bm.2540081

Hammel, E. M. (2006). *Iwo Jima*. Zenith Imprint.

Heinzen, J. (2014). A negotiated truce: The battle of Waterloo in European memory since the Second World War. *History & Memory*, *26*(1), 39–74. doi:10.2979/histmemo.26.1.39

Henderson, J. (2000). War as a tourist attraction: The case of Vietnam. *International Journal of Tourism Research*, *2*(4), 269–280. doi:10.1002/1522-1970(200007/08)2:4<269::AID-JTR219>3.0.CO;2-A

Hertzog, A. (2012). War Battlefields, tourism and imagination. *Tourism Review*, (1), 1–13. doi:10.4000/viatourism.1283

Jankowski, P. (2014). Verdun: The longest battle of the great war. Oxford University Press.

Jung, K., Nguyen, V. T., Piscarac, D., & Yoo, S. C. (2020). Meet the virtual Jeju dol Haru bang—The mixed VR/Ar application for cultural immersion in Korea's main heritage. *ISPRS International Journal of Geo-Information*, *9*(6), 367. doi:10.3390/ijgi9060367

Keegan, J. (2011). *Six armies in Normandy: From D-day to the liberation of Paris June 6th-August 25th, 1944*. Random House.

Keenan, J. (2013). Death by bullet, fire or vapor: Examining the decision to use the atomic bomb to end World War II in the Pacific theatre. *The Exposition*, *1*(2), 1–7.

Kler, B. K., & Forsythe, C. P. (2022). Battlefield tourism: the legacy of Sandakan in Malaysian Borneo. In Handbook of niche tourism (pp. 236-249). Edward Elgar Publishing. doi:10.4337/9781839100185.00031

Lee, Y. J. (2016). The relationships amongst emotional experience, cognition, and behavioral intention in battlefield tourism. *Asia Pacific Journal of Tourism Research*, *21*(6), 697–715. doi:10.1080/10941665.2015.1068195

Lelo, L., & Jamal, T. (2013). 3 African Americans at sites of darkness: Roots-seeking, diasporic identities and place making. In Dark tourism and place identity managing and interpreting dark places (pp. 28-45). Routledge. doi:10.4324/9780203134900

Leopold, T. (2007). A proposed code of conduct for war heritage sites. In Battlefield tourism: History, place, and ınterpretation (pp. 49-57). Elsevier. doi:10.4324/9780080548340

Light, D. (2017). Progress in dark tourism and thanatourism research: An uneasy relationship with heritage tourism. *Tourism Management*, *61*, 275–301. doi:10.1016/j.tourman.2017.01.011

Lloyd, D. W. (1994). *Tourism, Pilgrimage, and the Commemoration of the Great War in Great Britain, Australia and Canada, 1919–1939* [Unpublished PhD Thesis]. Cambridge University.

Lloyd, D. W. (2014). *Battlefield tourism: Pilgrimage and the commemoration of the Great War in Britain, Australia and Canada, 1919-1939*. A&C Black.

Lockstone-Binney, L., Hall, J., & Atay, L. (2013). Exploring the conceptual boundaries of diaspora and battlefield tourism: Australians' travel to the Gallipoli battlefield, Turkey, as a case study. *Tourism Analysis*, *18*(3), 297–311. doi:10.3727/108354213X13673398610736

Miles, S. (2014). Battlefield sites as dark tourism attractions: An analysis of experience. *Journal of Heritage Tourism*, *9*(2), 134–147. doi:10.1080/1743873X.2013.871017

Miles, S. T. (2012). *Battlefield tourism: Meanings and interpretations* [Unpublished PhD Thesis]. University of Glasgow.

Mileva, S. V. (2018). Potential of development of dark tourism in Bulgaria. *International Journal of Tourism Cities*, *4*(2), 144–153. doi:10.1108/IJTC-05-2017-0029

Mitchell, J. P. (2003). Looking forward to the past: National identity and history in Malta. *Identities (Yverdon)*, *10*(3), 377–398. doi:10.1080/10702890390228919

Mortensen, M. (2013). The making and remakings of an American icon: 'Raising the flag on ɪwo jima' from photojournalism to global, digital media. In A. Gjelsvik (Ed.), Eastwood's Iwo Jima: Critical engagements with flags of our fathers and letters from Iwo Jima (pp. 15-35). Columbia University Press. doi:10.7312/gjel16564-003

National Park Service. (2023). *Gettysburg NMP.* National Park Service. https://irma.nps.gov/Stats/SSRSReports/Park%20Specific%20Reports/Annual%20Park%20Recreation%20Visitation%20(1904%20-%20Last%20Calendar%20Year)?Park=GETT

Ngo, P. M., & Bui, H. T. (2019). Contested interpretation of Vietnam war heritage: Tour guides' mediating roles. *Journal of Tourism and Adventure*, *2*(1), 61–84. doi:10.3126/jota.v2i1.25933

Nguyen, V. T., Jung, K., Yoo, S., Kim, S., Park, S., & Currie, M. (2019, December 9-11). Civil War battlefield experience: Historical event simulation using augmented reality technology. In *2019 IEEE International Conference on Artificial Intelligence and Virtual Reality (AIVR)* (pp. 294-2943). 10.1109/AIVR46125.2019.00068

Olsen, D. H. (2003). Heritage, tourism, and the commodification of religion. *Tourism Recreation Research*, *28*(3), 99–104. doi:10.1080/02508281.2003.11081422

Packer, J., Ballantyne, R., & Uzzell, D. (2019). Interpreting war heritage: Impacts of Anzac Museum and battlefield visits on Australians' understanding of national identity. *Annals of Tourism Research*, *76*, 105–116. doi:10.1016/j.annals.2019.03.012

Panakera, C. (2007). World War II and Tourism Development in Solomon Islands. In C. Ryan (Ed.), Battlefield tourism: History, place, and interpretation (pp. 125-141). Routledge.

Parekh, B. (1995). The concept of national identity. *Journal of Ethnic and Migration Studies*, *21*(2), 255–268. doi:10.1080/1369183X.1995.9976489

Pennell, C. (2018). Taught to remember? British youth and First World War centenary battlefield tours. *Cultural Trends*, *27*(2), 83–98. doi:10.1080/09548963.2018.1453449

Piekarz, M. (2007). *Hot war tourism: The live Battlefield and the ultimate adventure holiday? In Battlefield tourism: History, place, and ɪnterpretation.* Routledge.

Poria, Y., Biran, A., & Reichel, A. (2009). Visitors' preferences for interpretation at heritage sites. *Journal of Travel Research*, *48*(1), 92–105. doi:10.1177/0047287508328657

Poria, Y., Reichel, A., & Biran, A. (2006). Heritage site management: Motivations and expectations. *Annals of Tourism Research, 33*(1), 162–178. doi:10.1016/j.annals.2005.08.001

Prideaux, B. (2007). Echoes of war: Battlefield tourism. In C. Ryan (Ed.), Battlefield tourism history-place and interpretation (pp. 81-95). Routledge.

Prost, A., & Krumeich, G. (2015). *Verdun 1916*. Tallandier.

Raynor, M. (2004). *English battlefields*. Tempus.

Rivera, L. A. (2008). Managing "spoiled" national identity: War, tourism, and memory in Croatia. *American Sociological Review, 73*(4), 613–634. doi:10.1177/000312240807300405

Rose, E. P. (2019). Geology and the Allied liberation of Normandy: Highlights to help mark the 75th anniversary of D-Day, 6 June 1944. *Geology Today, 35*(3), 107–114. doi:10.1111/gto.12269

Russ, J. A. (2001). VLR! VII Fighter Command Operations from Iwo Jima, April-August 1945. *Air Power History, 48*(3), 16–25.

Ryan, C. (Ed.). (2007). *Battlefield tourism history-place and interpretation*. Routledge.

Scates, B. (2002). In Gallipoli's shadow: Pilgrimage, memory, mourning and the Great War. *Australian Historical Studies, 33*(119), 1–21. doi:10.1080/10314610208596198

Sears, S. W. (2003). *Landscape Turned Red: The Battle of Antietam*. Houghton Mifflin Harcourt.

Seaton, A. V. (1996). Guided by the dark: From thanatopsis to thanatourism. *International Journal of Heritage Studies, 2*(1), 234–244. doi:10.1080/13527259608722178

Seaton, A. V. (1999). War and thanatourism: Waterloo 1815–1914. *Annals of Tourism Research, 26*(1), 130–158. doi:10.1016/S0160-7383(98)00057-7

Seaton, A. V. (2000). "Another weekend away looking for dead bodies ... ": Battlefield tourism on the Somme and in Flanders. *Tourism Recreation Research, 25*(3), 63–77. doi:10.1080/02508281.2000.11014926

Sharma, N. (2022). Acknowledging the shades of grey: The past, present and future of dark tourism in India. In R. Kumar & A. Parida (Eds.), Indian Tourism (pp. 125-142). Emerald Publishing. doi:10.1108/978-1-80262-937-820221009

Sharpley, R., & Stone, P. R. (Eds.). (2009). The darker side of travel: The theory and practice of dark tourism. Channel View Publications. doi:10.21832/9781845411169

Slade, P. (2003). Gallipoli tourism. *Annals of Tourism Research, 30*(4), 779–794. doi:10.1016/S0160-7383(03)00025-2

Smith, V. (1998). War and tourism. *Annals of Tourism Research, 25*(1), 202–227. doi:10.1016/S0160-7383(97)00086-8

Spillman, L. (1997). *Nation and commemoration: Creating national identities in the United States and Australia.* Cambridge University Press. doi:10.1017/CBO9780511520938

Stolcke, V. (1997). The 'Nature' of nationality. In Citizenship and exclusion (pp. 61-80). Palgrave Macmillan. doi:10.1057/9780230374591_4

Stone, P., & Sharpley, R. (2008). Consuming dark tourism: A thanatological perspective. *Annals of Tourism Research, 35*(2), 574–595. doi:10.1016/j.annals.2008.02.003

Stone, P. R. (2006). A dark tourism spectrum: Towards a typology of death and macabre related tourist sites, attractions and exhibitions. *Tourism: An International Interdisciplinary Journal, 54*(2), 145–160.

Stone, P. R. (2011). Dark tourism: Towards a new post-disciplinary research agenda. *International Journal of Tourism Anthropology, 1*(3-4), 318–332. doi:10.1504/IJTA.2011.043713

Stone, P. R. (2011). Dark tourism and the cadaveric carnival: Mediating life and death narratives at Gunther von Hagens' Body Worlds. *Current Issues in Tourism, 14*(7), 685–701. doi:10.1080/13683500.2011.563839

Tarlow, P. E. (2007). Dark tourism: The appealing 'dark' side of tourism and more. In D. Weaver (Ed.), Niche tourism (pp. 47-58). Routledge. doi:10.4324/9780080492926

Taru, P., Chingombe, W., & Mukwada, G. (2014). A bullet laden park: Potential for battlefield tourism in the Golden Gate Highlands National Park. *African Journal of Hospitality, Tourism and Leisure, 2*, 1–9.

Upton, A., Schänzel, H., & Lück, M. (2018). Reflections of battlefield tourist experiences associated with Vietnam war sites: An analysis of travel blogs. *Journal of Heritage Tourism, 13*(3), 197–210. doi:10.1080/1743873X.2017.1282491

Van der Merwe, C. D. (2014). Battlefields tourism: The status of heritage tourism in Dundee, South Africa. *Bulletin of Geography. Socio-Economic Series, 26*(26), 121–139. doi:10.2478/bog-2014-0049

Venter, D. (2011). Battlefield tourism in the South African context. *African Journal of Hospitality, Tourism and Leisure*, *1*(3), 1–5.

Virgili, S., Delacour, H., Bornarel, F., & Liarte, S. (2018). 'From the Flames to the Light': 100 years of the commodification of the dark tourist site around the Verdun battlefield. *Annals of Tourism Research*, *68*, 61–72. doi:10.1016/j.annals.2017.11.005

Wahlert, G. (2011). *Exploring Gallipoli: An Australian army battlefield guide* (Vol. 4). Simon and Schuster.

White, L., & Frew, E. (Eds.). (2013). *Dark tourism and place identity: Managing and interpreting dark places*. Routledge. doi:10.4324/9780203134900

Winter, C. (2009). Tourism social memory and the Great War. *Annals of Tourism Research*, *36*(4), 607–626. Advance online publication. doi:10.1016/j.annals.2009.05.002

Winter, C. (2011). Battlefield visitor motivations: Explorations in the Great War town of Ieper, Belgium. *International Journal of Tourism Research*, *13*(2), 164–176. doi:10.1080/14616688.2011.575075

Winter, C. (2012). Commemoration of the Great War on the Somme: Exploring personal connections. *Journal of Tourism and Cultural Change*, *10*(3), 248–263. doi:10.1080/14766825.2012.694450

Yoon, Y., & Uysal, M. (2005). An examination of the effects of motivation and satisfaction on destination loyalty: A structural model. *Tourism Management*, *26*(1), 45–56. doi:10.1016/j.tourman.2003.08.016

Zhang, C. X., Fong, L. H. N., Li, S., & Ly, T. P. (2019). National identity and cultural festivals in postcolonial destinations. *Tourism Management*, *73*, 94–104. doi:10.1016/j.tourman.2019.01.013

Zhang, J. J. (2010). Of Kaoliang, bullets and knives: Local entrepreneurs and the battlefield tourism enterprise in Kinmen (Quemoy), Taiwan. *Tourism Geographies*, *12*(3), 395–411. doi:10.1080/14616688.2010.494685

ADDITIONAL READING

Baldwin, F., & Sharpley, R. (2009). Battlefield tourism: Bringing organised violence back to life. In *The darker side of travel* (pp. 186–206). Channel View Publications. doi:10.21832/9781845411169-011

Butler, R., & Suntikul, W. (Eds.). (2013). *Tourism and war*. Routledge. doi:10.4324/9780203107706

Deckert, F. (2021). *Battlefield tourism on the Western Front of the Great War*. GRIN Verlag.

Dunkley, R., Morgan, N., & Westwood, S. (2011). Visiting the trenches: Exploring meanings and motivations in battlefield tourism. *Tourism Management, 32*(4), 860–868. doi:10.1016/j.tourman.2010.07.011

Eade, J., & Katić, M. (Eds.). (2017). *Military pilgrimage and battlefield tourism: Commemorating the dead*. Routledge. doi:10.4324/9781315595436

Lloyd, D. W. (2014). *Battlefield tourism: Pilgrimage and the commemoration of the Great War in Britain, Australia and Canada, 1919-1939*. A&C Black.

McKay, J. (2013). A critique of the militarisation of Australian history and culture thesis: The case of Anzac battlefield tourism. *Portal: Journal of Multidisciplinary International Studies, 10*(1), 1-25.

Miles, S. (2014). Battlefield sites as dark tourism attractions: An analysis of experience. *Journal of Heritage Tourism, 9*(2), 134–147. doi:10.1080/1743873X.2013.871017

Miles, S. T. (2012). *Battlefield tourism: Meanings and interpretations* [Doctoral dissertation]. University of Glasgow.

Ryan, C. (Ed.). (2007). *Battlefield Tourism: History, place and interpretation*. Routledge. doi:10.4324/9780080548340

Sharpley, R., & Stone, P. R. (Eds.). (2009). The darker side of travel: The theory and practice of dark tourism. Channel View Publications. doi:10.21832/9781845411169

KEY TERMS AND DEFINITIONS

Bushido: Bushido is a concept that defines the ethical and moral principles of the Japanese warrior class. The term, meaning "the way of the warrior," emphasizes values such as honor, loyalty, courage, and self-discipline that were expected to be followed by Japanese warriors. The origins of Bushido can be traced back to the behavior of the samurai, the feudal warrior class of Japan. Traditionally, these principles were considered disciplined rules followed in every aspect of a samurai's life. Today, Bushido still holds a significant place in Japanese culture, continuing to influence areas such as modern Japanese discipline, work ethics, and sports.

D-Day: D-Day is the term used for the start of the Normandy Invasion on June 6, 1944 during WWII. Allied forces launched the operation to liberate France from Nazi occupation. It was a significant amphibious invasion that involved land, air, and naval forces and marked a turning point in the war.

Dark Tourism: Dark tourism is a type of tourism that involves visiting places where significant but tragic historical and cultural events have occurred. It is popular for reasons such as understanding and remembering history, exploring cultural identity, and personal satisfaction.

Morbid Tourism: Blom (2000) defined morbid tourism as a form of tourism that involves visiting places where death events that affect a large number of people have occurred and places where the phenomenon of death is presented.

Thana Tourism: According to Seaton (1996), thanatourism is a tourism type that is derived from the Ancient Greek word "thanatos," which means to personalize death, and is more specifically focused on violent deaths than dark tourism. It involves visiting sites related to natural disasters, wars, crimes against humanity, tragic events, and death in general. Thanatourism aims to educate tourists about past events, suffering, and tragedy, but it has also been subject to ethical and moral debates.

Compilation of References

A brief history. (2016, May 3). FBI. https://www.fbi.gov/history/brief-history

Agapito, D., Mendes, J., & Oom do Valle, P. (2012). The Rural Village as an open door to nature-based tourism in Portugal: The Aldeia da Pedralva case. *Tourism (Zagreb)*, *60*(3), 325–338.

Agüera, O. F. (2013). El Turismo Comunitario como Herramienta para el Desarrollo Sostenible de Destinos Subdesarrollados. Nómadas. *Journal of Social and Juridical Sciencies, 38*(2). https://www.theoria.eu/nomadas/38/francisco_orgaz.pdf

Ahmad, Y. (2006). The scope and definitions of heritage: From tangible to intangible. *International Journal of Heritage Studies*, *12*(3), 292–300. doi:10.1080/13527250600604639

Akbulut, O., & Ekin, Y. (2018). Kültürel miras turizmi olarak savaş alanları turizmi: Türkiye'de yer alan savaş anıtlarının coğrafi bilgi sistemleri analizi. *Hitit Üniversitesi Sosyal Bilimler Enstitüsü Dergisi*, *11*(1), 395–420. doi:10.17218/hititsosbil.397914

Akbulut, O., & Ekin, Y. (2022). Yaşayan tarih: Savaş alanları turizmi ve yeniden canlandırma toplulukları örneği. *Kent Akademisi Dergisi*, *15*(2), 439–451. doi:10.35674/kent.988625

Akinci, Z. (2013). Management of accessible tourism and its market in Turkey. *International Journal of Business and Management Studies*, *2*(2), 413–426.

Al Muala, A. (2012). Assessing the relationship between marketing mix and loyalty through tourists satisfaction in Jordan curative tourism. *American Academic & Scholarly Research Journal*, *4*(2), 7-23.

Al-Ansi, A., & Han, H. (2019). Role of halal-friendly destination performances, value, satisfaction, and trust in generating destination image and loyalty. *Journal of Destination Marketing and Management, 13*(December), 51–60. doi:10.1016/j.jdmm.2019.05.007

Albacete-Sáez, C., Fuentes-Fuentes, M., & Lloréns-Montes, F. (2007). Service quality measurement in rural accommodation. *Annals of Tourism Research*, *34*(1), 45–65. doi:10.1016/j.annals.2006.06.010

Aldeias do Xisto. (2015). *Relatório de Avaliação EEC PROVERE rede das Aldeias do Xisto* [EEC PROVERE Assessment Report Schist Villages network]. ADXTUR.

Aldeias do Xisto. (2023). *Aldeias*. Available at: https://aldeiasdoxisto.pt/aldeias

Alén, E., Domínguez, T., & Losada, N. (2012). New opportunities for the tourism market: Senior tourism and accessible tourism. *Visions for global tourism industry: Creating and sustaining competitive strategies*, 139-166.

Al-Husseini, S., & Elbeltagi, I. (2012). The Impact of Leadership Style and Knowledge Sharing on Innovation in Iraqi Higher Education Institutions. In *Proceedings of the 4th European Conference on Intellectual Capital*. Arcada University of Applied Sciences.

Alipour, H., Fatemi, H., & Malazizi, N. (2020). Is edu-tourism a sustainable option? A case study of residents' perceptions. *Sustainability (Basel)*, *12*(15), 5937. Advance online publication. doi:10.3390u12155937

Alipour, H., & Kilic, H. (2005). An institutional appraisal of tourism development and planning: The case of the Turkish Republic of North Cyprus (TRNC). *Tourism Management*, *26*(1), 79–94. doi:10.1016/j.tourman.2003.08.017

Alipour, H., Vaziri, R. K., & Ligay, E. (2011). Governance as Catalyst to Sustainable Tourism Development: Evidence from North Cyprus. *Journal of Sustainable Development*, *4*(5). Advance online publication. doi:10.5539/jsd.v4n5p32

Alipour-Hafezi, M., Ashrafi-Rizi, H., Kazempour, Z., & Shahbazi, M. (2013). Using 4P marketing model in academic libraries: An experience. *International Journal of Information Science and Management, 11*(2), 45-58.

Alonso-Sañudo, A. I. (2002). La aplicación del concepto de turismo sostenible en los países desarrollados. In *Turismo sostenible* (pp. 25 – 36). Iepala Editorial, 2002.

Alturas, B., & Oliveira, L. S. (2016). Consumers using Social Media: Impact on Companies' Reputation. Radical Marketing. In *Academy of Marketing Annual Conference Proceedings*. Newcastle Business School at Northumbria University.

Amabile, T., Schatzel, E., Moneta, G., & Kramer, S. (2004). Leader behaviors and the work environment for creativity: Perceived leader support. *The Leadership Quarterly*, *15*(1), 5–32. doi:10.1016/j.leaqua.2003.12.003

Amankwaa, A., Gyensare, M. A., & Susomrith, P. (2019). Transformational leadership with innovative behaviour. *Leadership and Organization Development Journal*, *4*(4), 402–420. doi:10.1108/LODJ-10-2018-0358

Anderson, J. T. (2006). Snow day: Memories of growing up in the Bronx. *Gettysburg Review*, *19*(3), 257–311.

Andraz, J. M., Norte, N. M., & Gonçalves, H. S. (2015). Effects of tourism on regional asymmetries: Empirical evidence for Portugal. *Tourism Management*, *50*, 257–267. doi:10.1016/j.tourman.2015.03.004

Anton, C., Salvador, L., & Monné, R. (1996). Turismo rural, desarrollo local y preservación del ambiente. Elementos para un desarrollo sostenible del turismo en la zona de montaña Prades-Montsant, Cataluña. *Ería: Revista cuatrimestral de geografía, 41*, 227 – 238. Disponible en: https://dialnet.unirioja.es/servlet/articulo?codigo=34843

An, Y.-S., & Seo, K.-K. (2013). A study on service satisfaction factor analysis of an online secondhand bookstore. *Jouranl of Digital Convergence, 11*(11), 251–256. doi:10.14400/JDPM.2013.11.11.251

Araoz, G. F. (2011). Preserving heritage places under a new paradigm. *Journal of Cultural Heritage Management and Sustainable Development, 1*(1), 55–60. doi:10.1108/20441261111129933

Ariwibowo, A., Budimen, H., & Dwinanto, A. (2022). Development of Gastronomy Tourism of Cultural Heritage Spice Route in Kota Serang, Banten. *International Forum on Spice Route 2022*. https://ocs.brin.go.id/index.php/ifsr/ifsr2022/paper/view/81

Asgarnezhad Nouri, B., Nemati, V., & Abbasgholizadeh, N. (2019). The Effect of Perceived Value on the Destination Image, Satisfaction and Loyalty of Medical Tourists: A Case Study in Ardabil. *Journal of Health, 10*(1), 34–49. doi:10.29252/j.health.10.1.34

Ashworth, G. J., & Isaac, R. K. (2015). Have we illuminated the dark? Shifting perspectives on 'dark' tourism. *Tourism Recreation Research, 40*(3), 316–325. doi:10.1080/02508281.2015.1075726

Ashworth, G., & Hartmann, R. (Eds.). (2005). *Horror and Human Tragedy Revisited: The Management of Sites of Atrocities for Tourism*. Cognizant.

Atay, L. (2008). Alan kılavuzluğunun Çanakkale İli'ndeki uygulaması ve ilgili turizm mevzuatı açısından değerlendirilmesi. *Anatolia: Turizm Araştırmaları Dergisi, 19*(2), 169–176.

Aucancela, I., Betty, P., Velasco, S., & Víctor, M. (2021). Gestión turística como herramienta de desarrollo sostenible de la microcuenca del río Chimborazo, Cantón Riobamba. *Revista de Ciencias Sociales y Humanidades Chakiñan, 13*, 102-116. Disponible en: http://scielo.senescyt.gob.ec/pdf/rchakin/n13/2550-6722-rchakin-13-00102.pdf

Avolio, B. J., & Gardner, W. L. (2005). Authentic Leadership Development: Getting to the Root of Positive Forms of Leadership. *The Leadership Quarterly, 16*(3), 315–338. doi:10.1016/j.leaqua.2005.03.001

Avraham, E., & Ketter, E. (2017). Destination image repair while combatting crises: Tourism marketing in Africa. *Tourism Geographies, 19*(5), 780–800. doi:10.1080/14616688.2017.1357140

Axelrod, A. (2016). *The battle of Verdun*. Rowman & Littlefield.

Axtell, C. M., Holman, D. J., Unsworth, K. L., Wall, T. D., Waterson, P. E., & Harrington, E. (2000). Shop-floor innovation: Facilitating the suggestion and implementation of ideas. *Journal of Occupational and Organizational Psychology, 73*(3), 265–285. doi:10.1348/096317900167029

Bagozzi, R. P., & Yi, Y. (1988). On the evaluation of structural equation models. *Journal of the Academy of Marketing Science*, *16*(1), 74–94. doi:10.1007/BF02723327

Baldwin, F., & Sharpley, R. (2009). Battlefield tourism: bringing organised violence back to life. In The darker side of travel: the theory and practice of dark tourism (pp. 186-206). Channel View.

Bandyopadhyay, S. (2016). Factors affecting the adoption of electronic word-of-mouth in the tourism industry. *International Journal of Business and Social Science*, *7*(1), 10–18.

Barbuddhe, S. B., & Singh, N. P. (2014). *Agro-Eco-Tourism: A New Dimension to Agriculture*. ICAR Research.

Barcellini, S. (2009). *La Meuse face au défi du centenaire de la grande guerre 2014-2018. Propositions pour une refondation de la politique mémorielle*. Rapport de la Mission Histoire.

Barreto, A. M. (2013). Do Users Look at Banner Ads on Facebook? *Journal of Research in Interactive Marketing*, *7*(2), 119–139. doi:10.1108/JRIM-Mar-2012-0013

Bass, B. M. (2000). The Future of Leadership in Learning Organizations. *The Journal of Leadership Studies*, *7*(3), 19–40. doi:10.1177/107179190000700302

Bass, B. M. (2008). *The Bass handbook of leadership: Theory, research, & managerial applications* (4th ed.). Free Press.

Bass, B. M., & Avolio, B. J. (1990). *Transformational leadership development: Manual for the multifactor leadership questionnaire*. Consulting Psychologists Press.

Bass, B. M., & Riggio, R. E. (2006). *Transformational Leadership* (2nd ed.). Lawrence Erlbaum Associates, Inc. doi:10.4324/9781410617095

Bateman, T. S., & Crant, J. M. (1993). The proactive component of organizational behavior: A measure and correlates. *Journal of Organizational Behavior*, *14*(2), 103–118. doi:10.1002/job.4030140202

Bauman, Z. (1990). *Privatization of Ambivalence, Modernity and Ambivalence*. Polity Press.

Bauman, Z. (1992). Survival as a Social Construct. *Theory, Culture & Society*, *9*(1), 1–36. doi:10.1177/026327692009001002

Bec, A., Moyle, B., Timms, K., Schaffer, V., Skavronskaya, L., & Little, C. (2019). Management of immersive heritage tourism experiences: A conceptual model. *Tourism Management*, *72*, 117–120. doi:10.1016/j.tourman.2018.10.033

Belhassen, Y., Caton, K., & Stewart, W. P. (2008). The search for authenticity in the pilgrim experience. *Annals of Tourism Research*, *35*(3), 668–689. doi:10.1016/j.annals.2008.03.007

Bennett, T. (2009). A study of the management leadership style preferred by it subordinates. *Journal of Organizational Culture Communications and Conflict*, *13*, 1–15.

Bennis, W. G., & Thomas, R. J. (2002). Crucibles of leadership. *Harvard Business Review*, *80*(9), 39–45. PMID:12227145

Berger, P., & Luckmann, T. (1967). *The Social Construction of Reality: A Treatise in the Sociology of Knowledge*. Doubleday.

Bifulco, F. (2016). *Managing cultural heritage: innovation perspectives, customer experience, resources enhancement, performance management*. McGraw-Hill Education.

Bigley, J. D., Lee, C., Chon, J., & Yoon, Y. (2010). Motivations for war-related tourism:A case of DMZ visitors in Korea. *Tourism Geographies*, *12*(2), 371–394. doi:10.1080/14616688.2010.494687

Bigné Alcañiz, E., Font Aulet, X., & Andreu Simó, L. (2000). *Marketing de destinos turísticos. Análisis y estrategias de desarrollo*. Escuela Superior de Gestión Comercial y Marketing.

Bijlani, V. A. (2021, September). *Sustainable digital transformation of heritage tourism. In 2021 IoT Vertical and Topical Summit for Tourism*. IEEE.

Biran, A., Poria, Y., & Oren, G. (2011). Sought experiences at (dark) heritage sites. *Annals of Tourism Research*, *38*(3), 820–841. doi:10.1016/j.annals.2010.12.001

Bird, G., Westcott, M., & Thiesen, N. (2018). Marketing Dark Heritage: Building Brands, Myth-Making And Social Marketing. In The Palgrave Handbook of Dark Tourism. Palgrave Macmillan.

Bissell, L. (2009). *Understanding Motivation and Perception at Two Dark Tourism Attractions in Winnipeg, MB* [Master's Thesis]. Retrieved from https://mspace.lib.umanitoba.ca/bitstream/handle/1993/29742/Bissell_Understanding_motivation.pdf?sequence=1

Black, J. (2010). *The Battle of Waterloo*. Random House.

Blake, J. (2008). UNESCO's 2003 Convention on Intangible Cultural Heritage: the implications of community involvement in 'safeguarding'. In Intangible heritage (pp. 59-87). Routledge.

Blom, T. (2000). Morbid tourism - A Postmodern Market Niche with an Example from Althorp. *NorskGeografiskTidsskrift - Norsk Geografisk Tidsskrift*, *54*(1), 29–36. doi:10.1080/002919500423564

Boerner, S., Eisenbeiss, S., & Griesser, D. (2007). Followers behaviour and organizational performance: The impact of transformational leaders. *Journal of Leadership & Organizational Studies*, *13*(3), 15–26. doi:10.1177/10717919070130030201

Bommer, M., & Jalajas, D. S. (2004). Innovation sources of large and small technology-based firms. *IEEE Transactions on Engineering Management*, *51*(1), 13–18. doi:10.1109/TEM.2003.822462

Borpuzari, P. (2022). India's spice exports need to grow at 19.5% to meet $10 billion target by FY27. *The Economics Times.* https://economictimes.indiatimes.com/small-biz/trade/exports/insights/indias-spice-exports-need-to-grow-at-19-5-to-meet-10-billion-target-by-fy27/articleshow/92253030.cms

Böttger, T., Rudolph, T., Evanschitzky, H., & Pfrang, T. (2017). Customer inspiration: Conceptualization, scale development, and validation. *Journal of Marketing*, *81*(6), 116–131. doi:10.1509/jm.15.0007

Boukas, N. (2019). Rural tourism and residents' well-being in Cyprus: Towards a conceptualized framework of the appreciation of rural tourism for islands' sustainable development and competitiveness. *International Journal of Tourism Anthropology*, *7*(1), 60–86. doi:10.1504/IJTA.2019.098105

Bowman, M. S., & Pezzullo, P. C. (2009). What's so 'dark'about 'dark tourism'?: Death, tours, and performance. *Tourist Studies*, *9*(3), 187–202. doi:10.1177/1468797610382699

Brendon, P. (1991). *Thomas Cook: 150 years of popular tourism.* Secker (Martin) & Warburg Ltd.

Brogan, C., & Smith, J. (2009). *Trust agents: Using the web to build influence, improve reputation, and earn trust.* Wiley.

Brownell, J. (2010). Leadership in the service of hospitality. *Cornell Hospitality Quarterly*, *51*(3), 363–378. doi:10.1177/1938965510368651

Buekens, W. (2013). Coping with the innovation paradoxes: The challenge for a new game leadership. *Procedia Economics and Finance*, *6*, 205–212. doi:10.1016/S2212-5671(13)00133-0

Buil, I., Martínez, E., & Matute, J. (2019). Transformational leadership and employee performance: The role of identification, engagement and proactive personality. *International Journal of Hospitality Management*, *77*, 64–75. doi:10.1016/j.ijhm.2018.06.014

Bull, C., & Weed, M. (1999). Niche markets and small island tourism: The development of sports tourism in Malta. *Managing Leisure, 4*(3), 142-155.

Burns, J. M. (1978). *Leadership.* Harper Perennial.

Byran, A., & Hyde, K. F. (2013). Guest Editorial: New Perspectives On Dark Tourism. *International Journal of Culture, Tourism and Hospitality Research*, *7*(3), 191–198. doi:10.1108/IJCTHR-05-2013-0032

Caddick-Adams, P. (2005). II. footprints in the mud: The British Army's approach to the battlefield tour experience. *Defence Studies*, *5*(1), 15–26. doi:10.1080/14702430500096368

Çakar, K. (2019). Transnational tourism experiences at Gallipoli. *Tourism Management*, *74*, 411–412.

Calantone, R. J., Cavusgil, T. S., & Zhao, Y. (2002). Learning orientation, firm innovation capability, and firm performance. *Industrial Marketing Management, 31*(6), 515–524. doi:10.1016/S0019-8501(01)00203-6

Calheiros, A. C., Moro, S., & Rita, P. (2017). Sentiment Classification of Consumer-Generated Online Reviews Using Topic Modeling. *Journal of Hospitality Marketing & Management, 26*(7), 675–693. doi:10.1080/19368623.2017.1310075

Campelo, A. (2018). Immaterial heritage and sense of place. In *Cultural Heritage* (pp. 129–138). Routledge. doi:10.4324/9781315107264-10

Campos, A. C., Mendes, J., Valle, P. O. D., & Scott, N. (2018). Co-creation of tourist experiences: A literature review. *Current Issues in Tourism, 21*(4), 369–400. doi:10.1080/13683500.2015.1081158

Carbone, F. (2022). "Don't look back in anger". War museums' role in the post conflict tourism-peace nexus. *Journal of Sustainable Tourism, 30*(2-3), 565–583. doi:10.1080/09669582.2021.1901909

Carignani, F., Clemente, L., Iodice, G., & Bifulco, F. (2023). Digital Marketing in Cultural Heritage: An Approach to Metaverse. In Cultural Marketing and Metaverse for Consumer Engagement (pp. 142-163). IGI Global.

Carlson, E. D. (2021). War of great distances: Allied strategy in the Pacific War, 1941–1945. In The Routledge history of the Second World War (pp. 101-116). Routledge. doi:10.4324/9780429455353

Carmeli, A., & Waldman, D. (2010). Leadership, behavioral context, and the performance of work groups in a knowledge-intensive setting. *The Journal of Technology Transfer, 35*(4), 384–400. doi:10.100710961-009-9125-3

Caro-González, F., Javier, A., Guzmán, J., Alberto, O.-A., & Francisco, C.-V. Mario. (2015). Turismo, desarrollo sostenible y percepción de los stakeholders. Un estudio de caso en República Dominicana. *Revista de Economía del Caribe, 15*, 153 – 182. Disponible en: http://www.scielo.org.co/pdf/ecoca/n15/n15a06.pdf

Carrabine, E. (2011). Images of Torture: Culture, Politics and Power. *Crime, Media, Culture, 7*(1), 5–30. doi:10.1177/1741659011404418

Carrigan, A. (2014). Dark tourism and postcolonial studies: Critical intersections. *Postcolonial Studies, 17*(3), 236–250. doi:10.1080/13688790.2014.993425

Casanueva, R. C., & Gallego, A. A. (2012). *Empresas y organizaciones turísticas*. Pirámide.

Castro, C., Periñan, M., & Bueno, J. (2008). Transformational leadership and followers' attitudes: The mediating role of psychological empowerment. *International Journal of Human Resource Management, 19*(10), 1842–1863. doi:10.1080/09585190802324601

Causevic, S., & Lynch, P. (2011). Phoenix Tourism: Post-Conflict Tourism Role. *Annals of Tourism Research, 38*(3), 780–800. doi:10.1016/j.annals.2010.12.004

Cavaglieri, M., & Steindel, G. E. (2009). Um lugar para observar, conversar, ler, comprar-livros e outros suportes de informação e lazer: Uma análise dos sebos da cidade de Florianópolis [A place to observe, talk, read, buy-books and other information and leisure media: An analysis of the tallow of the city of Florianópolis]. *The Information Society*, *19*(3), 55–64.

Cave, J., & Buda, D. (2018). Souvenirs in dark tourism: Emotions and symbols. Palgrave Macmillan UK.

Cemetery and Serial Killers Walking Tour of Philadelphia – Philadelphia Pennsylvania . (2023). Tripadvisor. https://www.tripadvisor.com/AttractionProductReview-g60795-d12859723-or10-Cemetery_and_Serial_Killers_Walking_Tour_of_Philadelphia-Philadelphia_Pennsylv.html

Černe, M., Jaklič, M., & Škerlavaj, M. (2013). Authentic leadership, creativity, and innovation: A multilevel perspective. *Leadership*, *9*(1), 63–85. doi:10.1177/1742715012455130

Chang, L. H. (2017). Tourists' perception of dark tourism and its impact on their emotional experience and geopolitical knowledge: A comparative study of local and non-local tourists. *Journal of Tourism Research & Hospitality*, *6*(3), 1–5. doi:10.4172/2324-8807.1000169

Chang, S. Y., Tsaur, S. H., Yen, C. H., & Lai, H. R. (2020). Tour member fit and tour member–leader fit on group package tours: Influences on tourists' positive emotions, rapport, and satisfaction. *Journal of Hospitality and Tourism Management*, *42*, 235–243. doi:10.1016/j.jhtm.2020.01.016

Chang, S., Gong, Y., & Shum, C. (2011). Promoting innovation in hospitality companies through human resource management practices. *International Journal of Hospitality Management*, *30*(4), 812–818. doi:10.1016/j.ijhm.2011.01.001

Chan, M. K. Y., Chiu, D. K. W., & Lam, E. T. H. (2020). Effectiveness of overnight learning commons: A comparative study. *Journal of Academic Librarianship*, *46*(7), 102253. doi:10.1016/j.acalib.2020.102253 PMID:34173399

Chan, M. M. W., & Chiu, D. K. W. (2022). Alert driven customer relationship management in online travel agencies: event-condition-actions rules and key performance indicators. In A. Naim & S. Kautish (Eds.), *Building a brand image through electronic customer relationship management* (pp. 268–303). IGI Global. doi:10.4018/978-1-6684-5386-5.ch012

Chan, T. T. W., Lam, A. H. C., & Chiu, D. K. W. (2020). From Facebook to Instagram: Exploring user engagement in an academic library. *Journal of Academic Librarianship*, *46*(6), 102229. doi:10.1016/j.acalib.2020.102229 PMID:34173399

Chan, V. H. Y., Ho, K. K. W., & Chiu, D. K. W. (2022). Mediating effects on the relationship between perceived service quality and public library app loyalty during the COVID-19 era. *Journal of Retailing and Consumer Services*, *67*, 102960. doi:10.1016/j.jretconser.2022.102960

Charles Manson Biography. (2021, February 4). Biography. https://www.biography.com/crime-figure/charles-manson

Chatterjee, S., & Prasad, M. V. D. (2019). The Evolution of Agri-Tourism practices in India: Some Success Stories. *Madridge Journal of Agriculture and Environmental Sciences*, *1*(1), 19–25. doi:10.18689/mjaes-1000104

Chen, C. F., & Tsai, D. C. (2007). How destination image and evaluative factors affect behavioral intentions? *Tourism Management*, *28*(4), 1115–1122. doi:10.1016/j.tourman.2006.07.007

Chen, D. (2016). Intelligent application system of warehouse management based on RFID technology. *China Digital Cable TV*, *12*, 1372–1374.

Cheng, W., Tian, R., & Chiu, D.K.W. (2023). Travel vlogs influencing tourist decisions: Information preferences and gender differences. *Aslib Journal of Information Management*, ahead-of-print. doi:10.1108/AJIM-05-2022-0261

Cheng, J., Yuen, A. H., & Chiu, D. K. W. (2022). (in press). Systematic review of MOOC research in mainland China. *Library Hi Tech*. Advance online publication. doi:10.1108/LHT-02-2022-0099

Cheng, W. W. H., Lam, E. T. H., & Chiu, D. K. W. (2020). Social media as a platform in academic library marketing: A comparative study. *Journal of Academic Librarianship*, *46*(5), 102188. doi:10.1016/j.acalib.2020.102188

Chen, H. X., & Sun, Y. (2023). Applying war heritage in the national World War II history course for college students in China: An exploration of digitization strategies. *Sustainability (Basel)*, *15*(3), 2417–2439. doi:10.3390u15032417

Chen, K. Y., & Hsu, Y. L. (2021). Developing a model of backpackers' exploratory curiosity. *Tourism and Hospitality Management*, *27*(1), 1–23. doi:10.20867/thm.27.1.1

Chen, N., Dwyer, L., & Firth, T. (2014). Effect of dimensions of place attachment on residents' word-of-mouth behavior. *Tourism Geographies*, *16*(5), 826–843. doi:10.1080/14616688.2014.915877

Chen, S. (2017). The Relationship between Innovation and Firm Performance: A Literature Review. *Proceedings of the 7th International Conference on Social Network, Communication and Education, in Advances in Computer Science Research*. 10.2991nce-17.2017.132

Chen, W. J. (2011). Innovation in hotel services: Culture and personality. *International Journal of Hospitality Management*, *30*(1), 64–72. doi:10.1016/j.ijhm.2010.07.006 PMID:32287854

Chen, Y., Chiu, D. K. W., & Ho, K. K. W. (2018). Facilitating the learning of the art of Chinese painting and calligraphy at Chao Shao-an Gallery. *Micronesian Educators*, *26*, 45–58.

Cheung, L. S. N., Chiu, D. K. W., & Ho, K. K. W. (2022). A quantitative study on utilizing electronic resources to engage children's reading and learning: Parents' perspectives through the 5E instructional model. *The Electronic Library*, *40*(6), 662–679. doi:10.1108/EL-09-2021-0179

Cheung, M., & Wong, C. (2011). Transformational leadership, leader support, and employee creativity. *Leadership and Organization Development Journal*, *32*(7), 657–672. doi:10.1108/01437731111169988

Cheung, T. Y., Ye, Z., & Chiu, D. K. W. (2021). Value chain analysis of information services for the visually impaired: A case study of contemporary technological solutions. *Library Hi Tech*, *39*(2), 625–642. doi:10.1108/LHT-08-2020-0185

Cheung, V. S. Y., Lo, J. C. Y., Chiu, D. K. W., & Ho, K. K. W. (2023). Evaluating social media's communication effectiveness on travel product promotion: Facebook for college students in Hong Kong. *Information Discovery and Delivery*, *51*(1), 66–73. doi:10.1108/IDD-10-2021-0117

Chew, E. Y. T., & Jahari, S. A. (2014). Destination image as a mediator between perceived risks and revisit intention: A case of post-disaster Japan. *Tourism Management*, *40*, 382–393. doi:10.1016/j.tourman.2013.07.008

Che, X., & Ip, B. (2018). *Social networks in China*. Chandos Publishing.

Chiang, C. F., & Jang, S. (2008). The Antecedents and Consequences of Psychological Empowerment: The Case of Taiwan's Hotel Companies. *Journal of Hospitality & Tourism Research (Washington, D.C.)*, *32*(1), 40–61. doi:10.1177/1096348007309568

Chin, G. Y. L., & Chiu, D. K. W. (2023). RFID-based Robotic Process Automation for Smart Museums with an Alert-driven Approach. In R. Tailor (Ed.), *Application and Adoption of Robotic Process Automation for Smart Cities*. IGI Global.

Chiu, D. K. W., & Ho, K. K. W. (2022a). Special selection on contemporary digital culture and reading. *Library Hi Tech*, *40*(5), 1204–1209. doi:10.1108/LHT-10-2022-516

Chiu, D. K. W., & Ho, K. K. W. (2022b). Editorial: 40th anniversary: contemporary library research. *Library Hi Tech*, *40*(6), 1525–1531. doi:10.1108/LHT-12-2022-517

Cho, H. J., & Pucik, V. (2005). Relationship between innovativeness, quality, growth, profitability, and market value. *Strategic Management Journal*, *26*(6), 555–575. doi:10.1002mj.461

Christie, M. F., & Mason, P. A. (2003). Transformative tour guiding: Training tour guides to be critically reflective practitioners. *Journal of Ecotourism*, *2*(1), 1–16. doi:10.1080/14724040308668130

Chung, A. C. W., & Chiu, D. K. (2016). OPAC Usability Problems of Archives: A Case Study of the Hong Kong Film Archive. [IJSSOE]. *International Journal of Systems and Service-Oriented Engineering*, *6*(1), 54–70. doi:10.4018/IJSSOE.2016010104

Chung, C., Chiu, D. K. W., Ho, K. K. W., & Au, C. H. (2020). Applying social media to environmental education: Is it more impactful than traditional media? *Information Discovery and Delivery*, *48*(4), 255–266. doi:10.1108/IDD-04-2020-0047

Churchill, G. A. Jr. (1979). A paradigm for developing better measures of marketing constructs. *JMR, Journal of Marketing Research*, *16*(1), 64–73. doi:10.1177/002224377901600110

Churchill, N., & Barratt-Pugh, C. (2020). The digital entanglement of humanities, literacy, and storytelling. In K. W. Kung (Ed.), *Reconceptualizing the digital humanities in Asia* (pp. 141–154). Springer. doi:10.1007/978-981-15-4642-6_9

Cillo, V., Rialti, R., Del Giudice, M., & Usai, A. (2021). Niche tourism destinations' online reputation management and competitiveness in big data era: Evidence from three Italian cases. *Current Issues in Tourism, 24*(2), 177–191. doi:10.1080/13683500.2019.1608918

Clarke, P., & Eastgate, A. (2011). Cultural capital, life course perspectives and Western Front battlefield tours. *Journal of Tourism and Cultural Change, 9*(1), 31–44. doi:10.1080/14766825.2010.527346

Clark, R., Hartline, M., & Jones, K. (2009). The Effects of Leadership Style on Hotel Employees' Commitment to Service Quality. *Cornell Hospitality Quarterly, 50*(2), 209–231. doi:10.1177/1938965508315371

Cobb, H. H. (1995). *American battlefields: A complete guide to the historic conflicts in words, maps, and photos*. Macmillan.

Coghlan, A., & Weiler, B. (2018). Examining transformative processes in volunteer tourism. *Current Issues in Tourism, 21*(5), 567–582. doi:10.1080/13683500.2015.1102209

Cohen, E. (2018). Thanatourism: A comparative approach. In The Palgrave Handbook of Dark Tourism Studies. Palgrave Macmillan. doi:10.1057/978-1-137-47566-4_6

Cohen, E. (2018). Thanatourism: A comparative approach. The Palgrave handbook of dark tourism studies, 157-171.

Cohen, E. H. (2011). Educational dark tourism at an in populo site: The Holocaust Museum in Jerusalem. *Annals of Tourism Research, 38*(1), 193–209. doi:10.1016/j.annals.2010.08.003

Cohen, E., & Cohen, S. A. (2015). Beyond Eurocentrism in Tourism: A Paradigm Shift to Mobilities. *Tourism Recreation Research, 40*(2), 157–168. doi:10.1080/02508281.2015.1039331

Cole, S., & Eriksson, J. (2010). Tourism and human rights. In Tourism and inequality: Problems and prospects (pp. 107–125). doi:10.1079/9781845936624.0107

Collins-Kreiner, N. (2016). Dark tourism as/is pilgrimage. *Current Issues in Tourism, 19*(12), 1185–1189. doi:10.1080/13683500.2015.1078299

Comisión Mundial para el Medio Ambiente y el Desarrollo (CMMAD). (1987). *Reporte de la Comisión Mundial del Medio Ambiente y Desarrollo: Nuestro Futuro Común*. Organización de las Naciones Unidas. Disponible en: http://www.un-documents.net/our-common-future.pdf

Conferencia Mundial de Turismo Sostenible (WCST). (1995). *Carta del turismo sostenible*. Programa Hombre y Biosfera. Centro Mundial del Patrimonio. Programa de las Naciones Unidas para el Medio Ambiente. Organización Mundial del Turismo. Consejo Científico Internacional para el Desarrollo de las Islas. Comisión de las Comunidades Europeas. Decenio Mundial para el Desarrollo Cultural. Gobierno de Canarias. Cabildo de Lanzarote. Secretaría General de Turismo. Ministerio de Obras Públicas, Transportes y Medio Ambiente. Universidad de la Laguna.

Connell, J. (2006). Medical tourism: The newest of niches. *Tourism Recreation Research, 31*(1), 99–102. doi:10.1080/02508281.2006.11081252

Conti, E., & Cassel, H. S. (2020). Liminality in nature-based tourism experiences as mediated through social media. *Tourism Geographies*, *22*(2), 413–432. doi:10.1080/14616688.2019.1648544

Cooke, J., & Dickson, H. (2006). *Lonely Planet Bluelist: The Best in Travel 2007*. Lonely Planet Publications.

Corsane, G. (2005). *Heritage, Museums and Galleries: An Introductory Reader*. Routledge.

Costa, C. A., & Chalip, L. (2005). Adventure sport tourism in rural revitalisation: An ethnographic evaluation. *European Sport Management Quarterly*, *5*(3), 257–279. doi:10.1080/16184740500190595

Costa, J., Gomes, J., & Montenegro, M. (2014). Did the context of economic crisis affect the image of Portugal as a tourist destination: Strategic question overview. *Worldwide Hospitality and Tourism Themes*, *6*(5), 392–396. doi:10.1108/WHATT-09-2014-0025

Costa, J., Montenegro, M., & Gomes, J. (2018). What challenges and opportunities will lead to success? *Worldwide Hospitality and Tourism Themes*, *10*(6), 631–634. doi:10.1108/WHATT-08-2018-0053

Costa, S. M., Moro, S., Rita, P., & Alturas, B. (2023). Customer experience through online reviews from TripAdvisor: The case of Orlando theme parks. *International Journal of Technology Marketing*, *17*(1), 48–77. doi:10.1504/IJTMKT.2023.127352

Coyl-Shepherd, D. D., & Hanlon, C. (2013). Family play and leisure activities: Correlates of parents' and children's socio-emotional well-being. *International Journal of Play*, *2*(3), 254–272. doi:10.1080/21594937.2013.855376

Cream City Cannibal: Jeffery Dahmer Walking Tour. (2023). Tripadvisor. https://www.tripadvisor.com/AttractionProductReview-g60097-d23983034-Cream_City_Cannibal_Jeffery_Dahmer_Walking_Tour-Milwaukee_Wisconsin.html

Crespell, P., & Hansen, E. (2009). Antecedents to innovativeness in the forest product industry. *Journal of Forest Products Business Research*, *6*(1), 1–20.

Crespo Jareño, J. A. (2020). *Orígenes e impactos del ecoturismo en la Revista Kalpana, Edición Especial, Número 18*. Disponible en https://dialnet.unirioja.es/descarga/articulo/7834208.pdf

Creswell, J. W., & Creswell, J. D. (2017). *Research design: Qualitative, quantitative, and mixed methods approaches*. Sage publications.

Crime Museum Vienna - Vienna. (2023). Tripadvisor. https://www.tripadvisor.com/Attraction_Review-g190454-d591127-Reviews-Crime_Museum_Vienna-Vienna.html

Crosby, A., Prato, N., Solsona, J., & Gómez, O. A. (2009). *Re-inventando el turismo rural. Gestión y desarrollo*. Laertes.

Csikszentmihalyi, M., & Seligman, M. (2000). Positive psychology. *The American Psychologist*, *55*(1), 5–14. doi:10.1037/0003-066X.55.1.5 PMID:11392865

Cuccia, T. (2020). Intangible cultural heritage. In *Handbook of Cultural Economics* (3rd ed., pp. 294–303). Edward Elgar Publishing.

Cumming, G. (2008). Inference by eye: Reading the overlap of independent confidence intervals. *Statistics in Medicine*, 205–220. PMID:18991332

Cunha, C. R., Mendonça, V., Morais, E. P., & Carvalho, A. (2018). The role of gamification in material and immaterial cultural heritage. In *Proceedings of the 31st International Business Information Management Association Conference (IBIMA)* (pp. 6121-6129). International Business Information Management Association (IBIMA).

Daft, R. L. (2008). *The New Era of Management* (2nd ed.). Thomson South-Western Corporation.

Dahka, Z. Y., Hajiheydari, N., & Rouhani, S. (2020). User response to e-WOM in social networks: How to predict a content influence in Twitter. *International Journal of Internet Marketing and Advertising*, *14*(1), 91–111. doi:10.1504/IJIMA.2020.106041

Dai, C., & Chiu, D. K. W. (2023). Impact of COVID-19 on reading behaviors and preferences: Investigating high school students and parents with the 5E instructional model. *Library Hi Tech*. Advance online publication. doi:10.1108/LHT-10-2022-0472

Dai, F., Wang, D., & Kirillova, K. (2022). Travel inspiration in tourist decision making. *Tourism Management*, *90*, 104484. doi:10.1016/j.tourman.2021.104484

Dale, C., & Robinson, N. (2011). Dark Tourism. In P. Robinson, S. Heitmann, & P. Dieke (Eds.), *Research Themes for Tourism* (pp. 205–217). CABI. doi:10.1079/9781845936846.0205

Dancausa Millán, M. G., Perez Naranjo, L. M., Hernandez Rojas, R. D., & Millan Vazquez de la Torre, M. G. (2019). Cemetery tourism in southern Spain: An analysis of demand. *Tourism and Hospitality Management*, *25*(1), 37–52. doi:10.20867/thm.25.1.1

Dann, G. (1998). The Dark Side of Tourism. Centre International de Recherches et d'Etudes Touristiques.

Dann, G., & Seaton, A. (Eds.). (2001). Slavery, Contested Heritage and Thanatourism. Haworth Hospitality Press.

Dann, G. (1994). Tourism: The Nostalgia Industry of the Future. In W. Theobald (Ed.), *Global Tourism: The Next Decade* (pp. 55–67). Butterworth Heinemann.

Darban, G., Karatepe, O. M., & Rezapouraghdam, H. (2022). Does work engagement mediate the impact of green human resource management on absenteeism and green recovery performance? *Employee Relations*, *44*(5), 1092–1108. doi:10.1108/ER-05-2021-0215

Darcy, S., & Dickson, T. J. (2009). A whole-of-life approach to tourism: The case for accessible tourism experiences. *Journal of Hospitality and Tourism Management, 16*(1), 32–44. doi:10.1375/jhtm.16.1.32

Dark Philly Adult Night Tour - Philadelphia Pennsylvania . (2023). Tripadvisor. https://www.tripadvisor.com.sg/AttractionProductReview-g60795-d11854370-or10-Dark_Philly_Adult_Night_Tour-Philadelphia_Pennsylvania.html

Dashper, K. (2014). *Rural Tourism: An International Perspective*. Cambridge Scholars Publishing.

Davis, P. (2011). Total war. *The Metropolitan Review*, (Spring), 18–25.

DC Insider Tours – Washington DC District of Columbia . (2023). Tripadvisor. https://www.tripadvisor.com/ShowUserReviews-g28970-d3410803-r292929652-DC_Insider_Tours-Washington_DC_District_of_Columbia.html

De Esteban, Curiel, Javier, Antonovica, & Arta. (2010). El ecoturismo como modelo internacional de desarrollo sostenible del turismo cultural. *Revista Teoría y praxis, 6*(8), 43 – 53. Disponible en: http://risisbi.uqroo.mx/handle/20.500.12249/622

De Jong, J. P. J., & Den Hartog, D. (2010). Measuring innovative work behaviour. *Creativity and Innovation Management, 19*, 23–36. DOI:.00547.x doi:10.1111/j.1467-8691.2010

Deng, Q., Allard, B., Lo, P., Chiu, D. K., See-To, E. W., & Bao, A. Z. (2019). The role of the library café as a learning space: A comparative analysis of three universities. *Journal of Librarianship and Information Science, 51*(3), 823–842. doi:10.1177/0961000617742469

Deng, S., & Chiu, D. K. W. (2023). Analyzing Hong Kong Philharmonic Orchestra's Facebook Community Engagement with the Honeycomb Model. In M. Dennis & J. Halbert (Eds.), *Community Engagement in the Online Space* (pp. 31–47). IGI. Global. doi:10.4018/978-1-6684-5190-8.ch003

Deng, W., Chin, G. Y.-l., Chiu, D. K. W., & Ho, K. K. W. (2022). Contribution of Literature Thematic Exhibition to Cultural Education: A Case Study of Jin Yong's Gallery. *Micronesian Educators, 32*, 14–26.

Derrien, M. M., & Stokowski, P. A. (2020). Discursive constructions of night sky experiences: Imagination and imaginaries in national park visitor narratives. *Annals of Tourism Research, 85*, 103038. doi:10.1016/j.annals.2020.103038

Desjardin, T. A. (2003). *These honored dead: How the story of Gettysburg shaped American memory*. Da Capo Press.

Devaraja, T. S., & Deepak, K. (2014). Role of Tour Operator in Sustainable Supply Chain Management of Tourism - A Case Study on Bharat International Travels (Bit) in Mysore City. *Global Journal for Research Analysis, 3*(8). Advance online publication. doi:10.15373/22778160/August2014/8

Devesa, M., Laguna, M., & Palacios, A. (2010). The role of motivation in visitor satisfaction: Empirical evidence in rural tourism. *Tourism Management*, *31*(4), 547–552. doi:10.1016/j.tourman.2009.06.006

Dhoke, Perke, & Karanjalkar. (2020). Economic analysis in marketing of turmeric in Sangli District. *Journal of Pharmacognosy and Phytochemistry*, *9*(5), 618–620. www.phytojournal.com

Dias, Á., Patuleia, M., Silva, R., Estêvão, J., & González-Rodríguez, M. R. (2022). Post-pandemic recovery strategies: Revitalizing lifestyle entrepreneurship. *Journal of Policy Research in Tourism, Leisure & Events*, *14*(2), 97–114. doi:10.1080/19407963.2021.1892124

Dicks, B. (2004). *Culture on display: The production of contemporary visitability*. McGraw-Hill Education.

Dillette, A. K., Douglas, A. C., & Andrzejewski, C. (2021). Dimensions of holistic wellness as a result of international wellness tourism experiences. *Current Issues in Tourism*, *24*(6), 794–810. doi:10.1080/13683500.2020.1746247

Director of Economics and statistics GoI. (2022). *Economic Survey of Maharashtra 2021-22*. Author.

Directorate of Arecanut and Spices Development (DASD). (2023). *About DASD*. DASD. https://www.dasd.gov.in/index.php/content/index/home

Doğaner, S. (2006). Savaş ve turizm: Troya ve Gelibolu savaş alanları. *Türk Coğrafya Dergisi*, (46), 1–21.

Drăgulănescu, I.-V., & Druţu, M. (2012). Rural Tourism for Local Economic Development. International. *Journal of Academic Research in Accounting, Finance and Management Sciences*, *2*(1), 196-203.

Driessen, S., Grever, M., & Reijnders, S. (2022). Lessons of war. The significance of battlefield tours for the Dutch military. *Critical Military Studies*, *8*(2), 214–232. doi:10.1080/23337486.2019.1651044

Drvenkar, N., Banožić, M., & Živić, D. (2015). Development of memorial tourism as a new concept-possibilities and restrictions. *Tourism and Hospitality Management*, *21*(1), 63–77. doi:10.20867/thm.21.1.5

DTLA Murder Mystery Ghost Tour – Los Angeles California . (2023). Tripadvisor. https://www.tripadvisor.com/AttractionProductReview-g32655-d16753250-DTLA_Murder_Mystery_Ghost_Tour-Los_Angeles_California.html

Duarte, P. (2010). Evolution of rural tourism in Portugal: A 25 years analysis. *e-Review of Tourism Research (eRTR)*, *8*(3), 41-56.

Duarte, M., Dias, Á., Sousa, B., & Pereira, L. (2023). Lifestyle Entrepreneurship as a Vehicle for Leisure and Sustainable Tourism. *International Journal of Environmental Research and Public Health, 20*(4), 3241. doi:10.3390/ijerph20043241 PMID:36833935

Dubois, A., & Gadde, L.-E. (2002). Systematic combining: An abductive approach to case research. *Journal of Business Research, 55*(7), 553–560. doi:10.1016/S0148-2963(00)00195-8

Du, J., & Sun, X. (2015, November). Research on Holographic and Digital Protection of Cultural Heritage Information. In *2015 International Conference on Social Science, Education Management and Sports Education* (pp. 1148-1151). Atlantis Press. 10.2991semse-15.2015.295

Dunkle, R. (2008). *Gladiators Violence and Spectacle in Ancient Rome*. Pearson Education Limited.

Dunkley, R., Morgan, N., & Westwood, S. (2011). Visiting the trenches: Exploring meanings and motivations in battlefield tourism. *Tourism Management, 32*(4), 860–868. doi:10.1016/j.tourman.2010.07.011

Durán-Vázquez, R., Lorenzo-Valdés, A., & Moreno-Quezada, G. (2012). Innovation and CSR Impact on Financial Performance of Selected Companies in Mexico. *Journal of Entrepreneurship. Management and Innovation, 8*(3), 5–20.

Eastern State Penitentiary - Philadelphia Pennsylvania. (2023). Tripadvisor. https://www.tripadvisor.com/Attraction_Review-g60795-d102763-Reviews-Eastern_State_Penitentiary-Philadelphia_Pennsylvania.html

Eastern State Penitentiary. (2023). Eastern State Penitentiary. https://www.easternstate.org/

Ecoturixtlan. (2023). Recuperado de www.oaxaca-mio.com/ecoturixtlan.htm

Edwards, J. L., & Winkler, C. K. (1997). Representative form and the visual ideograph: The Iwo Jima image in editorial cartoons. *The Quarterly Journal of Speech, 83*(3), 289–310. doi:10.1080/00335639709384187

Eisenbeiss, S. A., van Knippenberg, D., & Boerner, S. (2008). Transformational leadership and team innovation: Integrating team climate principles. *The Journal of Applied Psychology, 93*(6), 1438–1446. doi:10.1037/a0012716 PMID:19025260

Elkins, T., & Keller, R. T. (2003). Leadership in research and development organizations: A literature review and conceptual framework. *The Leadership Quarterly, 14*(4), 587–606. doi:10.1016/S1048-9843(03)00053-5

Ellemers, N., De Gilder, D., & Haslam, S. A. (2004). Motivating individuals and groups at work: A social identity perspective on leadership and group performance. *Academy of Management Review, 29*(3), 459–478. doi:10.2307/20159054

Elliot, S., & Lever, M. W. (2022). You Want to go where? Shifts in social media behaviour during the COVID-19 pandemic. *Annals of Leisure Research*, 1–15. doi:10.1080/11745398.2022.2041448

Enz, C., & Siguaw, J. (2003). Innovations in hotel practice. *The Cornell Hotel and Restaurant Administration Quarterly*, *44*(5–6), 115–123. doi:10.1177/001088040304400516

Enz, C., Verma, R., Walsh, K., Kimes, S. E., & Siguaw, J. (2010). Cases in innovative practices in hospitality and related services. *Cornell Hospitality Report*, *10*(10), 4–26.

Eppich, R., & Grinda, J. L. G. (2019). Sustainable financial management of tangible cultural heritage sites. *Journal of Cultural Heritage Management and Sustainable Development*, *9*(3), 282–299. doi:10.1108/JCHMSD-11-2018-0081

Erkutlu, H. (2008). The impact of transformational leadership on organizational and leadership effectiveness: The Turkish case. *Journal of Management Development*, *27*(7), 708–726. Advance online publication. doi:10.1108/02621710810883616

Esfehani, M. H., & Albrecht, J. N. (2019). Planning for intangible cultural heritage in tourism: Challenges and implications. *Journal of Hospitality & Tourism Research (Washington, D.C.)*, *43*(7), 980–1001. doi:10.1177/1096348019840789

Evans, G. (2002). Living in a World Heritage City: Stakeholders in the dialectic of the universal and particular. *International Journal of Heritage Studies*, *8*(2), 117–135. doi:10.1080/13527250220143913

Ezeuduji, I. O. (2015). Strategic event-based rural tourism development for sub-Saharan Africa. *Current Issues in Tourism*, *18*(3), 212–228. doi:10.1080/13683500.2013.787049

Fabio, M., & Elena, D. (2014). *Handbook of Research on Retailer-Consumer Relationship Development*. Business Science Reference.

Fairley, S., Gibson, H., & Lamont, M. (2018). Temporal manifestations of nostalgia: Le Tour de France. *Annals of Tourism Research*, *70*, 120–130. doi:10.1016/j.annals.2017.09.004

Faraldo, J. M., & Rodríguez-López, C. (2014). *Introducción a la historia del turismo*. Alianza Editorial.

Farmaki, A. (2013). Dark tourism revisited: A supply/demand conceptualisation. *International Journal of Culture, Tourism and Hospitality Research*, *7*(3), 281–292. doi:10.1108/IJCTHR-05-2012-0030

Farmaki, A., & Antoniou, K. (2017). Politicising dark tourism sites: Evidence from Cyprus. *Worldwide Hospitality and Tourism Themes*, *9*(2), 175–186. doi:10.1108/WHATT-08-2016-0041

Fathi, R. (2021). *Centenary (Battlefield Tourism). In 1914-1918-online: International Encyclopedia of the First World War*. Freie Universitat Berlin.

Febrian, F. (2022). Targeting Niche Tourism Using Niche Media: Expanding the Media in a Digital Era towards the Hashtag Generation. *Multidisciplinary Digital Publishing Institute Proceedings*, *83*(1), 39. doi:10.3390/proceedings2022083039

Fernández García, R. (2011). *La dimensión económica. Desarrollo sostenible*. Editorial Club Universitario.

Feuchtwang, S. (2012). *Making place: State projects, globalisation and local responses in China*. Taylor and Francis. doi:10.4324/9781843147671

Fewster, K., Başarin, V., & Başarin, H. H. (2003). Gallipoli: The Turkish story. Allen & Unwin.

Figgins, S. G., Smith, M. J., Sellars, C. N., Greenlees, I. A., & Knight, C. J. (2016). "You really could be something quite special": A qualitative exploration of athletes' experiences of being inspired in sport. *Psychology of Sport and Exercise*, *24*, 82–91. doi:10.1016/j.psychsport.2016.01.011

Figueiredo, E., Kastenholz, E., & Pinho, C. (2014). Living in a rural tourism destination - exploring the views of local communities. *Revista Portuguesa de Estudos Regionais*, *36*, 3–12. doi:10.59072/rper.vi36.417

Firmino, A. (2010). New challenges for the organic farmers in India-Tourism, spices and herbs. *Journal for Geography*. www.fao.org/nr/giahs/en/

Fleischer, A., & Felsenstein, D. (2000). Support for rural tourism: Does it make a difference? *Annals of Tourism Research*, *27*(4), 1007–1024. doi:10.1016/S0160-7383(99)00126-7

Flood, A. (2020, February 10). *Whitechapel Mural Will Celebrate the Lives of Jack the Ripper's Victims*. The Guardian. https://www.theguardian.com/books/2020/feb/10/whitechapel-mural-jack-the-ripper-victims-the-five-hallie-rubenhold

Foard, G., & Partida, T. (2005). *Scotland's historic fields of conflict: An assessment for historic Scotland* by *the battlefields trust*. Retrieved: https://www.battlefieldstrust.com/media/660.pdf

Foley, M., & Lennon, J. (2021). Dark Tourism: Understanding visitor motivations and site interpretation. In Tourism Management: Theory and Practice. CABI Publishing.

Foley, M., & Lennon, J. J. (1997). Dark tourism: an ethical dilemma. In Hospitality, tourism and leisure management: Issues in strategy and culture (pp. 153-164). Cassell.

Foley, M., & Lennon, J. (1996). Editorial: Heart of darkness. *International Journal of Heritage Studies*, *2*(4), 195–197. doi:10.1080/13527259608722174

Foley, M., & Lennon, J. J. (1996). JFK and dark tourism: A fascination with assassination. *International Journal of Heritage Studies*, *2*(4), 198–211. doi:10.1080/13527259608722175

Fonia, R. (n.d.). *India's link with silk trade roads*. UNESCO. Retrieved March 13, 2023, from https://en.unesco.org/silkroad/countries-alongside-silk-road-routes/india

Fonseca, A. P., Seabra, C., & Silva, C. (2016). Dark Tourism: Concepts, Typologies And Sites. *Journal of Research Hospital*, *2*, 1–6.

Foote, K. E. (2003). *Shadowed ground: America's landscapes of violence and tragedy*. University of Texas Press. doi:10.7560/705258

Ford, C. M. (2002). The futurity of decisions as a facilitator of organizational creativity and change. *Journal of Organizational Change Management*, *15*(6), 635–646. doi:10.1108/09534810210449541

Foreman, M. (2022). *From the battlefield to the big screen: Audie murphy, laurence olivier, vivien leigh and dirk bogarde in WW2*. Frontline Books.

Fornell, C., & Larcker, D. F. (1981). Evaluating structural equation models with unobservable variables and measurement error. *JMR, Journal of Marketing Research*, *18*(1), 39–50. doi:10.1177/002224378101800104

Foulk, D. (2016). The impact of the "economy of history": The example of battlefield tourism in France. *Mondes du Tourisme*, (12), 41–57. doi:10.4000/tourisme.1338

Fraj, E., Matute, J., & Melero, I. (2015). Environmental strategies and organizational competitiveness in the hotel industry: The role of learning and innovation as determinants of environmental success. *Tourism Management*, *46*, 30–42. doi:10.1016/j.tourman.2014.05.009

François, P. (2013). 'The best way to see Waterloo is with your eyes shut': British'histourism', authenticity and commercialisation in the mid-nineteenth century. *Anthropological Journal on European Cultures*, *22*(1), 26–41. doi:10.3167/ajec.2013.220103

Fredrickson, L. M., & Anderson, D. H. (1999). A qualitative exploration of the wilderness experience as a source of spiritual inspiration. *Journal of Environmental Psychology*, *19*(1), 21–39. doi:10.1006/jevp.1998.0110

Freire-Medeiros, B. (2014). *Touring poverty*. Routledge. doi:10.4324/9780203840719

Fremantle Prison Torchlight Tour . (2023). Ghost Tours. https://www.ghosttourbookings.com.au/ghost-tours/fremantle-prison-torchlight-tour/

Frey, B. S., & Frey, B. S. (2021). *Overcoming Overtourism*. Springer International Publishing. doi:10.1007/978-3-030-63814-6

Fridrich, M., Stone, P. R., & Rukesha, P. (2018). Dark Tourism, Difficult Heritage, and Memorialization: A Case of the Rwandan Genocide. In The Palgrave Handbook of Dark Tourism Studies. Palgrave Macmillan.

Fritz Haarmann Murder Tour Through Hanover's Old Town . (2023). Viator. https://www.viator.com/tours/Hannover/Fritz-Haarmann-murder-tour-through-Hanovers-old-town/d28614-279650P1

Frochot, I. (2005). A benefit segmentation of tourists in rural areas: A Scottish perspective. *Tourism Management*, *26*(3), 335–346. doi:10.1016/j.tourman.2003.11.016

Frost, W. (2007). Refighting the Eureka Stockade: Managing a dissonant battlefield. In C. Ryan (Ed.), *Battlefield tourism: History, place and interpretation* (pp. 187–194). Elsevier. doi:10.1016/B978-0-08-045362-0.50022-1

Gao, X., & Chiu, D. K. W. (2023). Integration of cultural retailing with new technologies and media: A case study of the Suzhou Antique Bookstore. In J. D. Santos & I. V. Pereira (Eds.), *Management and Marketing for Improved Retail Competitiveness and Performance*. IGI Global.

Gargiulo, R., & Carignani, F. (2022). Novel Approaches in Profiling in Museums. In *Handbook of Research on Museum Management in the Digital Era* (pp. 228–247). IGI Global. doi:10.4018/978-1-7998-9656-2.ch012

Garg, V. K., Walters, B. A., & Priem, R. L. (2003). Chief executive scanning emphases, environmental dynamism, and manufacturing firm performance. *Strategic Management Journal*, *24*(8), 725–744. doi:10.1002mj.335

Gatewood, J. B., & Cameron, C. M. (2004). Battlefield pilgrims at Gettysburg National Military Park. *Ethnology*, *43*(3), 193–216. doi:10.2307/3774062

Ghorbanzadeh, D., Shabbir, M. S., Mahmood, A., & Kazemi, E. (2021). Investigating the role of experience quality in predicting destination image, perceived value, satisfaction, and behavioural intentions: A case of war tourism. *Current Issues in Tourism*, *24*(21), 3090–3106. doi:10.1080/13683500.2020.1863924

Gibson, D. C. (2008). The Relationship Between Serial Murder and the American Tourism Industry. *Journal of Travel & Tourism Marketing*, *20*(1), 45–60. doi:10.1300/J073v20n01_04

Gibson, D., Yai, E., & Pratt, S. (2022). Journeying into the past to discover the potential for WWII dark tourism in the Solomon Islands. *Current Issues in Tourism*, *25*(14), 2285–2302. doi:10.1080/13683500.2021.1957787

Gieling, J., & Ong, C. E. (2016). Warfare tourism experiences and national identity: The case of Airborne Museum 'Hartenstein' in Oosterbeek, the Netherlands. *Tourism Management*, *57*, 45–55. doi:10.1016/j.tourman.2016.05.017

Gill, A. S., Flaschner, A. B., & Shachar, M. (2006). Mitigating stress and burnout by implementing transformational leadership. *International Journal of Contemporary Hospitality Management*, *18*(6/7), 469–481. doi:10.1108/09596110610681511

Gill, F., Rico, R., Alcover, C. M., & Barrasa, A. (2005). Change-oriented leadership satisfaction and performance in work groups: Effects of team climate and group potency. *Journal of Managerial Psychology*, *20*(3/4), 312–328. doi:10.1108/02683940510589073

Gilson, J. F. (2018). Inspiring change in heritage interpretation. In S. Pulla & B. Schissel (Eds.), *Applied interdisciplinarity in scholar practitioner programs* (pp. 69–86). Palgrave Macmillan. doi:10.1007/978-3-319-64453-0_5

Ginzarly, M., & Teller, J. (2021). Online communities and their contribution to local heritage knowledge. *Journal of Cultural Heritage Management and Sustainable Development*, *11*(4), 361–380. doi:10.1108/JCHMSD-02-2020-0023

Gisbert-López, M. C., Verdú-Jover, A. J., & Gómez-Gras, J. M. (2014). The moderating effect of relationship conflict on the creative climate – innovation association: The case of traditional sectors in Spain. *International Journal of Human Resource Management, 25*(1), 47–67. doi:10.1080/09585192.2013.781525

Goldsmith, R. E. (1999). The personalised marketplace: Beyond the 4Ps. *Marketing Intelligence & Planning, 17*(4), 178–185. doi:10.1108/02634509910275917

Gong, J. Y., Schumann, F., Chiu, D. K. W., & Ho, K. K. W. (2017). Tourists' mobile information seeking behavior: An investigation on China's youth. *International Journal of Systems and Service-Oriented Engineering, 7*(1), 58–76. doi:10.4018/IJSSOE.2017010104

González, M. V. (2008). Intangible heritage tourism and identity. *Tourism Management, 29*(4), 807–810. doi:10.1016/j.tourman.2007.07.003

Gorokhova, T. (2021). The influence of the development of digital technologies on consumer behaviour. *Ukrainian Black Sea Region Agrarian Science, 112*(4), 45–54. doi:10.31521/2313-092X/2021-4(112)-5

Gössling, S., & Lane, B. (2015). Rural tourism and the development of Internet-based accommodation booking platforms: A study in the advantages, dangers and implications of innovation. *Journal of Sustainable Tourism, 23*(8-9), 1386–1403. doi:10.1080/09669582.2014.909448

Govindarajan, V., & Trimble, C. (2005). *Ten Rules for Strategic Innovators; from idea to execution.* Harvard Business School Press.

Gravari-Barbas, M., & Guinand, S. (Eds.). (2017). *Tourism and gentrification in contemporary metropolises: International perspectives.* Taylor & Francis. doi:10.4324/9781315629759

Gretzel, U., & Yoo, K. H. (2008). Use and impact of online travel reviews. In P. O'Connor, W. Hopken, & U. Gretzel (Eds.), *Information and communication technologies in tourism 2008, 35-46.* Springer. doi:10.1007/978-3-211-77280-5_4

Guerrero González, P., & Ramos Mendoza, J. R. (2014). *Introducción al turismo.* Grupo Editorial Patria.

Gummesson, E. (2017). Case theory in business and management: Reinventing case study research. *Sage (Atlanta, Ga.).* Advance online publication. doi:10.4135/9781473920811

Gunday, G., Ulusoy, G., Kilic, K., & Alpkan, L. (2011). Effects of innovation types on firm performance. *International Journal of Production Economics, 133*(2), 662–676. doi:10.1016/j.ijpe.2011.05.014

Guo, Y., Lam, A. H. C., Chiu, D. K. W., & Ho, K. K. W. (2022). Perceived quality of WhatsApp reference service: A quantitative study from user perspectives. *Information Technology and Libraries, 41*(3). Advance online publication. doi:10.6017/ital.v41i3.14325

Gupta, V., & Singh, S. (2013). How leaders impact employee creativity: A study of Indian R&D laboratories. *Management Research Review, 36*(1), 66–88. doi:10.1108/01409171311284594

Hakim, L., & Yanuwiadi, B. (2016). Identifying and Assessing Spice and Herbs Resources for Tourism Attractions: A Case Study from Banyuwangi, East Java. *Research Journal of Life Science, 3*(3), 154–159. http://rjls.ub.ac.id

Hall, D. (1994). Civil War reenactors and the postmodern sense of history. *The Journal of American Culture, 17*(3), 7–11. doi:10.1111/j.1542-734X.1994.00007.x

Hall, D. (2002). Brand development, tourism and national identity: The re-imaging of former Yugoslavia. *Journal of Brand Management, 9*(4), 323–334. doi:10.1057/palgrave.bm.2540081

Hall, J., Johnson, S., Wysocki, A., & Kepner, K. (2008). *Transformational Leadership: The Transformational of Managers and Associates*. University of Florida.

Hamel, G., & Labarre, P. (2011). Improving our capacity to manage. *The Wall Street Journal*.

Hammel, E. M. (2006). *Iwo Jima*. Zenith Imprint.

Handayani, B. (2018). Going to the Dark Sites With Intention: Construction of Niche Tourism. In M. Korstanje & B. George (Eds.), Virtual Traumascapes and Exploring the Roots of Dark Tourism (pp. 50-66). IGI Global. doi:10.4018/978-1-5225-2750-3.ch003

Haq, S., & Chandio, J. (2017). Transactional Leadership and its Impact on the Organizational Performance: A Critical Analysis. *International Journal of Trend in Scientific Research and Development*, 135-139.

Harrison, D. (2003). *The Sociology of Modernization and Development*. Routledge. doi:10.4324/9780203359587

Harrison, M. A., & Frederick, E. J. (2020, July 1). Interested in Serial Killers? Morbid Curiosity in College Students. *Current Psychology (New Brunswick, N.J.)*. Advance online publication. doi:10.100712144-020-00896-w

Harter, J. K., Schmidt, F. L., & Hayes, T. L. (2002). Business-unit-level relationship between employee satisfaction, employee engagement, and business outcomes: A meta-analysis. *The Journal of Applied Psychology, 87*(2), 268–279. doi:10.1037/0021-9010.87.2.268 PMID:12002955

Hartmann, R. (2014). Dark tourism, thanatourism and dissonance in heritage tourism management: New directions in contemporary tourism research. *Journal of Heritage Tourism, 9*(2), 166–182. doi:10.1080/1743873X.2013.807266

Hartmann, R., Lennon, J., Reynolds, D. P., Rice, A., Rosenbaum, A. T., & Stone, P. R. (2018). The history of dark tourism. *Journal of Tourism History, 10*(3), 269–295. doi:10.1080/1755182X.2018.1545394

Hart, T. (1998). Inspiration: Exploring the experience and its meaning. *Journal of Humanistic Psychology, 38*(3), 7–35. doi:10.1177/00221678980383002

Harwell, J. (2017). Rhyme or reason? Patterns in book pricing by format. *The Journal of Electronic Publishing : JEP*, *20*(1). Advance online publication. doi:10.3998/3336451.0020.104

Haunted Hollywood Walking Tour: True Crime, Creepy Tales – Los Angeles California . (2023). Tripadvisor. https://www.tripadvisor.com/AttractionProductReview-g32655-d12148364-Haunted_Hollywood_Walking_Tour_True_Crime_Creepy_Tales-Los_Angeles_California.html

Heidegger, M. (1996). *Being and time: A translation of Sein und Zeit.* SUNY press.

Heinzen, J. (2014). A negotiated truce: The battle of Waterloo in European memory since the Second World War. *History & Memory*, *26*(1), 39–74. doi:10.2979/histmemo.26.1.39

Hellier, P. K., Geursen, G. M., Carr, R. A., & Rickard, J. A. (2003). Customer repurchase intention. *European Journal of Marketing*, *37*(11/12), 1762–1800. doi:10.1108/03090560310495456

Helm, R., & Mark, A. (2007). Implications from cue utilisation theory and signalling theory for firm reputation and the marketing of new products. *International Journal of Product Development*, *4*(3/4), 396–411. doi:10.1504/IJPD.2007.012504

He, M., Liu, B., & Li, Y. (2021). Tourist inspiration: How the wellness tourism experience inspires tourist engagement. *Journal of Hospitality & Tourism Research (Washington, D.C.)*. Advance online publication. doi:10.1177/10963480211026376

Henderson, J. (2000). War as a Tourist Attraction: The Case of Vietnam. *International Journal of Tourism Research*, *2*(4), 269–280. doi:10.1002/1522-1970(200007/08)2:4<269::AID-JTR219>3.0.CO;2-A

Heneghan, M., Caslin, B., Ryan, M., & O'Donoghue, C. (2016). *Rural Tourism: Rural Economy & Development Programme*. Teagasc.

Hennig-Thurau, T., Gwinner, K. P., Walsh, G., & Gremler, D. D. (2004). Electronic Word-of-Mouth via Consumer-Opinion Platforms: What Motivates Consumers to Articulate Themselves on the Internet? *Journal of Interactive Marketing*, *18*(1), 38–52. doi:10.1002/dir.10073

Henriques, M. (2019). *How spices changed the ancient world.* BBC. https://www.bbc.com/future/bespoke/made-on-earth/the-flavours-that-shaped-the-world/

Herrmann, D., & Felfe, J. (2013). Moderators of the relationship between leadership style and employee creativity: The role of task novelty and personal initiative. *Creativity Research Journal*, *25*(2), 172–181. doi:10.1080/10400419.2013.783743

Hertzog, A. (2012). War Battlefields, tourism and imagination. *Tourism Review*, (1), 1–13. doi:10.4000/viatourism.1283

Hettiarachchi, I. C., De Silva, D. A. M., and Esham, M. (2021). International tourists' perceptions on spice markets: A guideline to introduce spice market concept into Sri Lankan context. *Journal of Agricultural Sciences - Sri Lanka*, *16*(3), 410–430. doi:10.4038/jas.v16i03.9467

Hinkin, T. R., & Schriesheim, C. A. (2008). A theoretical and empirical examination of the transactional and non-leadership dimensions of the Multifactor Leadership Questionnaire (MLQ). *The Leadership Quarterly*, *19*(5), 501–513. doi:10.1016/j.leaqua.2008.07.001

Hinsch, C., Felix, R., & Rauschnabel, P. A. (2020). Nostalgia beats the wow-effect: Inspiration, awe and meaningful associations in augmented reality marketing. *Journal of Retailing and Consumer Services*, *53*, 101987. doi:10.1016/j.jretconser.2019.101987

Hjalager, A.-M. (2002). Repairing innovation defectiveness in tourism. *Tourism Management*, *23*(5), 465–474. doi:10.1016/S0261-5177(02)00013-4

Hodgkinson, S., Prins, H., & Stuart-Bennett, J. (2017). Monsters, Madmen... and Myths: A Critical Review of the Serial Killing Literature. *Aggression and Violent Behavior*, *34*, 282–289. doi:10.1016/j.avb.2016.11.006

Home . (2023) Littledean Jail. https://www.littledeanjail.com/

Hongdao, Q., Bibi, S., Khan, A., Ardito, L., & Nurunnabi, M. (2019). Does what goes around really comes around? The mediating effect of CSR on the relationship between transformational leadership and employee's job performance in law firms. *Sustainability (Basel)*, *11*(12), 3366. doi:10.3390u11123366

Hooper, G., & Lennon, J. (2016). *Dark tourism.* Taylor & Francis. doi:10.4324/9781315575865

Hu, Q., Dinev, P., Hart, T., Cooke, D. (2012). Managing Employee Compliance with Information Security Policies: The Critical Role of Top Management and Organizational Culture. *Decisions Science*, *43*(4).

Huang, F., & Rice, J. (2009). The role of absorptive capacity in facilitating open innovation outcomes: A study of Australian SMEs in the manufacturing sector. *International Journal of Innovation Management*, *13*(02), 201–220. doi:10.1142/S1363919609002261

Huang, P.-S., Paulino, Y. C., So, S., Chiu, D. K. W., & Ho, K. K. W. (2021). Editorial - COVID-19 Pandemic and Health Informatics (Part 1). *Library Hi Tech*, *39*(3), 693–695. doi:10.1108/LHT-09-2021-324

Huang, P.-S., Paulino, Y. C., So, S., Chiu, D. K. W., & Ho, K. K. W. (2022). Guest editorial: COVID-19 Pandemic and Health Informatics Part 2. *Library Hi Tech*, *40*(2), 281–285. doi:10.1108/LHT-04-2022-447

Huber, D., Milne, S., & Hyde, K. F. (2018). Constraints and facilitators for senior tourism. *Tourism Management Perspectives*, *27*(March), 55–67. doi:10.1016/j.tmp.2018.04.003

Huertas-Valdivia, I., Llorens-Montes, F. J., & Ruiz-Moreno, A. (2018). Achieving engagement among hospitality employees: A serial mediation model. *International Journal of Contemporary Hospitality Management*, *30*(1), 217–241. doi:10.1108/IJCHM-09-2016-0538

Huete-Alcocer, N. (2017). A Literature review of word of mouth and electronic word of mouth: Implications for consumer behaviour. *Frontiers in Psychology*, *8*(July), 1256. Advance online publication. doi:10.3389/fpsyg.2017.01256 PMID:28790950

Humphreys, J. H., & Einstein, W. O. (2003). Nothing new under the sun: Transformational leadership from a historical perspective. *Management Decision*, *41*(1/2), 85–95. doi:10.1108/00251740310452934

Iacobucci, D. (2010). Structural equations modeling: Fit indices, sample size, and advanced topics. *Journal of Consumer Psychology*, *20*(1), 90–98. doi:10.1016/j.jcps.2009.09.003

Ilban, M. O., & Bezirgan, M. (2015). Effects of Destination Image and Total Perceived Value on Tourists' Behavioral Intentions: An Investigation of Domestic Festival Tourists. *Tourism Analysis*, *20*(November), 499–510. Advance online publication. doi:10.3727/108354215X14411980111370

Iliev, D. (2020). The evolution of religious tourism: Concept, segmentation and development of new identities. *Journal of Hospitality and Tourism Management*, *45*, 131–140. doi:10.1016/j.jhtm.2020.07.012

INEGI (Instituto Nacional de Estadística y Geografía). (2021). *Comunicado de prensa Numero 539/21 con fecha 24 de septiembre de 2021.* Estadísticas a propósito del día mundial del turismo. Disponible en: https://www.inegi.org.mx/contenidos/saladeprensa/aproposito/2021/EAP_Turismo21.pdf

Isaac, R. K. (2021). Editorial special issue in Dark Tourism. *Journal of Heritage Tourism*, *16*(4), 363–366. doi:10.1080/1743873X.2021.1920963

Isaac, R. K., Çakmak, E., & Butler, R. (Eds.). (2019). *Tourism and hospitality in conflict-ridden destinations*. Routledge. doi:10.4324/9780429463235

Išoraitė, M. (2016). Marketing mix theoretical aspects. *International Journal of Research - Granthaalayah, 4*(6), 25-37.

Ispas, A. (2012). The perceived leadership style and employee performance in the hotel industry: a dual approach. *A Review of International Comparative Management*, *13*(2), 294–304.

Ivanova, A., & Ibáñez, R. (2012). *Medio ambiente y política turística en México. Tomo I: Ecología, biodiversidad y desarrollo turístico*. D.F. Secretaría de Medio Ambiente y Recursos Naturales. Instituto Nacional de Ecología. Universidad Autónoma de Baja California Sur.

Ivanova, P., & Light, D. (2018). 'It's not that we like death or anything': Exploring the motivations and experiences of visitors to a lighter dark tourism attraction. *Journal of Heritage Tourism*, *13*(4), 356–369. doi:10.1080/1743873X.2017.1371181

Jack the Ripper Museum - London England. (2023). Tripadvisor. https://www.tripadvisor.com/Attraction_Review-g186338-d86386 94-Reviews-Jack_the_Ripper_Museum-London_England.html

Jack the Ripper Museum. (2023). Jack the Ripper Museum. https://www.jacktherippermuseum.com/

Jack the Ripper. (2023). Prater. https://www.praterwien.com/nc/en/attractions/details/a/jack-the-ripper/

JacobS. (2008). Understanding Culturally Sustainable Tourism An Observed Comparison of the Models followed by Kerala and Goa. *Conference on Tourism in India - Challenges Ahead*, 75–90. https://ssrn.com/abstract=1480374

Jacobs, J. M. (1962). *The death and life of great American cities*. Taylor & Francis.

Jacobs, T. P., & McConnell, A. R. (2022). Self-transcendent emotion dispositions: Greater connections with nature and more sustainable behavior. *Journal of Environmental Psychology*, *81*, 101797. doi:10.1016/j.jenvp.2022.101797

Jankowski, P. (2014). Verdun: The longest battle of the great war. Oxford University Press.

Janssen, O., & Van Yperen, N. W. (2004). Employees' goal orientations, the quality of leader member exchange, and the outcomes of job performance and job satisfaction. *Academy of Management Journal*, *47*(3), 368–384. doi:10.2307/20159587

Jászberényi, M. (2020). *A kulturális turizmus sokszínűsége*. Akadémiai Kiadó. doi:10.1556/9789634545224

Javier, S., & Bign, J. E. (2001). *Tourism image, evaluation variables and after purchase behaviour*. Academic Press.

Jenatabadi, H. (2015). *An Overview of Organizational Performance Index: Definitions and Measurements*. University of Malaya. doi:10.2139srn.2599439

Jeong, M., & Jeon, M. M. (2008). Customer reviews of hotel experiences through consumer generated media (CGM). *Journal of Hospitality & Leisure Marketing*, *17*(1-2), 121–138. doi:10.1080/10507050801978265

Jeong, Y., & Kim, S. (2020). A study of event quality, destination image, perceived value, tourist satisfaction, and destination loyalty among sport tourists. *Asia Pacific Journal of Marketing and Logistics*, *32*(4), 940–960. doi:10.1108/APJML-02-2019-0101

Jiang, J., Zhang, J., Zhang, H., & Yan, B. (2018). Natural soundscapes and tourist loyalty to nature-based tourism destinations: The mediating effect of tourist satisfaction. *Journal of Travel & Tourism Marketing*, *35*(2), 218–230. doi:10.1080/10548408.2017.1351415

Jiang, X., Chiu, D. K. W., & Chan, C. T. (2023). Application of the AIDA model in social media promotion and community engagement for small cultural organizations: A case study of the Choi Chang Sau Qin Society. In M. Dennis & J. Halbert (Eds.), *Community Engagement in the Online Space* (pp. 48–70). IGI Global. doi:10.4018/978-1-6684-5190-8.ch004

Jiang, Y., Ramkissoon, H., Mavondo, F. T., & Feng, S. (2017). Authenticity: The link between destination image and place attachment. *Journal of Hospitality Marketing & Management*, *26*(2), 105–124. doi:10.1080/19368623.2016.1185988

Jiménez-Zafra, S., Martín-Valdivia, M., Martínez-Cámara, E., & Ureña-López, L. (2015). Combining resources to improve unsupervised sentiment analysis at aspect level. *Journal of Information Science*, *42*(2), 1–19. doi:10.1177/0165551515593686

Jolliffe, L. (2014). Recognition of spices and cuisine as intangible heritage. In L. Jolliffe (Ed.), Spices and Tourism: Destinations, Attractions and Cuisines (Vol. 38, pp. 183–200). Channel View Publisher. doi:10.21832/9781845414443-013

Jones, S., Dodd, A., & Gruber, J. (2014). Development and validation of a new multidimensional measure of inspiration: Associations with risk for bipolar disorder. *PLoS One*, *9*(3), e91669. doi:10.1371/journal.pone.0091669 PMID:24670894

Jong, J. D., & Hartog, D. N. (2007). How leaders influence employees' innovative behaviour'. *European Journal of European Management*, *10*(1), 41–64.

Jora, O. D., Iacob, M., & Apăvăloaei, M. A. (2019). Public Cultural Heritage and Private Property Rights: Building Sustainable Community Through Individuality. In *Caring and Sharing: The Cultural Heritage Environment as an Agent for Change: 2016 ALECTOR Conference, Istanbul, Turkey* (pp. 213-224). Springer International Publishing.

Jordan, E. J., & Prayag, G. (2022). Residents' cognitive appraisals, emotions, and coping strategies at local dark tourism sites. *Journal of Travel Research*, *61*(4), 887–902. doi:10.1177/00472875211004761

Judd, C. (2015). Moderator Variable: Methodology. International Encyclopedia of the Social & Behavioral Sciences, 672-674.

Judge, T. A., & Piccolo, R. F. (2004). Transformational and transactional leadership: A meta-analytic test of their relative validity. *The Journal of Applied Psychology*, *89*(5), 755–768. doi:10.1037/0021-9010.89.5.755 PMID:15506858

Jung, D., Chow, C., & Wu, A. (2003). The role of transformational leadership in enhancing organizational innovation: Hypotheses and some preliminary findings. *The Leadership Quarterly*, *14*(4-5), 525–544. doi:10.1016/S1048-9843(03)00050-X

Jung, D., Wu, A., & Chow, C. W. (2008). Towards understanding the direct and indirect effects of CEOs' transformational leadership on firm innovation. *The Leadership Quarterly*, *119*(5), 582–594. doi:10.1016/j.leaqua.2008.07.007

Jung, K., Nguyen, V. T., Piscarac, D., & Yoo, S. C. (2020). Meet the virtual Jeju dol Haru bang—The mixed VR/Ar application for cultural immersion in Korea's main heritage. *ISPRS International Journal of Geo-Information*, *9*(6), 367. doi:10.3390/ijgi9060367

Junping, Q., & Huixia, Z. (2022). *Research on the protection and inheritance of intangible cultural heritage under the background of rural revitalisation*. Applied Mathematics and Nonlinear Sciences. doi:10.2478/amns.2021.2.00250

Kadam, A. S., Suryawanshi, R. R., Dhengale, S. S., & Shinde, H. R. (2019). Economic analysis of production of ginger in Sangli district of Maharashtra. *International Journal of Chemical Studies*, *7*(6), 1456–1460. http://www.chemijournal.com

Kahai, S. S., Sosik, J. J., & Avolio, B. J. (2003). Effects of leadership style, anonymity, and rewards on creativity-relevant processes and outcomes in an electronic meeting system context. *The Leadership Quarterly*, *14*(4-5), 499–524. doi:10.1016/S1048-9843(03)00049-3

Kalmuk, G., & Acar, A. (2015). The Mediating Role of Organizational Learning Capability on The Relationship Between Innovation and Firm's Performance: A Conceptual Framework. *Procedia: Social and Behavioral Sciences*, *210*, 164–169. doi:10.1016/j.sbspro.2015.11.355

Kamyabi, M., & Alipour, H. (2022). An Investigation of the Challenges Faced by the Disabled Population and the Implications for Accessible Tourism: Evidence from a Mediterranean Destination. *Sustainability (Basel)*, *14*(8), 4702. Advance online publication. doi:10.3390u14084702

Kani, Y., Aziz, Y. A., Sambasivan, M., & Bojei, J. (2017). Antecedents and outcomes of destination image of Malaysia. *Journal of Hospitality and Tourism Management*, *32*, 89–98. doi:10.1016/j.jhtm.2017.05.001

Kara, D., Uysal, M., Sirgy, M., & Leed, G. (2013). The effects of leadership style on employee well-being in hospitality. *International Journal of Hospitality Management*, *34*, 9–18. doi:10.1016/j.ijhm.2013.02.001

Karatepe, O. M., Rezapouraghdam, H., & Hassannia, R. (2020). Job insecurity, work engagement and their effects on hotel employees' non-green and nonattendance behaviors. *International Journal of Hospitality Management*, *87*, 102472. Advance online publication. doi:10.1016/j.ijhm.2020.102472

Karatepe, O. M., Rezapouraghdam, H., & Hassannia, R. (2021). Does employee engagement mediate the influence of psychological contract breach on pro-environmental behaviors and intent to remain with the organization in the hotel industry? *Journal of Hospitality Marketing & Management*, *30*(3), 326–353. doi:10.1080/19368623.2020.1812142

Kassarjian, H. H. (1977). Content analysis in consumer research. *The Journal of Consumer Research*, *4*(1), 8–18. doi:10.1086/208674

Kastenholz, E. (2004). 'Management of demand' as a tool in sustainable tourist destination development. *Journal of Sustainable Tourism*, *12*(5), 388–408. doi:10.1080/09669580408667246

Kastenholz, E., Carneiro, M. J., Marques, C. P., & Oliveira, E. (2020). Dark tourism and well-being: The role of emotional experiences and cultural learning. *Journal of Destination Marketing & Management*, *17*, 100423.

Kastenholz, E., Carneiro, M., Eusébio, C., & Figueiredo, E. (2013). Host-guest relationships in rural tourism: Evidence from two Portuguese villages. Anatolia. *Anatolia*, *24*(3), 367–380. doi:10.1080/13032917.2013.769016

Kastenholz, E., Carneiro, M., Marques, C., & Lima, J. (2012). Understanding and managing the rural tourism experience - The case of a historial village in Portugal. *Tourism Management Perspectives*, *4*(October), 207–214. doi:10.1016/j.tmp.2012.08.009

Kastenholz, E., Davis, D., & Paul, G. (1999). Segmenting tourism in rural areas: The case of north and central Portugal. *Journal of Travel Research*, *37*(4), 353–363. doi:10.1177/004728759903700405

Kattara, H., & El-Said, O. (2013). Innovation strategies: The implementation of creativity principles in Egyptian hotels. *Tourism and Hospitality Research*, *13*(3), 140–148. doi:10.1177/1467358414522053

Kavoura, A., & Bitsani, E. (2013). E-branding of rural tourism in Carinthia, Austria. *Tourism Review*, *61*(3), 289–312.

Keegan, J. (2011). *Six armies in Normandy: From D-day to the liberation of Paris June 6th-August 25th, 1944*. Random House.

Keenan, J. (2013). Death by bullet, fire or vapor: Examining the decision to use the atomic bomb to end World War II in the Pacific theatre. *The Exposition*, *1*(2), 1–7.

Kendle, A. (2008). *Dark Tourism: A fine line between curiosity and exploitation*. Vagabondish. Retrieved from: https://www.vagabondish.com/dark-tourism-travel-tours/

Khalili, A. (2016). Linking transformational leadership, creativity, innovation, and innovation-supportive climate. *Management Decision*, *54*(9), 2277–2293. doi:10.1108/MD-03-2016-0196

Khan, A., Bibi, S., Lyu, J., Garavelli, A. C., Pontrandolfo, P., & Perez Sanchez, M. A. (2020). Uncovering Innovativeness in Spanish Tourism Firms: The Role of Transformational Leadership, OCB, Firm Size, and Age. *Sustainability (Basel)*, *12*(10), 3989. doi:10.3390u12103989

Khandekar, R. G., Salvi, B. R., Sawant, V. S., Malshe, K. V., Rathod, R. R., & Parulekar, Y. R. (2016). Recent advances in propagation of black pepper and major tree spices in Maharashtra. *Indian Journal of Arecanut, Spices and Medicinal Plants*, *18*(3), 37–40.

Khan, M. J., Chelliah, S., Khan, F., & Amin, S. (2019). Perceived risks, travel constraints and visit intention of young women travelers: The moderating role of travel motivation. *Tourism Review*, *74*(3), 721–738. doi:10.1108/TR-08-2018-0116

Khan, M. T. (2014). The concept of 'marketing mix'and its elements. *International Journal of Information, Business and Management*, *6*(2), 95–107.

Khan, N. A., Shafi, S. M., & Ahangar, H. (2018). Digitization of cultural heritage: Global initiatives, opportunities and challenges. *Journal of Cases on Information Technology*, *20*(4), 1–16. doi:10.4018/JCIT.2018100101

Khoi, N. H., Le, A. N. H., & Tran, M. D. (2021). Tourist inspiration and its consequences: The moderating role of neuroticism. *International Journal of Tourism Research*, *23*(5), 901–913. doi:10.1002/jtr.2452

Khoi, N. H., Phong, N. D., & Le, A. N. H. (2020). Customer inspiration in a tourism context: An investigation of driving and moderating factors. *Current Issues in Tourism, 23*(21), 2699–2715. doi:10.1080/13683500.2019.1666092

Kim, S. H., Holland, S., & Han, H. S. (2013). A structural model for examining how destination image, perceived value, and service quality affect destination loyalty: A case study of Orlando. *International Journal of Tourism Research, 15*(4), 313–328. doi:10.1002/jtr.1877

Kim, S., Whitford, M., & Arcodia, C. (2019). Development of intangible cultural heritage as a sustainable tourism resource: The intangible cultural heritage practitioners' perspectives. *Journal of Heritage Tourism, 14*(5-6), 422–435. doi:10.1080/1743873X.2018.1561703

King, E. B., De Chermont, K., West, M. A., Dawson, J. F., & Hebl, M. R. (2007). How innovation can alleviate negative consequences of demanding work contexts: The influence of climate for innovation on organizational outcomes. *Journal of Occupational and Organizational Psychology, 80*(4), 631–645. doi:10.1348/096317906X171145

Kler, B. K., & Forsythe, C. P. (2022). Battlefield tourism: the legacy of Sandakan in Malaysian Borneo. In Handbook of niche tourism (pp. 236-249). Edward Elgar Publishing. doi:10.4337/9781839100185.00031

Kline, R. B. (2015). *Principles and practice of structural equation modeling*. Guilford publications.

Knight, A. (2019). Why Can't We Look Away from the Worst of Humanity? *Literary Hub.* https://lithub.com/on-dark-tourism-murder-hauntings-and-the-serial-killer-capital-of-australia/

Korstanje, M. E. (2020). The dark tourist: Consuming dark spaces in the periphery. In Tourism, Terrorism and Security. Emerald Publishing Limited.

Korstanje, M. E. (2016). *The rise of thana-capitalism and tourism.* Routledge. doi:10.4324/9781315457482

Korstanje, M. E. (2020). The dark tourist: Consuming dark spaces in the periphery. In M. Korstanje & H. Seraphin (Eds.), *Tourism, terrorism and security*. Emerald Group. doi:10.1108/978-1-83867-905-720201009

Korstanje, M. E., & Olsen, D. H. (2020). Negotiating the intersections between dark tourism and pilgrimage. In D. Olsen & M. Korstanje (Eds.), *Dark tourism and pilgrimage* (pp. 1–15). CABI. doi:10.1079/9781789241877.0001

Korstanje, M., & Seraphin, H. (2017). Revisiting the sociology of consumption in tourism. In S. K. Dixit (Ed.), *The Routledge handbook of consumer behaviour in hospitality and tourism* (pp. 16–25). Routledge.

Kotler, P., Armstrong, G., & Opresnik, M. O. (2021). *Principles of marketing* (18th ed.). Pearson.

Kozak, M., & Uca, S. (2008). Effective factors in the constitution of leadership styles: A study of Turkish hotel managers. *Anatolia, 19*(1), 117–130. doi:10.1080/13032917.2008.9687057

Krause, D. (2004). Influence-based leadership as a determinant of the inclination to innovate and of innovation-related behaviours: An empirical investigation. *The Leadership Quarterly*, *15*(1), 79–102. doi:10.1016/j.leaqua.2003.12.006

Krause, D., Gebert, D., & Kearney, E. (2007). Implementing process innovations the benefits of combining delegative-participative with consultative-advisory leadership. *Journal of Leadership & Organizational Studies*, *14*(1), 16–25. doi:10.1177/1071791907304224

Kumar, S., Valeri, M., & Shekhar. (2022). Understanding the relationship among factors influencing rural tourism: A hierarchical approach. *Journal of Organizational Change Management*, *35*(2), 385–407. doi:10.1108/JOCM-01-2021-0006

Kwon, J., & Boger, C. A. (2021). Influence of brand experience on customer inspiration and pro-environmental intention. *Current Issues in Tourism*, *24*(8), 1154–1168. doi:10.1080/13683500.2020.1769571

Ladkin, A., Fyall, A., Fletcher, J., & Shipway, R. (2008). London Tourism: A 'Post-Disaster' Marketing Response. *Journal of Travel & Tourism Marketing*, *23*(2–4), 95–111. doi:10.1300/J073v23n02_08

Lai, A. (2011). *Transformational-Transactional Leadership Theory*. 2011 AHS Capstone Projects. Paper 17. http://digitalcommons.olin.edu/ahs_capstone_2011/17

Lalangui, J., Espinoza Carrión, C. R., & Pérez Espinoza, M. J. (2017). Turismo sostenible, un aporte a la responsabilidad social empresarial: Sus inicios, características y desarrollo. *Universidad y Sociedad*, *9*(1), 148-153. Disponible en: http://scielo.sld.cu/pdf/rus/v9n1/rus21117.pdf

Lam, A. H. C., Ho, K. K. W., & Chiu, D. K. W. (2023). Instagram for student learning and library promotions? A quantitative study using the 5E Instructional Model. *Aslib Journal of Information Management*, *75*(1), 112–130. doi:10.1108/AJIM-12-2021-0389

Latham, K. F., Narayan, B., & Gorichanaz, T. (2019). Encountering the muse: An exploration of the relationship between inspiration and information in the museum context. *Journal of Librarianship and Information Science*, *51*(4), 1067–1076. doi:10.1177/0961000618769976

Law, R., Leung, R., Lo, A., Leung, D., & Fong, L. H. N. (2015). Distribution channel in hospitality and tourism: Revisiting disintermediation from the perspectives of hotels and travel agencies. *International Journal of Contemporary Hospitality Management*, *27*(3), 431–452. doi:10.1108/IJCHM-11-2013-0498

Lee, H. A., Law, R., & Murphy, J. (2011). Helpful reviewers in TripAdvisor, an online travel community. *Journal of Travel & Tourism Marketing*, *28*(7), 675–688. doi:10.1080/10548408.2011.611739

Lee, J. (2008). Effects of leadership and leader–member exchange on innovativeness. *Journal of Managerial Psychology*, *23*(6), 670–687. doi:10.1108/02683940810894747

Lee, Y. J. (2016). The relationships amongst emotional experience, cognition, and behavioral intention in battlefield tourism. *Asia Pacific Journal of Tourism Research, 21*(6), 697–715. doi:10.1080/10941665.2015.1068195

Leff, E. (2004). *Saber ambiental. Sustenibilidad, racionalidad, complejidad, poder*. Siglo XXI Editores.

Lehman, A., O'Rourke, N., Hatcher, L., & Stepanski, E. (2013). *JMP for Basic Univariate and Multivariate Statistics: Methods for Researchers and Social Scientists* (2nd ed.). SAS Institute.

Lehto, X. Y., Brown, S., Chen, Y., & Morrison, A. M. (2006). Yoga tourism as a niche within the wellness tourism market. *Tourism Recreation Research, 31*(1), 25–35. doi:10.1080/02508281.2006.11081244

Lei, S. Y., Chiu, D. K. W., Lung, M. M., & Chan, C. T. (2021). Exploring the aids of social media for musical instrument education. *International Journal of Music Education, 39*(2), 187–201. doi:10.1177/0255761420986217

Lengkeek, J. (2001). Leisure experience and imagination: Rethinking Cohen's modes of tourist experience. *International Sociology, 16*(2), 173–184. doi:10.1177/0268580901016002003

Lennon, J. (2005). Journeys in Understanding: What Is Dark Tourism? *The Sunday Observer*. Retrieved from http://www.guardian.co.uk/travel/2005/oct/23/darktourism.observerescapesection.

Lennon, J. (2005). Journeys Into Understanding. *The Guardian*. https://www.theguardian.com/travel/2005/oct/23/darktourism.observerescapesection

Lennon, J. J. (2018). Dark Tourism Visualisation: Some Reflections on the Role of Photograpy. In P. R. Stone, R. Hartmann, T. Seaton, R. Sharpley, & L. White (Eds.), *The Palgrave Handbook of Dark Tourism Studies*. Palgrave Macmillan. doi:10.1057/978-1-137-47566-4_24

Lennon, J., & Foley, M. (2000). *Dark Tourism. The Attraction of Death and Disaster*. Cengage Learning EMEA.

Lennon, J., & Foley, M. (2002). *Dark Tourism: The Attraction of Death and Disaster*. Continuum.

Leopold, T. (2007). A proposed code of conduct for war heritage sites. In Battlefield tourism: History, place, and ınterpretation (pp. 49-57). Elsevier. doi:10.4324/9780080548340

Leung, D., Law, R., Hoof, H., & Buhalis, D. (2013). Social Media in Tourism and Hospitality: A Literature Review. *Journal of Travel & Tourism Marketing, 30*(1-2), 3–22. doi:10.1080/10548408.2013.750919

Leung, T. N., Luk, C. K. L., Chiu, D. K. W., & Ho, K. K. W. (2022). User perceptions, academic library usage, and social capital: A correlation analysis under COVID-19 after library renovation. *Library Hi Tech, 40*(2), 304–322. doi:10.1108/LHT-04-2021-0122

Levy, S., Duan, W., & Boo, S. (2013). An analysis of one-star online reviews and responses in the Washington, D.C., lodging market. *Cornell Hospitality Quarterly, 54*(1), 49–63. doi:10.1177/1938965512464513

Li, M. (2021). Analysis on the scene construction of urban physical bookstore based on the perspective of knowledge communication. *Chinese Editors Journal,* (10), 43-46+52.

Liang, T. L., Chang, H. F., Ko, M. H., & Lin, C. W. (2017). Transformational leadership and employee voices in the hospitality industry. *International Journal of Contemporary Hospitality Management, 29*(1), 374–392. doi:10.1108/IJCHM-07-2015-0364

Light, D. (2017). Progress in dark tourism and thanatourism research: An uneasy relationship with heritage tourism. *Tourism Management, 61*, 275–301. doi:10.1016/j.tourman.2017.01.011

Light, D., & Ivanova, P. (2022). Thanatopsis and mortality mediation within "lightest" dark tourism. *Tourism Review, 77*(2), 622–635. doi:10.1108/TR-03-2021-0106

Li, K. K., & Chiu, D. K. W. (2022). A Worldwide Quantitative Review of the iSchools' Archival Education. *Library Hi Tech, 40*(5), 1497–1518. doi:10.1108/LHT-09-2021-0311

Lilienfeld, S. O., Lynn, S. J., Ruscio, J., & Beyerstein, B. L. (2010). *50 pszichológiai tévhit.* Partvonal Könyvkiadó.

Li, N., Liang, J., & Crant, J. M. (2010). The role of proactive personality in job satisfaction and organizational citizenship behavior: A relational perspective. *The Journal of Applied Psychology, 95*(2), 395–404. doi:10.1037/a0018079 PMID:20230079

Lin, C. H., & Kuo, B. Z. L. (2016). The behavioral consequences of tourist experience. *Tourism Management Perspectives, 18*, 84–91. doi:10.1016/j.tmp.2015.12.017

Lindberg, F., Hansen, A. H., & Eide, D. (2014). A multirelational approach for understanding consumer experiences within tourism. *Journal of Hospitality Marketing & Management, 23*(5), 487–512. doi:10.1080/19368623.2013.827609

Lingala, S., & Deshmukh, P. (2019). *A guidebook for integration of biodiversity spice sector in the Western Ghats.* https://sustainabledevelopment.in/wp-content/uploads/2020/06/1572605133IBBI-_Spice-publication_final-copy-10092019-2.pdf

Li, S., Xie, Z., Chiu, D. K. W., & Ho, K. K. W. (2023c). Sentiment Analysis and Topic Modeling Regarding Online Classes on the Reddit Platform: Educators versus Learners. *Applied Sciences (Basel, Switzerland), 13*(4), 2250. doi:10.3390/app13042250

Lisle, D. (2004). Gazing at Ground Zero: Tourism, Voyeurism and Spectacle. *Journal for Cultural Research, 8*(1), 3–21. doi:10.1080/1479758042000797015

Litvin, S., Goldsmith, R., & Pan, B. (2008). Electronic Word-of-Mouth in Hospitality and Tourism Management. *Tourism Management, 29*(3), 458–468. doi:10.1016/j.tourman.2007.05.011

Liu, Q., & Wu, X. (2013). Research on Chengdu vegetables marketing based on STP Model. *Asian Social Science*, *9*(4), 221–226. doi:10.5539/ass.v9n4p221

Liu, Y. (2020). The cultural persistence and future of antique bookstores under the context of "Internet plus": A case analysis of bookstores in Nanjing. *Renwen Tianxia*, (17), 56–60.

Liu, Y., Chiu, D. K. W., & Ho, K. K. W. (2023). Short-Form Videos for Public Library Marketing: Performance Analytics of Douyin in China. *Applied Sciences (Basel, Switzerland)*, *13*(6), 3386. doi:10.3390/app13063386

Li, X. (2021). Urban public cultural space as a dialogue: An analysis of the role of physical bookstores. *Journal of News Research*, *12*(04), 251–252.

Lloyd, D. W. (1994). *Tourism, Pilgrimage, and the Commemoration of the Great War in Great Britain, Australia and Canada, 1919–1939* [Unpublished PhD Thesis]. Cambridge University.

Lloyd, D. W. (2014). *Battlefield tourism: Pilgrimage and the commemoration of the Great War in Britain, Australia and Canada, 1919-1939*. A&C Black.

Lloyd-Jones, A., & Davies, A. J. (2006). An investigation of eMarketing within the second hand book trade. *BLED 2006 Proceedings*, Article 33. https://aisel.aisnet.org/bled2006/33

Lockstone-Binney, L., Hall, J., & Atay, L. (2013). Exploring the conceptual boundaries of diaspora and battlefield tourism: Australians' travel to the Gallipoli battlefield, Turkey, as a case study. *Tourism Analysis*, *18*(3), 297–311. doi:10.3727/108354213X13673398610736

Lo, P., Chiu, D. K. W., & Chu, W. (2014). Modeling Your College Library after a Commercial Bookstore? The Hong Kong Design Institute Library Experience. *Community & Junior College Libraries*, *19*(3-4), 59–76. doi:10.1080/02763915.2014.915186

Lo, P., Hsu, W.-E., Wu, S. H. S., Travis, J., & Chiu, D. K. W. (2021b). *Creating a Global Cultural City via Public Participation in the Arts: Conversations with Hong Kong's Leading Arts and Cultural Administrators*. Nova Science Publishers.

Lorenzo Linares, H., & Morales Garrido, G. (2014). Del desarrollo turístico sostenible al desarrollo local. Su comportamiento complejo. *Pasos (El Sauzal)*, *12*(2), 453–466. doi:10.25145/j.pasos.2014.12.033

Lowe, R., & Marriott, S. (2007). *Enterprise: Entrepreneurship and Innovation*. Elsevier.

Lu, S.S., Tian, R., & Chiu, D.K.W. (2023). Why do people not attend public library programs in the current digital age? *Library Hi Tech*, ahead-of-print. doi:10.1108/LHT-04-2022-0217

Lv, H. (2016). Introduce to the G20 the proverb "There is heaven above, and Suzhou and Hangzhou below". *Hangzhou Quality Life*, (04), 39–40.

Lv, X., Luo, H., Xu, S., Sun, J., Lu, R., & Hu, Y. (2022a). Dark tourism spectrum: Visual expression of dark experience. *Tourism Management*, *93*, 104580. doi:10.1016/j.tourman.2022.104580

Lv, X., Lu, R., Xu, S., Sun, J., & Yang, Y. (2022b). Exploring visual embodiment effect in dark tourism: The influence of visual darkness on dark experience. *Tourism Management, 89*, 104438. doi:10.1016/j.tourman.2021.104438

Macey, W. H., & Schneider, B. (2008). The meaning of employee engagement. *Industrial and Organizational Psychology: Perspectives on Science and Practice, 1*(1), 3–30. doi:10.1111/j.1754-9434.2007.0002.x

Machado Carvalho, M. A. (2023). Bone Chapels: Who might be interested in visiting and why? *Tourism Recreation Research*, 1–12. doi:10.1080/02508281.2023.2167908

Madanaguli, A., Kaur, P., Mazzoleni, A., & Dhir, A. (2022). The innovation ecosystem in rural tourism and hospitality – a systematic review of innovation in rural tourism. *Journal of Knowledge Management, 26*(7), 1732–1762. doi:10.1108/JKM-01-2021-0050

Madureira, L., & Alturas, B. (2022). Impact of Social Media Influencers on the Portuguese Tourism and Travel Industry in a Covid-19 Era. In Marketing and Smart Technologies. Proceedings of ICMarkTech 2021. Springer. doi:10.1007/978-981-16-9272-7_32

Maerk, J., y Boxill, I. (2000). *CDMX, México.* Turismo en el Caribe. Plaza y Valdés, S.A. de C.V.

Malkanthi, S. H. P., Ishana, A. S. F., Sivashankar, P., & Weeralal, J. L. K. (2015). Willingness to initiate spice-tourism in the Kolonna District Secretariat of Ratnapura district in Sri Lanka: Farmers' perspective. *Sri Lanka Journal of Food and Agriculture, 1*(1), 35. doi:10.4038ljfa.v1i1.5

Maltz, A. C., Shenhar, A. J., & Reilly, R. R. (2003). Beyond the balanced scorecard: Refining the search for organizational success measures. *Long Range Planning, 36*(2), 187–204. doi:10.1016/S0024-6301(02)00165-6

Manasseh, T., Müller-Sarmiento, P., Reuter, H., von Faber-Castell, C., & Pallua, C. (2012). Customer inspiration–a key lever for growth in European retail. *Marketing Review St. Gallen, 29*(5), 16–21. doi:10.136511621-012-0159-9

Marcheggiani, D., Täckström, O., Esuli, A., & Sebastiani, F. (2014). Hierarchical multi-label conditional random fields for aspect-oriented opinion mining. In *Advances in information retrieval* (pp. 273–285). Springer. doi:10.1007/978-3-319-06028-6_23

Markezic, E. (2022). True Crime Tourism: The Good, the Bad, and the Bundy. *Escape*. https://www.escape.com.au/destinations/north-america/usa/true-crime-tourism-the-good-the-bad-and-the-bundy/news-story/f8fd8859f0ede988126e1f66f36feaa0

Marpakwar, P. (2022). 1,800+ deaths in 8 months_ Sharp rise in farm suicides in Maharashtra _ Mumbai News. *Times of India*. https://timesofindia.indiatimes.com/city/mumbai/1800-deaths-in-8-months-sharp-rise-in-farm-suicides-in-maharashtra/articleshow/94820572.cms

Marsh, I., & Melville, G. (2019). *Crime, Justice and the Media* (3rd ed.). Routledge., doi:10.4324/9780429432194

Marson, D. (2011). From mass tourism to niche tourism. In *Research themes for tourism* (pp. 1–15). CABI. doi:10.1079/9781845936846.0001

Martínez Herrera, H. (2005). *El marco ético de la responsabilidad social empresarial*. Editorial Pontificia Universidad Javeriana.

Martínez Quintana, V. (2017). El turismo de naturaleza: Un producto turístico sostenible. *Arbor*, *193*(785), a396. doi:10.3989/arbor.2017.785n3002

Martínez Salvador, S. (2020). *Dirección de alojamientos turísticos*. Editorial Paraninfo.

Martínez Salvador, S. (2022). *Animación Turística*. Editorial Parainfo.

Martini, A., & Buda, D. M. (2020). Dark tourism and affect: Framing places of death and disaster. *Current Issues in Tourism*, *23*(6), 679–692. doi:10.1080/13683500.2018.1518972

Marzo-Navarro, M., Pedraja-Iglesias, M., & Vinzón, L. (2015). Sustainability indicators of rural tourism from the perspective of the residents. *Tourism Geographies*, *17*(4), 586–602. doi:10.1080/14616688.2015.1062909

Matravers, D. (2018). Is It Ever Morally Acceptable to Visit a Mass Murder Site? *Open.edu*. https://www.open.edu/openlearn/history-the-arts/it-ever-morally-acceptable-visit-mass-murder-site

Mauri, A. G., & Minazzi, R. (2013). Web reviews influence on expectations and purchasing intentions of hotel potential customer. *International Journal of Hospitality Management*, *34*(September), 99–107. doi:10.1016/j.ijhm.2013.02.012

McFarland, F. (2022). Strange Attraction. Inside the Morbid World of 'Dark Tourists' Who Love Serial Killers Like Jeffrey Dahmer & Flock to Areas Where They Lived. *The U.S. Sun*. https://www.the-sun.com/news/6447113/dark-tourists-visit-jeffrey-dahmer-home-serial-killer/

McManus, L. (2013). Customer accounting and marketing performance measures in the hotel industry: Evidence from Australia. *International Journal of Hospitality Management*, *33*, 140–152. doi:10.1016/j.ijhm.2012.07.007

Mead, T. (2019). *Tourists Are Flocking to Ted Bundy's Old House, and the Current Tenants Are Sick of It*. House Beautiful. https://www.housebeautiful.com/design-inspiration/real-estate/a27090576/ted-bundy-house/

Melian, A. G. (2016). *Accessible tourism: An integrated model of the behavior of tourists with disabilities in a destination*. Academic Press.

Melo, A., Hernández-Maestro, R., & Muñoz-Gallego, P. (2017). Service quality perceptions, online visibility, and business performance in rural lodging establishments. *Journal of Travel Research*, *56*(2), 250–262. doi:10.1177/0047287516635822

Meneses, T. Á. (2013). La planificación turística: Un aspecto clave para el desarrollo sostenible y regional de Boyacá. Revista de Investigación. *Desarrollo e Innovación, 3*(2), 101–110. doi:10.19053/20278306.2169

Meng, F., & Uysal, M. (2008). Effects of Gender Differences on Perceptions of Destination Attributes, Motivations, and Travel Values: An Examination of a Nature-Based Resort Destination. *Journal of Sustainable Tourism, 16*(4), 445–466. doi:10.1080/09669580802154231

Meng, Y., Chu, M. Y., & Chiu, D. K. W. (2022). The impact of COVID-19 on museums in the digital era: Practices and challenges in Hong Kong. *Library Hi Tech, 41*(1), 130–151. doi:10.1108/LHT-05-2022-0273

Merriam-Webster Inc. (n.d.). Retrieved from https://www.merriam-webster.com/dictionary/darkness

Metcalf, M., & Morelli, C. (2015). *The Art of Leading Change: Innovative Leaders Transformation Model.* Articles from Integral Leadership Review. http://integralleadershipreview.com/author/maureen-metcalf-and-carla-morelli/

Meyers-Levy, J., & Sternthal, B. (1991). Gender differences in the use of message cues and judgments. *JMR, Journal of Marketing Research, 28*(February), 84–96. doi:10.1177/002224379102800107

Michaelis, B., Stegmaier, R., & Sonntag, K. (2009). Shedding light on followers' innovation implementation behaviour: The role of transformational leadership, commitment to change, and climate for initiative. *Journal of Managerial Psychology, 25*(4), 408–429. doi:10.1108/02683941011035304

Miles, S. T. (2012). *Battlefield tourism: Meanings and interpretations* [Unpublished PhD Thesis]. University of Glasgow.

Miles, S. (2014). Battlefield sites as dark tourism attractions: An analysis of experience. *Journal of Heritage Tourism, 9*(2), 134–147. doi:10.1080/1743873X.2013.871017

Miles, W. F. (2002). Auschwitz: Museum Interpretation and Darker Tourism. *Annals of Tourism Research, 29*(4), 1175–1178. doi:10.1016/S0160-7383(02)00054-3

Mileva, S. V. (2018). Potential of development of dark tourism in Bulgaria. *International Journal of Tourism Cities, 4*(2), 144–153. doi:10.1108/IJTC-05-2017-0029

Millán, G. D., Rojas, R. D. H., & García, J. S. R. (2019). Analysis of the demand of dark tourism: A case study in Córdoba (Spain). *Mediterranean Journal of Social Sciences, 10*(1), 161–176. doi:10.2478/mjss-2019-0015

MindTools. (2022). *The Segmentation, Targeting and Positioning (STP) marketing model.* https://www.mindtools.com/a5llt9t/segmentation-targeting-and-positioning-model

Ministry of Agriculture and Cooperation. (2012). *State Agricultural Portal Software Requirement Specifications*. http://www.msamb.com

Ministry of Commerce and Industry. (2023). 14th Edition of the World Spice Congress (WSC) to be held in Mumbai from 16-18 February 2023. *PIB GOI*. https://pib.gov.in/PressReleasePage.aspx?PRID=1890034

Min, J., Yang, K., & Thapa-Magar, A. (2021). Dark tourism segmentation by tourists' motivations for visiting earthquake sites in Nepal: Implications for dark tourism. *Asia Pacific Journal of Tourism Research*, *26*(8), 866–878. doi:10.1080/10941665.2021.1925315

Mionel, V. (2019). Dark tourism and thanatourism: Distinct tourism typologies or simple analytical tools? *Tourism: An International Interdisciplinary Journal*, *67*(4), 423–437.

Mionel, V. (2020). (Not so) Dark tourism: The Merry Cemetery in Săpânţa (Romania)–an expression of folk culture. *Tourism Management Perspectives*, *34*, 100656. doi:10.1016/j.tmp.2020.100656

Mitchell, J. P. (2003). Looking forward to the past: National identity and history in Malta. *Identities (Yverdon)*, *10*(3), 377–398. doi:10.1080/10702890390228919

Mody, M., & Hanks, L. (2020). Consumption authenticity in the accommodations industry: The keys to brand love and brand loyalty for hotels and Airbnb. *Journal of Travel Research*, *59*(1), 173–189. doi:10.1177/0047287519826233

Molera, L., & Albaladejo, P. (2007). Profiling segments of tourists in rural areas of south-eastern Spain. *Tourism Management*, *28*(3), 757–767. doi:10.1016/j.tourman.2006.05.006

Morales, V., Barrionuevo, M., & Gutiérrez, L. (2010). Transformational leadership influence on organizational performance through organizational learning and innovation. *Journal of Business Research*, *65*(7), 1040–1050. doi:10.1016/j.jbusres.2011.03.005

Moro, S., & Rita, P. (2018). Brand strategies in social media in hospitality and tourism. *International Journal of Contemporary Hospitality Management*, *30*(1), 343–364. doi:10.1108/IJCHM-07-2016-0340

Moro, S., Rita, P., & Coelho, J. (2017a). Stripping customers' feedback on hotels through data mining: The case of Las Vegas Strip. *Tourism Management Perspectives*, *23*(July), 41–52. doi:10.1016/j.tmp.2017.04.003

Moro, S., Rita, P., & Cortez, P. (2017b). A text mining approach to analyzing Annals literature. *Annals of Tourism Research*, *66*(September), 208–210. doi:10.1016/j.annals.2017.07.011

Moro, S., Rita, P., & Oliveira, C. (2018). Factors influencing hotels' online prices. *Journal of Hospitality Marketing & Management*, *27*(4), 443–464. doi:10.1080/19368623.2018.1395379

Morris, M. H., Kuratko, D. F., & Covin, J. G. (2011). *Corporate entrepreneurship and innovation*. South-Western Cencage Learning.

Mortensen, M. (2013). The making and remakings of an American icon: 'Raising the flag on ıwo jima' from photojournalism to global, digital media. In A. Gjelsvik (Ed.), Eastwood's Iwo Jima: Critical engagements with flags of our fathers and letters from Iwo Jima (pp. 15-35). Columbia University Press. doi:10.7312/gjel16564-003

Morton, R. J. (2008). *Serial Murder. Multi-Disciplinary Perspectives for Investigators*. FBI Academy.

Mostra Serial Killer - Jesolo Veneto . (2023). Tripadvisor. https://www.tripadvisor.co.uk/Attraction_Review-g580246-d982 3089-Reviews-Mostra_Serial_Killer-Jesolo_Veneto.html

MoT GoI. (2021). *India tourism statistics at a glance 2021*. https://tourism.gov.in/sites/default/files/2021-09/English T ourisum 2021.pdf

Moufakkir, O., & Selmi, N. (2018). Examining the spirituality of spiritual tourists: A Sahara desert experience. *Annals of Tourism Research*, *70*, 108–119. doi:10.1016/j.annals.2017.09.003

Moura, A., Eusébio, C., & Devile, E. (2022). The 'why' and 'what for' of participation in tourism activities: Travel motivations of people with disabilities. *Current Issues in Tourism*, 1–17. doi:10.1080/13683500.2022.2044292

Mourtazina, E. (2020). Beyond the horizon of words: Silent landscape experience within spiritual retreat tourism. *International Journal of Culture, Tourism and Hospitality Research*, *14*(3), 349–360. doi:10.1108/IJCTHR-10-2019-0185

Mowatt, R. A., & Chancellor, C. H. (2011). Visiting death and life: Dark tourism and slave castles. *Annals of Tourism Research*, *38*(4), 1410–1434. doi:10.1016/j.annals.2011.03.012

Moyano Bonilla, C. (1995). Derecho a un Medio Ambiente Sano. *Boletín del Instituto de Investigaciones Jurídicas de la UNAM*, *84*. Disponible en http://biblio.juridicas.unam.mx/revista/DerechoComparado/ind ice.htm?n=82

Mumford, M. D., & Licuanan, B. (2004). Leading for innovation: Conclusions, issues, and directions. *The Leadership Quarterly*, *15*(1), 163–171. doi:10.1016/j.leaqua.2003.12.010

Mumford, M. D., Scott, G. M., Gaddis, B., & Strange, J. M. (2002). Leading creative people: Orchestrating expertise and relationships. *The Leadership Quarterly*, *13*(6), 705–750. doi:10.1016/S1048-9843(02)00158-3

Munar, A., & Jacobsen, J. (2013). Trust and Involvement in Tourism Social Media and Web-Based Travel Information Sources. *Scandinavian Journal of Hospitality and Tourism*, *13*(1), 1–19. doi:10.1080/15022250.2013.764511

Munar, A., & Jacobsen, J. (2014). Motivations for sharing tourism experiences through social media. *Tourism Management*, *43*(August), 46–54. doi:10.1016/j.tourman.2014.01.012

Muñoz, F. (2003). *El turismo explicado con claridad.* Amertown International.

Murder – The Exhibition – Budapest Central Hungary . (2023). Tripadvisor. https://www.tripadvisor.com/Attraction_Review-g274887-d3752902-Reviews-Murder_The_Exhibition-Budapest_Central_Hungary.html

Murder Wall . (2023). POMC. https://pomc.org/murder-wall/

Museum of Death - Los Angeles California . (2023). Tripadvisor. https://www.tripadvisor.com/Attraction_Review-g32655-d2018939-Reviews-Museum_of_Death-Los_Angeles_California.html

Museum of Death - New Orleand Louisiana. (2023). Tripadvisor. https://www.tripadvisor.com/Attraction_Review-g60864-d7951659-Reviews-Museum_of_Death-New_Orleans_Louisiana.html

Museum of Death. (2023). Museum of Death: http://www.museumofdeath.net/

Museum of Perth – Perth Greater Perth Western Australia . (2023). Tripadvisor. https://www.tripadvisor.com/Attraction_Review-g255103-d8813929-Reviews-Museum_of_Perth-Perth_Greater_Perth_Western_Australia.html

Museum of Perth. (2023). Museum of Perth. https://www.museumofperth.com.au/

Nagy, A. (2014). The orientation towards innovation of spa hotel management: The case of Romanian spa industry. *Procedia: Social and Behavioral Sciences*, *124*, 425–431. doi:10.1016/j.sbspro.2014.02.504

Nahavandi, A. (2015). *The Art and Science of Leadership*. Pearson Education Limited.

Naikwadi, V. A., & Chaskar, P. M. (2012). Plan: For marketing library and information services. *Indian Streams Research Journal*, *2*(10), 1–6.

Nair, B. B., & Mohanty, P. P. (2021). Positioning spice tourism as an emerging form of special interest tourism: Perspectives and strategies. *Journal of Ethnic Foods*, *8*(1), 10. Advance online publication. doi:10.118642779-021-00086-4

National Museum of Crime & Punishment - Washington DC District of Columbia . (2023). Tripadvisor. https://www.tripadvisor.com/Attraction_Review-g28970-d941651-Reviews-National_Museum_of_Crime_Punishment-Washington_DC_District_of_Columbia.html

National Park Service. (2023). *Gettysburg NMP.* National Park Service. https://irma.nps.gov/Stats/SSRSReports/Park%20Specific%20Reports/Annual%20Park%20Recreation%20Visitation%20(1904%20-%20Last%20Calendar%20Year)?Park=GETT

Nazir, M. U., Yasin, I., & Tat, H. H. (2021). Destination image's mediating role between perceived risks, perceived constraints, and behavioral intention. *Heliyon*, *7*(7), e07613. doi:10.1016/j.heliyon.2021.e07613 PMID:34368481

NCERT. (2022). Distribution, Density, Growth and Composition. In *India and Economy*. NCERT. https://ncert.nic.in/textbook/pdf/legy201.pdf

Nebel, B. J., & Wright, R. T. (1999). *Ciencias ambientales: ecología y desarrollo sostenible*. D.F. Pearson Educación.

Nejad, H., & Rowe, G. (2009). Strategic leadership: Short-term stability and long-term viability. *Ivey Business Journal*, *73*(5), 2–6.

Netease. (2021). *Renmin Road, the main axis of Suzhou, witnesses the development and changes of Suzhou*. https://www.163.com/dy/article/G37N4E3C0515ESFT.html

New Orleans Killers and Thrillers Tour – New Orleans Louisiana . (2023). Tripadvisor. https://www.tripadvisor.com/AttractionProductReview-g60864-d11469169-New_Orleans_Killers_and_Thrillers_Tour-New_Orleans_Louisiana.html

New Orleans True Murder Tour: Sinister Criminal Intentions – New Orleans Louisiana. (2023). Tripadvisor. https://www.tripadvisor.com/AttractionProductReview-g60864-d12955117-New_Orleans_True_Murder_Tour_Sinister_Criminal_Intentions-New_Orleans_Louisiana.html

Newton, M. (2006). *The Encyclopedia of Serial Killers* (2nd ed.). Infobase Publishing.

Ngo, P. M., & Bui, H. T. (2019). Contested interpretation of Vietnam war heritage: Tour guides' mediating roles. *Journal of Tourism and Adventure*, *2*(1), 61–84. doi:10.3126/jota.v2i1.25933

Nguyen, V. T., Jung, K., Yoo, S., Kim, S., Park, S., & Currie, M. (2019, December 9-11). Civil War battlefield experience: Historical event simulation using augmented reality technology. In *2019 IEEE International Conference on Artificial Intelligence and Virtual Reality (AIVR)* (pp. 294-2943). 10.1109/AIVR46125.2019.00068

Ni, Y., Lam, A. H. C., & Chiu, D. K. W. (2023). Leveraging Online Communities for Building Social Capital in University Libraries: A Case Study of Fudan University Medical Library. In Balance and Boundaries in Creating Meaningful Relationships in Online Higher Education. IGI Global.

Nicolau, J., & Santa-María, M. (2013). The effect of innovation on hotel market value. *International Journal of Hospitality Management*, *32*, 71–79. doi:10.1016/j.ijhm.2012.04.005

Night-Time Dark History and Vampire Walking Tour in Buda Castle District - Budapest . (2023). Tripadvisor. https://www.tripadvisor.com/AttractionProductReview-g274887-d17162384-Night_Time_Dark_History_and_Vampire_Walking_Tour_in_Buda_Castle_District-Budapest_.html

Ni, J., Chiu, D. K. W., & Ho, K. K. W. (2022). Information search behavior among Chinese self-drive tourists in the smartphone era. *Information Discovery and Delivery*, *50*(3), 285–296. doi:10.1108/IDD-05-2020-0054

Novakovic, J. (2021). The Role of Museums in a Digital World: Attracting Youth and Overcoming COVID19 Obstacles. *Cultural Management: Science and Education*, 59-68.

Novelli, M. (2018). Niche tourism: Past, present and future. The SAGE handbook of tourism management: Applications of theories and concepts of tourism, 1, 344-359.

Nunnally, J. C., & Bernstein, I. H. (1994). *Psychological theory*. MacGraw-Hill.

Nyanjom, J., Boxall, K., & Slaven, J. (2018). Towards inclusive tourism? Stakeholder collaboration in the development of accessible tourism. *Tourism Geographies*, *20*(4), 675–697. doi:10.1080/14616688.2018.1477828

Nybakk, E., & Jenssen, J. I. (2012). Innovation strategy, working climate, and financial performance in traditional manufacturing firms: An empirical analysis. *International Journal of Innovation Management*, *16*(2), 1–30. doi:10.1142/S1363919611003374

Nyomozás az első magyar sorozatgyilkos után – séta . (2023). Kertvárosi Időutazó. https://stayhappening.com/e/nyomoz%C3%A1s-az-els%C5%91-magyar-sorozatgyilkos-ut%C3%A1n-s%C3%A9ta-E2ISTS5714A

O'Neill, S. (2002). Soham Pleads with Trippers to Stay Away. *Daily Telegraph*. http:// www.telegraph.co.uk/news/main.jhtml?xml=/news/2002/08/26/nfen26.xml&sSheet=/portal/2002/08/26/ixport.html

OECD. (1994). *Tourism strategies and rural development: organisation for economic co-operation and development*. OECD.

Ogretmenoglu, M., Mavric, B., & Dincer, F. I. (2022). Using a bibliometric approach to shed light on dark tourism. *Podium (São Paulo)*, *11*(2), 328–352. doi:10.5585/podium.v11i2.19902

Ojeda García, C., & Mármol Sinclair, P. (2016). *Marketing Turístico*. Parainfo.

Oke, A., Munshi, N., & Walumbwa, F. (2009). The Influence of Leadership on Innovation Processes and Activities. *Organizational Dynamics*, *38*(1), 64–72. doi:10.1016/j.orgdyn.2008.10.005

Oke, A., Walumbwa, F. O., & Myers, A. (2012). Innovation Strategy, Human Resource Policy, and Firms' Revenue Growth: The Roles of Environmental Uncertainty and Innovation Performance. *Decision Sciences*, *43*(2), 273–302. doi:10.1111/j.1540-5915.2011.00350.x

Oleynick, V. C., Thrash, T. M., LeFew, M. C., Moldovan, E. G., & Kieffaber, P. D. (2014). The scientific study of inspiration in the creative process: Challenges and opportunities. *Frontiers in Human Neuroscience*, *8*, 436. doi:10.3389/fnhum.2014.00436 PMID:25009483

Oliver, M., & Barnes, C. (2012). Back to the future: The World Report on Disability. *Disability & Society*, *27*(4), 575–579. doi:10.1080/09687599.2012.686781

Oliver, R. L. (1980). A cognitive model of the antecedents and consequences of satisfaction decisions. *JMR, Journal of Marketing Research*, *17*(4), 460–469. doi:10.1177/002224378001700405

Olsen, D. H. (2003). Heritage, tourism, and the commodification of religion. *Tourism Recreation Research*, *28*(3), 99–104. doi:10.1080/02508281.2003.11081422

Olsen, D. H., & Korstanje, M. E. (Eds.). (2019). *Dark tourism and pilgrimage*. CABI.

OpenBook. (2022). *2021 China book retail market report*. http://www.cnfaxie.org/detail.html?id=25&contentId=535

Orfila-Sintes, F., Crespi-Cladera, R., & Martinez-Ros, E. (2005). Innovation activity in the hotel industry: Evidence from Balearic Islands. *Tourism Management*, *26*(6), 851–865. doi:10.1016/j.tourman.2004.05.005

Organización Mundial del Turismo. (1999). *Código Ético Mundial para el Turismo. Por un turismo responsable*. Asamblea General de la OMT. Disponible en https://webunwto.s3.eu-west-1.amazonaws.com/s3fs-public/2019-10/gcetpassportglobalcodees.pdf

Organización Mundial del Turismo. (2022). *Comunicado de prensa. La recuperación del turismo gana impulso tras la disminución de las restricciones y el aumento de la confianza*. Disponible en: https://webunwto.s3.eu-west-1.amazonaws.com/s3fs-public/2022-06/220606-unwto-barometer-es.pdf?VersionId=qi2kpKlUpSD4j_AJ Qilm9kIlpZ_tiHMt

Osaze, B. E., & Anao, A. R. (2000). *Managerial Finance*. Uniben Press.

Ottenbacher, M., & Shaw, V. (2002). The Role of Employee Management in NSD: Preliminary Results from a Study of the Hospitality SectorIn Proceedings of the 2002 Product Development and Management Association (PDMA) Research Conference, 109-133.

Ottenbacher, M. (2007). Innovation management in hospitality industry: Different strategies for achieving success. *Journal of Hospitality & Tourism Research (Washington, D.C.)*, *31*(4), 431–454. doi:10.1177/1096348007302352

Ottenbacher, M., & Gnoth, J. (2005). How to Develop Successful Hospitality Innovation. *The Cornell Hotel and Restaurant Administration Quarterly*, *46*(2), 205–222. doi:10.1177/0010880404271097

Ottenbacher, M., Gnoth, J., & Jones, P. (2006). Identifying determinants of success in development of new high-contact services: Insights from hospitality industry. *International Journal of Service Industry Management*, *17*(4), 344–363. doi:10.1108/09564230610680659

Ousby, I. (1990). *The Englishman's England: taste, travel and the rise of tourism*. Cambridge University Press.

Packer, J., & Ballantyne, R. (2016). Conceptualizing the visitor experience: A review of literature and development of a multifaceted model. *Visitor Studies*, *19*(2), 128–143. doi:10.1080/10645578.2016.1144023

Packer, J., Ballantyne, R., & Uzzell, D. (2019). Interpreting war heritage: Impacts of Anzac Museum and battlefield visits on Australians' understanding of national identity. *Annals of Tourism Research*, *76*, 105–116. doi:10.1016/j.annals.2019.03.012

Page, S. J., & Connell, J. (2009). *Tourism: a modern synthesis* (3rd ed.). Cengage Learning.

Pallardy, R. (2021, August 17). *Elizabeth Báthory*. https://www.britannica.com/biography/Elizabeth-Bathory

Panakera, C. (2007). World War II and Tourism Development in Solomon Islands. In C. Ryan (Ed.), Battlefield tourism: History, place, and interpretation (pp. 125-141). Routledge.

Pan, J., Liu, S., Ma, B., & Qu, Z. (2018). How does proactive personality promote creativity? A multilevel examination of the interplay between formal and informal leadership. *Journal of Occupational and Organizational Psychology*, *91*(4), 852–874. doi:10.1111/joop.12221

Parekh, B. (1995). The concept of national identity. *Journal of Ethnic and Migration Studies*, *21*(2), 255–268. doi:10.1080/1369183X.1995.9976489

Park, D., & Yoon, Y. (2009). Segmentation by motivation in rural tourism: A Korean case study. *Tourism Management*, *30*(1), 99–108. doi:10.1016/j.tourman.2008.03.011

Park, E., Choi, B. K., & Lee, T. J. (2019). The role and dimensions of authenticity in heritage tourism. *Tourism Management*, *74*, 99–109. doi:10.1016/j.tourman.2019.03.001

Parks-Leduc, L., Feldman, G., & Bardi, A. (2015). Personality traits and personal values: A meta-analysis. *Personality and Social Psychology Review*, *19*(1), 3–29. doi:10.1177/1088868314538548 PMID:24963077

Parrey, S. H., Hakim, I. A., & Rather, R. A. (2019). Mediating role of government initiatives and media influence between perceived risks and destination image: A study of conflict zone. *International Journal of Tourism Cities*, *5*(1), 90–106. doi:10.1108/IJTC-02-2018-0019

Patiar, A., & Mia, L. (2009). Transformational leadership style, market competition and departmental performance: Evidence from luxury hotels in Australia. *International Journal of Hospitality Management*, *28*(2), 254–262. doi:10.1016/j.ijhm.2008.09.003

Pennell, C. (2018). Taught to remember? British youth and First World War centenary battlefield tours. *Cultural Trends*, *27*(2), 83–98. doi:10.1080/09548963.2018.1453449

Pérez de las Heras, M. (2003). *La guía del ecoturismo. O como conservar la naturaleza a través del turismo*. Mundi-Prensa Libros, S.A.

Peterson, W., Gijsbers, G., & Wilks, M. (2003). *An organizational performance assessment system for agricultural research organizations: Concepts, methods, and procedures*. ISNAR Research Management Guidelines.

Peterson, S. J., Walumbwa, F. O., Byron, K., & Myrowitz, J. (2009). CEO positive psychological traits, transformational leadership, and firm performance in high-technology start-up and established firms. *Journal of Management, 35*(2), 348–368. doi:10.1177/0149206307312512

Philbin, T., & Philbin, M. (2009). *Killer Book of Serial Killers*. Sourcebooks, Inc.

Phillips, W. J., Wolfe, K., Hodur, N., Leistritz, F. L., Management, H., Dakota, N., Leadership, H., & Dakota, N. (2013). *Tourist Word of Mouth and Revisit. 104*(November 2011), 93–104. https://doi.org/ doi:10.1002/jtr

Phillips, W. J., Wolfe, K., Hodur, N., & Leistritz, F. L. (2013). Tourist word of mouth and revisit intentions to rural tourism destinations: A case of North Dakota, USA. *International Journal of Tourism Research, 15*(1), 93–104. doi:10.1002/jtr.879

Piekarz, M. (2007). *Hot war tourism: The live Battlefield and the ultimate adventure holiday? In Battlefield tourism: History, place, and ınterpretation*. Routledge.

Podoshen, J. S. (2013). Dark tourism motivations: Simulation, emotional contagion and topographic comparison. *Tourism Management, 35*, 263–271. doi:10.1016/j.tourman.2012.08.002

Podsakoff, P. M., MacKenzie, S. B., & Podsakoff, N. P. (2012). Sources of method bias in social science research and recommendations on how to control it. *Annual Review of Psychology, 63*(1), 539–569. doi:10.1146/annurev-psych-120710-100452 PMID:21838546

Polat, N., & Hermans, E. (2016). A model proposed for sustainable accessible tourism (SAT). *Tékhne (Instituto Politécnico do Cávado e do Ave), 14*(2), 125–133. doi:10.1016/j.tekhne.2016.11.002

Police Museum. (2023). Google. https://www.google.com/search?q=budapest+police+museum&biw=1462&bih=794&sxsrf=AOaemvIfi_deSrtmttn0rpuWDbwUbL7hVw%3A1634149352659&ei=6CNnYa-9J5GK9u8Ppcia2As&ved=0ahUKEwiv0bLdgMjzAhURhf0HHSWkBrsQ4dUDCA4&uact=5&oq=budapest+police+museum&gs_lcp=Cgdnd3Mtd2l6EA

Popiel, M. (2016). Barriers in Undertaking Tourist Activity by Disabled People. *Prace Naukowe Akademii Im. Jana Długosza w Częstochowie. Kultura Fizyczna, 15*(3), 103–110.

Poria, Y., Biran, A., & Reichel, A. (2009). Visitors' preferences for interpretation at heritage sites. *Journal of Travel Research, 48*(1), 92–105. doi:10.1177/0047287508328657

Poria, Y., Reichel, A., & Biran, A. (2006). Heritage site management: Motivations and expectations. *Annals of Tourism Research, 33*(1), 162–178. doi:10.1016/j.annals.2005.08.001

Potts, T. J. (2012). 'Dark tourism' and the 'kitschification' of 9/11. *Tourist Studies, 12*(3), 232–249. doi:10.1177/1468797612461083

Powell, R., Kennell, J., & Barton, C. (2018). Dark Cities: A dark tourism index for Europe's tourism cities, based on the analysis of DMO websites. *International Journal of Tourism Cities, 4*(1), 4–21. doi:10.1108/IJTC-09-2017-0046

Pratt, M. (2022). *What is the 4P marketing matrix?* https://www.business.org/marketing/sales/marketing-101-4p-matrix/

Prayag, G. (2009). Tourists' evaluations of destination image, satisfaction, and future behavioral intentions-the case of mauritius. *Journal of Travel & Tourism Marketing*, *26*(8), 836–853. doi:10.1080/10548400903358729

Preko, A., Mohammed, I., Gyepi-Garbrah, T. F., & Allaberganov, A. (2021). Islamic tourism: Travel motivations, satisfaction and word of mouth, Ghana. *Journal of Islamic Marketing*, *12*(1), 124–144. doi:10.1108/JIMA-04-2019-0082

Prideaux, B. (2007). Echoes of war: Battlefield tourism. In C. Ryan (Ed.), Battlefield tourism history-place and interpretation (pp. 81-95). Routledge.

Proença, S., & Souziakis, E. (2008). Tourism as an economic growth factor: A case study for Southern European countries. *Tourism Economics*, *14*(4), 791–806. doi:10.5367/000000008786440175

Prost, A., & Krumeich, G. (2015). *Verdun 1916*. Tallandier.

Qiao, G., Cao, Y., & Zhang, J. (2022). Accessible Tourism – understanding blind and vision-impaired tourists' behaviour towards inclusion. *Tourism Review*. Advance online publication. doi:10.1108/TR-03-2022-0129

Qiao, G., Ding, L., Zhang, L., & Yan, H. (2021). Accessible tourism: a bibliometric review (2008–2020). In *Tourism Review*. Emerald Group Holdings Ltd. doi:10.1108/TR-12-2020-0619

Qiu, M. (2018). Antique bookstores as media: A study based on Wuhan Bookstores. *Press Outpost*, (2), 15–17.

Quesada Castro, R. (2000). *Elementos del turismo*. Editorial Universidad Estatal a Distancia.

Rafferty, A., & Griffin, M. (2004). Dimensions of transformational leadership: Conceptual and empirical extensions. *The Leadership Quarterly*, *15*(3), 329–354. doi:10.1016/j.leaqua.2004.02.009

Raine, R. (2013). A dark tourist spectrum. *International Journal of Culture, Tourism and Hospitality Research*, *7*(3), 242–256. doi:10.1108/IJCTHR-05-2012-0037

Rajanbabu, R., & Ganesan, S. (2015). Growth and Instability in Production of Indian Spices. *International Journal of Management and Social Development*, *2*(9), 42–50.

Rajasekaram, K., Hewege, C. R., & Perera, C. R. (2022). "Tourists' experience" in dark tourism: A systematic literature review and future research directions. *Asia Pacific Journal of Tourism Research*, *27*(2), 206–224. doi:10.1080/10941665.2022.2046118

Rama, S., Wulung, P., Yuliawati, A. K., Abdullah, U., & Fitriyani, E. (2021). Spice cultural heritage in geotourism trail. *Journal of Engineering Science and Technology*, 76–87.

Rancho La Bellota. (2023). Recuperado de: www.rancholabellota.com

Raynor, M. (2004). *English battlefields*. Tempus.

Regalado Pezúa, O., & Arias Valencia, J. (2006). Desarrollo Sostenible en Turismo: Una propuesta para Machu Picchu. *Journal of Economics, Finance and Administrative Science, 11*(20), 63-73. Disponible en: https://www.redalyc.org/pdf/3607/360735259003.pdf

Reisinger, Y. (2015). *Transformational tourism: Host perspectives*. SABI. doi:10.1079/9781780643922.0000

Renaud, L. (2020). Reconsidering global mobility–distancing from mass cruise tourism in the aftermath of COVID-19. *Tourism Geographies*, *22*(3), 679–689. doi:10.1080/14616688.2020.1762116

Rendőrmúzeum. (2023). http://www.rendormuzeum.com/fooldal

Resnick, P., Zeckhauser, R., Friedman, E., & Kuwabara, K. (2000). Reputation systems. *Communications of the ACM*, *43*(12), 45–48. doi:10.1145/355112.355122

Reyno Momberg, M. (2007). *Responsabilidad Social Empresarial (RSE) como Ventaja Competitiva*. Universidad Técnica Federico Santa María.

Rezapouraghdam, H., Akhshik, A., & Ramkissoon, H. (2021). Application of machine learning to predict visitors' green behavior in marine protected areas: Evidence from Cyprus. *Journal of Sustainable Tourism*, 1–25. Advance online publication. doi:10.1080/09669582.2021.1887878

Ribeiro, M., & Marques, C. (2002). Rural tourism and the development of less favoured areas - Between rhetoric and practice. *International Journal of Tourism Research*, *4*(3), 211–220. doi:10.1002/jtr.377

Richard Ramirez. (n.d.). In *Encyclopedia Britannica online*. https://www.britannica.com/biography/Richard-Ramirez

Richards, G. (2018). Cultural tourism: A review of recent research and trends. *Journal of Hospitality and Tourism Management*, *36*, 12–21. doi:10.1016/j.jhtm.2018.03.005

Richards, S., Aziz, N., Bale, S., Bick, D., Das, S., Gastier-Foster, J., Grody, W. W., Hegde, M., Lyon, E., Spector, E., Voelkerding, K., & Rehm, H. L. (2015). Standards and guidelines for the interpretation of sequence variants: A joint consensus recommendation of the American College of Medical Genetics and Genomics and the Association for Molecular Pathology. *Genetics in Medicine*, *17*(5), 405–423. doi:10.1038/gim.2015.30 PMID:25741868

Rivas García, J., & Magadán Díaz, M. (2012). *Planificación turística y desarrollo sostenible*. Septem Ediciones.

Rivas, García, & Jesús. (2004). *Estructura y Economía del Mercado Turístico*. Septem Ediciones, S. L.

Rivera, L. A. (2008). Managing "spoiled" national identity: War, tourism, and memory in Croatia. *American Sociological Review*, *73*(4), 613–634. doi:10.1177/000312240807300405

Roberts, C. (2018). Education the (dark) masses: dark tourism and sensemaking. In P. R. Stone, R. Hartmann, T. Seaton, R. Sharpley, & L. White (Eds.), *The Palgrave Handbook of Dark Tourism Studies* (pp. 606–637). doi:10.1057/978-1-137-47566-4_25

Roberts, C., & Stone, P. (2014). *Dark tourism and dark heritage: Emergent themes, issues and consequences* (I. Convery, G. Corsane, & P. Davis, Eds.). Newcastle University Press.

Robinson, N. (2015). *Dark tourism motivations: an investigation into the motivations of visitors to sites associated with dark tourism* [Doctoral dissertation]. University of Salford.

Robinson, H. S., Carrillo, P., Anumba, C., & Ghassani, A. M. A. (2005). Review and implementation of performance management models in construction engineering organizations. *Construction Innovation*, *5*(4), 203–217. doi:10.1108/14714170510815258

Robinson, M., & Novelli, M. (2007). Niche tourism: an introduction. In *Niche tourism* (pp. 1–10). Routledge.

Robledo, M. A., & Batle, J. (2017). Transformational tourism as a hero's journey. *Current Issues in Tourism*, *20*(16), 1736–1748. doi:10.1080/13683500.2015.1054270

Rodríguez Sánchez-Escalonilla, N. (2016). *Organización y prestación del servicio de recepción en alojamientos UF0052*. Ediciones Paraninfo, S.A.

Rojek, C. (1993). Ways of Escape. Basingstoke: Macmillan. 1997 Indexing, Dragging and the Social Construction of Tourist Sights. In C. Rojek & J. Urry (Eds.), *Touring Cultures: Transformations of Travel and Theory* (pp. 52–74). Routledge.

Rosalina, P. D., Dupre, K., & Wang, Y. (2021). Rural tourism: A systematic literature review on definitions and challenges. *Journal of Hospitality and Tourism Management*, *47*(June), 134–149. doi:10.1016/j.jhtm.2021.03.001

Rose, E. P. (2019). Geology and the Allied liberation of Normandy: Highlights to help mark the 75th anniversary of D-Day, 6 June 1944. *Geology Today*, *35*(3), 107–114. doi:10.1111/gto.12269

Rosewood, J. (2015). *Edmund Kemper: The True Story of the Co-ed Killer*. CreateSpace Independent Publishing Platform.

Rosing, K., Frese, M., & Bausch, A. (2011). Explaining the heterogeneity of the leadership–innovation relationship: Ambidextrous leadership. *The Leadership Quarterly*, *22*(5), 956–974. doi:10.1016/j.leaqua.2011.07.014

Rowley, J. E. (2018). *Information Marketing*. Routledge.

Rubio-Escuderos, L., García-Andreu, H., & Ullán De La Rosa, J. (2021). *Accessible tourism: origins, state of the art and future lines of research*. Academic Press.

Russ, J. A. (2001). VLR! VII Fighter Command Operations from Iwo Jima, April-August 1945. *Air Power History*, *48*(3), 16–25.

Ryan, C. (Ed.). (2007). *Battlefield tourism history-place and interpretation*. Routledge.

Ryan, C., & Kohli, R. (2006). The Buried Village, New Zealand—An Example of Dark Tourism? *Asia Pacific Journal of Tourism Research, 11*(3), 211–226. doi:10.1080/10941660600753240

Ryan, R. M., & Deci, E. L. (2000). Self-determination theory and the facilitation of intrinsic motivation, social development, and well-being. *The American Psychologist, 55*(1), 68–78. doi:10.1037/0003-066X.55.1.68 PMID:11392867

Sánchez Pérez, M., Terán, Yépez, E., Marín-Carrillo, M. B., Marín-Carrillo, G. M. (2021). *La COVID-19 y el sector turístico en España: Impacto sobre el comportamiento del consumidor turístico*. Editorial Universidad de Almería.

Sandaruwani, J. A. R. C., & Athulagnanapala, W. K. (n.d.). *Impact of Tourism Operations in Spice Gardens for the Sustainable Tourism Development: A Case Study in Sri Lanka*. Academic Press.

Sandvik, I. L., Duhan, D. F., & Sandvik, K. (2014). Innovativeness and profitability: An empirical investigation in the Norwegian hotel industry. *Cornell Hospitality Quarterly, 55*(2), 165–185. doi:10.1177/1938965514520963

Satghare, H., & Sawant, M. (2018). SWOT Analysis of Marketing Strategies Applied by MTDC for Promotion of Maharashtra Tourism. *Atna: Journal of Tourism Studies, 13*(2), 79–95. doi:10.12727/ajts.20.6

Sather-Wagstaff, J. (2016). *Heritage that hurts: Tourists in the memoryscapes of September 11*. Routledge. doi:10.4324/9781315427539

Scates, B. (2002). In Gallipoli's shadow: Pilgrimage, memory, mourning and the Great War. *Australian Historical Studies, 33*(119), 1–21. doi:10.1080/10314610208596198

Schechter, H., & Everitt, D. (1997). *The A to Z Encyclopedia of Serial Killers*. Pocket Books.

Schmalbruch, S. (2018). US Trips for Fans of True Crime. *Insider*. https://www.insider.com/best-true-crime-vacations-2018-5

Schouten, J. W., McAlexander, J. H., & Koenig, H. F. (2007). Transcendent customer experience and brand community. *Journal of the Academy of Marketing Science, 35*(3), 357–368. doi:10.100711747-007-0034-4

Scott-Halsell, S., Shumate, S. R., & Blum, S. (2008). Using a Model of Emotional Intelligence Domains to Indicate Transformational Leaders in the Hospitality Industry. *Journal of Human Resources in Hospitality & Tourism, 7*(1), 99–113. doi:10.1300/J171v07n01_06

Sears, S. W. (2003). *Landscape Turned Red: The Battle of Antietam*. Houghton Mifflin Harcourt.

Seaton, T. (2009). Purposeful otherness: Approaches to the management of thanatourism. In The Darker Side of Travel Bristol: Channel View Publications.

Seaton, A. V. (1996). Guided by the dark: From thanatopsis to thanatourism. *International Journal of Heritage Studies, 2*(4), 234–244. doi:10.1080/13527259608722178

Seaton, A. V. (1999). War and thanatourism: Waterloo 1815–1914. *Annals of Tourism Research, 26*(1), 130–158. doi:10.1016/S0160-7383(98)00057-7

Seaton, A. V. (2000). "Another weekend away looking for dead bodies ... ": Battlefield tourism on the Somme and in Flanders. *Tourism Recreation Research, 25*(3), 63–77. doi:10.1080/02508281.2000.11014926

Seaton, A. V., & Lennon, J. (2004). Moral Panics, Ulterior Motives and Alterior Desires: Thanatourism in the Early 21st Century. In T. V. Singh (Ed.), *New Horizons in Tourism: Strange Experiences and Stranger Practices* (pp. 63–82). CABI. doi:10.1079/9780851998633.0063

Seaton, A. V., & Lennon, J. J. (2004). *New Horizons in Tourism: Strange Experiences And Stranger Practices.* CABI.

Seaton, T. (2018). Dark tourism history. In *The Palgrave handbook of dark tourism studies* (pp. 1–2). Palgrave Macmillan. doi:10.1057/978-1-137-47566-4

Secretaría de Turismo. (2022a). *Unidad de Información y Seguimiento.* Disponible en https://www.datatur.sectur.gob.mx/SitePages/versionesRAT.aspx

Secretaría de Turismo. (2022b). *Resultados de la Actividad Turística junio 2022.* Disponible en: https://datatur.sectur.gob.mx/RAT/RAT-2022-06(ES).pdf

Seethalekshmy, C., & Subramaniyan, M. (2021). Sustainable tourism development in spice tourism. In *Sustainable tourism development: An in-depth analysis* (Vol. 1, pp. 1–19). Multi Spectrum Publication.

Seibert, S. E., & Kraimer, M. L. (2001). The five-factor model of personality and career success. *Journal of Vocational Behavior, 58*(1), 1–21. doi:10.1006/jvbe.2000.1757

Seraphin, H., & Korstanje, M. E. (2021). *Dark tourism tribes: social capital as a variable. In Consumer Tribes in Tourism: Contemporary Perspectives on Special-Interest Tourism.* Springer-Nature.

Shanker, R., Bhanugopan, R., Van der Heijden, B. I., & Farrell, M. (2017). Organizational climate for innovation and organizational performance: The mediating effect of innovative work behavior. *Journal of Vocational Behavior, 100*, 67–77. doi:10.1016/j.jvb.2017.02.004

Sharma, N. (2022). Acknowledging the shades of grey: The past, present and future of dark tourism in India. In R. Kumar & A. Parida (Eds.), Indian Tourism (pp. 125-142). Emerald Publishing. doi:10.1108/978-1-80262-937-820221009

Sharma, P., & Nayak, J. K. (2020). Examining experience quality as the determinant of tourist behavior in niche tourism: An analytical approach. *Journal of Heritage Tourism, 15*(1), 76–92. doi:10.1080/1743873X.2019.1608212

Sharpley, R., & Wright, D. (2018). Disasters and Disaster Tourism; The Role of the Media. In The Palgrave Handbook of Dark Tourism Studies. Palgrave Macmillan.

Sharpley, R. (2020). Coronavirus and dark tourism: Managing thanatourism in the face of new risks. *Tourism Geographies*, *22*(3), 610–623.

Sharpley, R., & Stone, P. (2009). *The Darker Side of Travel: The Theory and Practice of Dark Tourism.* Channel View Publications. doi:10.21832/9781845411169

Shekhar, S., & Valeri, M. (2022). Evolving Themes in Dark Tourism Research: A Review Study. *Tourism (Zagreb)*, *70*(4), 624–641. doi:10.37741/t.70.4.6

Shenoy, J. (2020). Indian spices in great demand in post Covid 19 times; exports up 34%_ ASSOCHAM. *Times of India*. https://timesofindia.indiatimes.com/business/india-business/indian-spices-in-great-demand-in-post-covid-19-times-exports-up-34-assocham/articleshow/77048023.cms

Sheth, K. (2020). Countries that have produced the most serial killers. *WorldAtlas*. https://www.worldatlas.com/articles/countries-that-have-produced-the-most-serial-killers.html

Shim, C., & Santos, C. A. (2014). Tourism, place and placelessness in the phenomenological experience of shopping malls in Seoul. *Tourism Management*, *45*, 106–114. doi:10.1016/j.tourman.2014.03.001

Shin, S. J., & Zhou, J. (2007). When is educational specialization heterogeneity related to creativity in research and development teams? Transformational leadership as a moderator. *The Journal of Applied Psychology*, *92*(6), 1709–1721. doi:10.1037/0021-9010.92.6.1709 PMID:18020807

Shi, Y., & Bu, H. (2021). A preliminary study on the marketing mode of antique bookstores conforming to the trend of The Times. *Time-Honored Brand Marketing*, (8), 25–26.

Shukla, A., & Yadav, N. (2018). Role of Indian spices in Indian in history. *International Journal of Management Research and Review*, *8*(11), 1–6. www.ijmrr.com

Sidali, K. L., Kastenholz, E., & Bianchi, R. (2018). Food tourism, niche markets and products in rural tourism: Combining the intimacy model and the experience economy as a rural development strategy. In *Rural Tourism* (pp. 47–65). Routledge. doi:10.4324/9781315111865-3

Sinclair, D. (2018). What justification is there for including the mass suicide of Jonestown as part of a Guyana dark tourism narrative in 2025? *Worldwide Hospitality and Tourism Themes*, *10*(5), 592–604. doi:10.1108/WHATT-05-2018-0035

Singh, A., Hossain, F., & Burman, R. (2022). Crop husbandry in Pre-independent India. In *Indian agriculture after independence*. www.icar.org.in

Sintobin, T. (2021). Traveller, tourist and the 'lost art of travelling': The debate continues. In *Routledge Handbook of the Tourist Experience* (pp. 215–234). Routledge. doi:10.4324/9781003219866-20

Siriraj Medical Museum . (2023). Siriraj Medical Museum. https://www.si.mahidol.ac.th/sirirajmuseum/siriraj-museum-en.html

Sisto, R., Cappelletti, G. M., Bianchi, P., & Sica, E. (2022). Sustainable and accessible tourism in natural areas: A participatory approach. *Current Issues in Tourism, 25*(8), 1307–1324. doi:10.1080/13683500.2021.1920002

Skarzynski, P., & Gibson, R. (2008). *Innovation to the core: a blueprint for transforming the way your company innovates*. Harvard Business Press.

Skinner, J. (2016). Walking the Falls: Dark tourism and the significance of movement on the political tour of West Belfast. *Tourist Studies, 16*(1), 23–39. doi:10.1177/1468797615588427

Skotis, A., & Livas, C. (2022). A data-driven analysis of experience in urban historic districts. *Annals of Tourism Research Empirical Insights, 3*(2), 100052. doi:10.1016/j.annale.2022.100052

Sky. (2020). *Special plan for the protection of the historical and cultural city of Suzhou (2035)*. Suzhou Natural Resources and Planning Bureau. http://suzhou.bendibao.com/news/20201017/81860.shtm

Slade, P. (2003). Gallipoli tourism. *Annals of Tourism Research, 30*(4), 779–794. doi:10.1016/S0160-7383(03)00025-2

Slåtten, T., & Mehmetoglu, M. (2011). Antecedents and effects of engaged frontline employees: A study from the hospitality industry. *Managing Service Quality, 21*(1), 88–107. doi:10.1108/09604521111100261

Slåtten, T., Svensson, G., & Sværi, S. (2011). Empowering leadership and the influence of a humorous work climate on service employees' creativity and innovative behaviour in frontline service jobs. *International Journal of Quality and Service Sciences, 3*(3), 267–284. doi:10.1108/17566691111182834

Smith, V. (1998). War and tourism. *Annals of Tourism Research, 25*(1), 202–227. doi:10.1016/S0160-7383(97)00086-8

Snyder, N., & Duarte, D. (2003). *Strategic innovation: embedding innovation as a core competency in your organization*. John Wiley & Sons.

Soava, G. (2015). Development prospects of the tourism industry in the digital age. *Revista tinerilor economişti*, (25), 101-116.

Sokol, A., Gozdek, A., Figurska, I., & Blaskova, M. (2015). Organizational climate of higher education institutions and its implications for the development of creativity. *Procedia: Social and Behavioral Sciences, 182*, 279–288. doi:10.1016/j.sbspro.2015.04.767

Soligo, M., & Dickens, D. R. (2020). Rest in fame: Celebrity tourism in Hollywood cemeteries. *Tourism, Culture & Communication, 20*(2-3), 141–150. doi:10.3727/109830420X15894802540214

Somech, A. (2006). The effects of leadership style and team process on performance and innovation in functionally heterogeneous teams. *Journal of Management, 32*(1), 132–157. doi:10.1177/0149206305277799

Sontakke, Syed, & Sawate. (2018). Studies on extraction of essential oils from spices (Cardamom and Cinnamon). *International Journal of Chemical Studies, 6*(2), 2787–2789. https://www.ncdex.com

Sparks, B., & Browning, V. (2010). Complaining in cyberspace: The motives and forms of hotel guests' complaints online. *Journal of Hospitality Marketing & Management, 19*(7), 797–818. doi:10.1080/19368623.2010.508010

Spenceley, A. (2021). *The future of nature-based tourism: Impacts of COVID-19 and paths to sustainability*. Academic Press.

Spice Board of India. (2021). *Major Spice per State wise area and production of spices*. http://www.indianspices.com/sites/default/files/majorspiceStatewise2022_v2.pdf

Spice board of India. (n.d.). *India_Spice_Map*. Spice board of India. Retrieved February 20, 2023, from http://www.indianspices.com/sites/default/files/INDIA_SPICE_MAP.jpg

Spillman, L. (1997). *Nation and commemoration: Creating national identities in the United States and Australia*. Cambridge University Press. doi:10.1017/CBO9780511520938

Sprengel Museum Hannover - Hannover Lower Saxony . (2023). Tripadvisor. https://www.tripadvisor.de/Attraction_Review-g187351-d266726-Reviews-Sprengel_Museum_Hannover-Hannover_Lower_Saxony.html

Sprengel Museum. (2023). Sprengel Museum. https://www.sprengel-museum.de/en

Stolcke, V. (1997). The 'Nature' of nationality. In Citizenship and exclusion (pp. 61-80). Palgrave Macmillan. doi:10.1057/9780230374591_4

Stone, P. (2006). A Dark Tourism Spectrum: Towards a Typology of Death and Macabre Related Tourist Sites, Attractions and Exhibitions. *Tourism: An Interdisciplinary International Journal, 54*(2), 145–160. http://clok.uclan.ac.uk/27720/1/27720%20fulltext_stamped.pdf

Stone, A., Russell, R., & Patterson, K. (2004). Transformational versus servant leadership: A difference in leader focus. *Leadership and Organization Development Journal, 25*(3/4), 349–361. doi:10.1108/01437730410538671

Stone, P. (2013). Dark tourism scholarship: A critical review. *International Journal of Culture, Tourism and Hospitality Research, 7*(3), 307–318. doi:10.1108/IJCTHR-06-2013-0039

Stone, P. R. (2006). A dark tourism spectrum: Towards a typology of death and macabre related tourist sites, attractions and exhibitions. *Tourism: An International Interdisciplinary Journal, 54*(2), 145–160.

Stone, P. R. (2011). Dark tourism and the cadaveric carnival: Mediating life and death narratives at Gunther von Hagens' Body Worlds. *Current Issues in Tourism, 14*(7), 685–701. doi:10.1080/13683500.2011.563839

Stone, P. R. (2011). Dark tourism: Towards a new post-disciplinary research agenda. *International Journal of Tourism Anthropology, 1*(3-4), 318–332. doi:10.1504/IJTA.2011.043713

Stone, P. R. (2012). Dark tourism and significant other death: Towards a model of mortality mediation. *Annals of Tourism Research, 39*(3), 1565–1587. doi:10.1016/j.annals.2012.04.007

Stone, P. R. (2021). Dark tourism and place identity: The Role of Visitor Experience in Shaping Perceptions of Contested Places. *Tourism Geographies, 23*(1), 52–72.

Stone, P. R., & Sharpley, R. (2021). Dark tourism: A Review of The Literature. *Journal of Hospitality and Tourism Management, 46*, 70–83.

Stone, P., & Sharpley, R. (2008). Consuming dark tourism: A thanatological perspective. *Annals of Tourism Research, 35*(2), 574–595. doi:10.1016/j.annals.2008.02.003

Strange, C., & Kempa, M. (2003). Shades of dark tourism: Alcatraz and Robben Island. *Annals of Tourism Research, 30*(2), 386–405. doi:10.1016/S0160-7383(02)00102-0

Streiner, D. L. (2003). Being inconsistent about consistency: When coefficient alpha does and doesn't matter. *Journal of Personality Assessment, 80*(3), 17–22. doi:10.1207/S15327752JPA8003_01 PMID:12763696

Struan, R. (n.d.). *The Trade in Spices*. Retrieved March 13, 2023, from https://en.unesco.org/silkroad/sites/default/files/knowledge-bank-article/the%20trade%20in%20spices.pdf

Subramaniam, M., & Youndt, M. (2005). The influence of intellectual capital on the types of innovative capabilities. *Academy of Management Journal, 48*(3), 450–463. doi:10.5465/amj.2005.17407911

Su, L., Tang, B., & Nawijn, J. (2021). How tourism activity shapes travel experience sharing: Tourist well-being and social context. *Annals of Tourism Research, 91*, 103316. doi:10.1016/j.annals.2021.103316

Sundbo, J., Orfila-Sintes, F., & Sørensen, F. (2007). The innovative behaviour of tourism firms-Comparative studies of Denmark and Spain. *Research Policy, 36*(1), 88–106. doi:10.1016/j.respol.2006.08.004

Sung, Y. Y. C., & Chiu, D. K. W. (2022). E-book or print book: Parents' current view in Hong Kong. *Library Hi Tech, 40*(5), 1289–1304. doi:10.1108/LHT-09-2020-0230

Sun, X., Chiu, D. K. W., & Chan, C. T. (2022). Recent Digitalization Development of Buddhist Libraries: A Comparative Case Study. In S. Papadakis & A. Kapaniaris (Eds.), *The Digital Folklore of Cyberculture and Digital Humanities* (pp. 251–266). IGI Global. doi:10.4018/978-1-6684-4461-0.ch014

Su, S., Zhao, Q., & Tian, H. (2019). Saving the antique book Industry in Jiangnan Area: A case analysis of bookstores in Nanjing, Suzhou, Yangzhou and Hangzhou. *Market Weekly*, (6), 143–145.

Suzhou Antique Bookstore. (2015). *An ancient city, a time-honored bookstore, and a cultural label of the city*. https://mp.weixin.qq.com/s/Nk1aX2bHkteR813hREm7ww

Suzhou Writers Association. (2017). Yearly Report. *Su Zhou Literary and Art Association Joint Report*, 68-71.

Szabo, M. R., Radu, D., Gavrilas, S., Chambre, D., & Iditoiu, C. (2010). Antioxidant and antimicrobial properties of selected spice extracts. *International Journal of Food Properties*, *13*(3), 535–545. doi:10.1080/10942910802713149

Szűcs, Gy. (2015). 100 éve halt meg a nőfaló cinkótai sorozatgyilkos. *Index*. https://index.hu/tudomany/tortenelem/anagyhaboru/2015/02/06/100_eve_halt_meg_a_hirhedt_cinkotai_sorozatgyilkos/

Tajeddini, K. (2010). Effect of customer orientation and entrepreneurial orientation on innovativeness: Evidence from the hotel industry in Switzerland. *Tourism Management*, *31*(2), 221–231. doi:10.1016/j.tourman.2009.02.013

Tarifa-Fernández, J., Carmona-Moreno, E., & Sánchez-Fernández, R. (2022). An attempt to clarify what deserves to remain dark: A long look back. *Tourism and Hospitality Research*. doi:10.1177/14673584221110358

Tarlow, P. (2005). Dark Tourism: The Appealing 'Dark' Side of Tourism and More. In Niche Tourism: Contemporary Issues, Trends and Cases. Elsevier.

Tarlow, P. E. (2007). Dark tourism: The appealing 'dark'side of tourism and more. In D. Weaver (Ed.), Niche tourism (pp. 47-58). Routledge. doi:10.4324/9780080492926

Tarlow, P. (2005). Dark Tourism: The Appealing 'Dark' Side of Tourism and More. In M. Noveilli (Ed.), *Niche Tourism: Contemporary Issues, Trends and Cases* (pp. 47–57). Elsevier. doi:10.1016/B978-0-7506-6133-1.50012-3

Tarlow, P. E. (2007). Dark tourism: The appealing 'dark'side of tourism and more. In M. Novelli (Ed.), *Niche tourism* (pp. 47–58). Routledge.

Taru, P., Chingombe, W., & Mukwada, G. (2014). A bullet laden park: Potential for battlefield tourism in the Golden Gate Highlands National Park. *African Journal of Hospitality, Tourism and Leisure*, *2*, 1–9.

TaselotzinH. (2023). Recuperado de: http://taselotzin.mex.tl/frameset.php?url=/intro.html

Teicher, M. H. (2000, October 1). *Wounds That Time Won't Heal: The Neurobiology of Child Abuse*. Cerberum. https://www.dana.org/article/wounds-that-time-wont-heal/

Ten Bells. (2023). https://www.tenbells.com/

Terrible Crimes in Vienna. Hidden Secrets of the Inner City – Vienna . (2023). Tripadvisor. https://www.tripadvisor.com/AttractionProductReview-g190454-d21193573-Terrible_crimes_in_Vienna_hidden_secrets_of_the_inner_city-Vienna.html

Tevely, T. V., & Smith, M. K. (2022). Blurring the Boundaries Between Fact and Fiction: Serial Killers in the Context of Dark Tourism. *Tourism & Heritage Journal*, *4*(4), 53–75. Advance online publication. doi:10.1344/THJ.2022.4.4

The Amsterdam Dungeon (2023). https://www.thedungeons.com/amsterdam/en/

The Blood and Tears Walk – London England . (2023). Tripadvisor. https://www.tripadvisor.com/Attraction_Review-g186338-d1383076-Reviews-The_Blood_And_Tears_Walk-London_England.html

The Issue of Child Abuse. (n.d.). Childhelp. https://www.childhelp.org/child-abuse/

The Jack the Ripper Walking Tour in London – London England . (2023). Tripadvisor. https://www.tripadvisor.com/AttractionProductReview-g186338-d18935383-The_Jack_The_Ripper_Walking_Tour_in_London-London_England.html

The London Dungeon. (2023). https://www.thedungeons.com/london/

The Memorial. (2021). Rose South LA. http://rosesouthla.org/the-memorial/

The Victims of Fritz Haarmann Memorial. (2023). Atlas Obscura. https://www.atlasobscura.com/places/the-victims-of-fritz-haarmann-memorial

Thevenot, G. (2007). Blogging as a social media. *Tourism and Hospitality Research*, *7*(3-4), 287–289. doi:10.1057/palgrave.thr.6050062

Thomson, R. (2022). Haunted Venice – Legends, Mysteries and Stories to Creep Yourself out about the Most Romantic Place on Earth. *Rossi Writes*. https://rossiwrites.com/italy/venice/haunted-venice-legends-mysteries-stories/

Thrash, T. M., & Elliot, A. J. (2003). Inspiration as a psychological construct. *Journal of Personality and Social Psychology*, *84*(4), 871–889. doi:10.1037/0022-3514.84.4.871 PMID:12703654

Thrash, T. M., & Elliot, A. J. (2004). Inspiration: Core characteristics, component processes, antecedents, and function. *Journal of Personality and Social Psychology*, *87*(6), 957–973. doi:10.1037/0022-3514.87.6.957 PMID:15598117

Thrash, T. M., Elliot, A. J., Maruskin, L. A., & Cassidy, S. E. (2010a). Inspiration and the promotion of well-being: Tests of causality and mediation. *Journal of Personality and Social Psychology*, *98*(3), 488–506. doi:10.1037/a0017906 PMID:20175626

Thrash, T. M., Maruskin, L. A., Cassidy, S. E., Fryer, J. W., & Ryan, R. M. (2010b). Mediating between the muse and the masses: Inspiration and the actualization of creative ideas. *Journal of Personality and Social Psychology, 98*(3), 469–487. doi:10.1037/a0017907 PMID:20175625

Thrash, T. M., Maruskin, L. A., Moldovan, E. G., Oleynick, V. C., & Belzak, W. C. (2017). Writer–reader contagion of inspiration and related states: Conditional process analyses within a cross-classified writer×reader framework. *Journal of Personality and Social Psychology, 113*(3), 466–491. doi:10.1037/pspp0000094 PMID:27124379

Thrash, T. M., Moldovan, E. G., Oleynick, V. C., & Maruskin, L. A. (2014). The psychology of inspiration. *Social and Personality Psychology Compass, 8*(9), 495–510. doi:10.1111pc3.12127

Times of India. (2022). Hot trade: How Indian earns billions from its spices. *The Times of India.* https://timesofindia.indiatimes.com/business/india-business/hot-trade-how-indian-earns-billions-from-its-spices/articles how/93558184.cms

Timothy, D. J. (2018). Geography: The substance of tourism. *Tourism Geographies, 20*(1), 166–169. doi:10.1080/14616688.2017.1402948

Tipu, S., Ryan, J., & Fantazy, K. (2012). Transformational leadership in Pakistan: An examination of the relationship. *Journal of Management & Organization, 18*(4), 461–480. doi:10.5172/jmo.2012.18.4.461

Titta, N. (2010). *Bachelor's Thesis in Nature and Soft Adventure*. Lahti University of Applied Sciences.

Todorova, G. (2015). Marketing communication mix. *Trakia Journal of Sciences, 13*(Suppl.1), 368–374. doi:10.15547/tjs.2015.s.01.063

Toivonen, A. (2022). Sustainability dimensions in space tourism: The case of Finland. *Journal of Sustainable Tourism, 30*(9), 2223–2239. doi:10.1080/09669582.2020.1783276

Torabi Farsani, N., Zeinali, H., & Moaiednia, M. (2018). Food heritage and promoting herbal medicine-based niche tourism in Isfahan, Iran. *Journal of Heritage Tourism, 13*(1), 77–87. doi:10.1080/1743873X.2016.1263307

Tøssebro, J. (2004). Introduction to the special issue: Understanding disability. *Scandinavian Journal of Disability Research, 6*(1), 3–7. doi:10.1080/15017410409512635

Tóth, T. Zs., & Papp-Váry, Á. F. (2022). From Nuclear Disaster to Film Tourism: The Impact of the Chernobyl Mini-Series on the Exclusion Zone. In R. Baleiro & R. Pereira (Eds.), Global Perspectives on Literary Tourism and Film-Induced Tourism (pp. 280-301). IGI Global. https://doi.org/ doi:10.4018/978-1-7998-8262-6

Tour Leader Venice . (2023). Serial Killers of Venice Tour. https://www.tourleadervenice.com/project/venice-and-the-seri al-killers-tour/

Townsend, L., Wallace, C., Smart, A., & Norman, T. (2016). Building Virtual Bridges: How Rural Micro-Enterprises Develop Social Capital in Online and Face-to-Face Settings. *Sociologia Ruralis*, *56*(1), 29–47. doi:10.1111oru.12068

Tracey, J., & Hinkin, T. (1996). How transformational leaders lead in the hospitality industry. *International Journal of Hospitality Management*, *15*(2), 165–176. doi:10.1016/0278-4319(95)00059-3

Trading Economics. (2016). *Portugal - Rural Population.* Available at: https://tradingeconomics.com/portugal/rural-population-percent-of-total-population-wb-data.html

TripAdvisor Insights. (2013). *How to optimize your attraction listing on TripAdvisor.* Available at: https://www.tripadvisor.com/TripAdvisorInsights/n710/how-optimize-your-attraction-listing-tripadvisor

TripAdvisor Insights. (2015). *All about your TripAdvisor bubble rating.* Available at: https://www.tripadvisor.com/TripAdvisorInsights/n2640/all-about-your-tripadvisor-bubble-rating

Troitiño, V., & Miguel, Á. (1998). Turismo y desarrollo sostenible en ciudades históricas. Ería, Revista cuatrimestral de geografía. *Ejemplar dedicado a: El turismo en las ciudades históricas, 47*, 211-227.

Troitiño, V., & Miguel, Á. (2002). *El patrimonio arquitectónico y urbanístico como recurso. In La función social del patrimonio histórico: El turismo cultural.* Ediciones de la Universidad de Castilla-La Mancha.

Troshin, A. S., Sokolova, A. P., Ermolaeva, E. O., Magomedov, R. M., & Fomicheva, T. L. (2020). Information technology in tourism: Effective strategies for communication with consumers. *Journal of Environmental Management & Tourism*, *11*(2 (42)), 322–330. doi:10.14505//jemt.v11.2(42).10

True Crimes of Georgetown. (2023). DC by foot. https://freetoursbyfoot.com/true-crimes-of-georgetown/

Trunfio, M., Petruzzellis, L., & Nigro, C. (2006). Tour operators and alternative tourism in Italy: Exploiting niche markets to increase international competitiveness. *International Journal of Contemporary Hospitality Management*, *18*(5), 426–438. doi:10.1108/09596110610673556

Tsang, A. L. Y., & Chiu, D. K. W. (2022). Effectiveness of virtual reference services in academic libraries: A qualitative study based on the 5E Learning Model. *Journal of Academic Librarianship*, *48*(4), 102533. doi:10.1016/j.acalib.2022.102533

Tsaur, S. H., Yen, C. H., & Lin, Y. S. (2022). Destination inspiration: Scale development and validation. *Journal of Travel & Tourism Marketing*, *39*(5), 484–500. doi:10.1080/10548408.2022.2148040

Tse, H. L., Chiu, D. K. W., & Lam, A. H. C. (2022). From reading promotion to digital literacy: An analysis of digitalizing mobile library services with the 5E Instructional Model. In A. Almeida & S. Esteves (Eds.), *Modern Reading Practices and Collaboration Between Schools, Family, and Community* (pp. 239–256). IGI Global. doi:10.4018/978-1-7998-9750-7.ch011

Tseng, S. (2010). The Correlation between Organizational Culture and Knowledge Conversion on Corporate Performance. *Journal of Knowledge Management, 14*(2), 269–284. doi:10.1108/13673271011032409

Tunbridge, J. E., & Ashworth, G. J. (1996). *Dissonant heritage. The Management of the Past as a Resource in Conflict.* Belhaven Press.

Tung, V., & Ritchie, J. (2011). Exploring the essence of memorable tourism experiences. *Annals of Tourism Research, 38*(4), 1367–1386. doi:10.1016/j.annals.2011.03.009

Turismo de Portugal, I. P. (2015). *Turismo 2020: cinco princípios para uma ambição* [Tourism 2020: five principles for an ambition]. Available at: https://www.turismodeportugal.pt/Portugu%C3%AAs/turismodeportugal/newsletter/2015/Documents/TURISMO2020-5Principios.pdf

Two Feet & a Heartbeat Tours – Perth Greater Perth Western Australia . (2023). Tripadvisor. https://www.tripadvisor.com/ShowUserReviews-g255103-d1992671-r761709116-Two_Feet_a_Heartbeat_Tours-Perth_Greater_Perth_Western_Australia.html

Tzanelli, R. (2016). *Thanatourism and cinematic representations of risk: Screening the end of tourism.* Routledge. doi:10.4324/9781315624105

UNESCO. (n.d.). *What are the spice routes?* UNESCO. Retrieved March 14, 2023, from https://en.unesco.org/silkroad/content/what-are-spice-routes#:~:text=The%20Spice%20Routes%2C%20also%20known,across%20the%20Mediterranean%20to%20Europe

Ungarisches Haus. (2023). Atlas Obscura. https://www.atlasobscura.com/places/ungarisches-haus-hungarian-house

Ungerer, M. P. (2011). *Viable business strategies; a field book for leaders.* Knowres publishing.

University of Minnesota. (n.d.). *4.6 SWOT Analysis – Mastering Strategic Management.* Retrieved March 19, 2023, from https://open.lib.umn.edu/strategicmanagement/chapter/4-6-swot-analysis/

Unsworth, K. L., & Parker, S. (2003). Proactivity and innovation: promoting a new workforce for the new workplace. In T. D. W. Holman, C. V. Clegg, P. Sparrow, & A. Howard (Eds.), *The new workplace: a guide to the human impact of modern working practices* (pp. 175–196). John Wiley.

UNWTO. (2018). *Tourism and culture synergies.* UNWTO.

UNWTO. (2022). *Accessible tourism identified as 'game changer' for destinations*. https://www.unwto.org/news/accessible-tourism-identified-as-game-changer-for-destinations

Upton, A., Schänzel, H., & Lück, M. (2018). Reflections of battlefield tourist experiences associated with Vietnam war sites: An analysis of travel blogs. *Journal of Heritage Tourism, 13*(3), 197–210. doi:10.1080/1743873X.2017.1282491

Vaccaro, I. G., Jansen, J. J., Van Den Bosch, F. A., & Volberda, H. W. (2012). Management innovation and leadership: The moderating role of organizational size. *Journal of Management Studies*, *49*(1), 28–51. doi:10.1111/j.1467-6486.2010.00976.x

Valente, S., & Figueiredo, E. (2003). O turismo que existe não é aquele que se quer... [The tourism that exists is not what you want...]. 1º Encontro de Turismo em Espaços Rurais.

Valentine, S., Godkin, L., Fleischman, G., & Kidwell, R. (2011). Corporate ethical values, group creativity, job satisfaction and turnover intention: The impact of work context on work response. *Journal of Business Ethics*, *98*(3), 353–372. doi:10.100710551-010-0554-6

Van der Merwe, C. D. (2014). Battlefields tourism: The status of heritage tourism in Dundee, South Africa. *Bulletin of Geography. Socio-Economic Series*, *26*(26), 121–139. doi:10.2478/bog-2014-0049

Vargas, M. I. R. (2015). Determinant factors for small business to achieve innovation, high performance and competitiveness: Organizational learning and leadership style. *Procedia: Social and Behavioral Sciences*, *169*, 43–52. doi:10.1016/j.sbspro.2015.01.284

Varghese, B., & Paul, N. (2014). A Literature Review on Destination Management Organization (DMO). *ZENITH International Journal of Multidisciplinary Research, 4*(12), 82–88. http://zenithresearch.org.in/images/stories/pdf/2014/DEC/ZIJMR/9_ZIJMR_VOL4_ISSUE12_DECEMBER2014.pdf

Venice Ghost & Legends Walking City Tour – Venice Veneto . (2023). Tripadvisor. https://www.tripadvisor.com/AttractionProductReview-g187870-d13075842-Venice_Ghost_Legends_walking_city_tour-Venice_Veneto.html

Venter, D. (2011). Battlefield tourism in the South African context. *African Journal of Hospitality, Tourism and Leisure*, *1*(3), 1–5.

Vermeulen, L. E., & Seegers, D. (2009). Tried and tested: The impact of online hotel reviews on consumer consideration. *Tourism Management*, *30*(1), 123–127. doi:10.1016/j.tourman.2008.04.008

Victorino, L., Verma, R., Plaschka, G., & Dev, C. (2005). Service innovation and customer choice in the hospitality industry. *Managing Service Quality*, *15*(6), 555–576. doi:10.1108/09604520510634023

Vienna Criminal History Tour in German. (2023). Get Your Guide. https://www.getyourguide.com/vienna-l7/vienna-criminal-history-tour-in-german-t51444/

Virgen Aguilar, C. R. (2014). *Turismo y desarrollo sustentable. Un acercamiento al estudio del turismo.* Asociación Mexicana de Centros de Enseñanza Superior en Turismo y Gastronomía, A.C., Universidad de Guadalajara. Universidade Federal do Parná.

Virgili, S., Delacour, H., Bornarel, F., & Liarte, S. (2018). 'From the Flames to the Light': 100 years of the commodification of the dark tourist site around the Verdun battlefield. *Annals of Tourism Research, 68*, 61–72. doi:10.1016/j.annals.2017.11.005

Visitare Serial Killer. A Jesolo Venezia Mostra. (2023). Serial Killer. http://www.mostraserialkiller.it/serial-killer/visitare-serial-killer-a-jesolo-venezia-mostra.html

Vogeler Ruiz, C., & Hernández Armand, E. (2018). *Introducción al turismo. Análisis y estructura. Editorial Universitaria.* Ramón Areces.

Volmer, J., Spurk, D., & Niessen, C. (2012). Leader–member exchange (LMX), job autonomy, and creative work involvement. *The Leadership Quarterly, 23*(3), 456–465. doi:10.1016/j.leaqua.2011.10.005

Wahlert, G. (2011). *Exploring Gallipoli: An Australian army battlefield guide* (Vol. 4). Simon and Schuster.

Walby, K., & Piché, J. (2011). The polysemy of punishment memorialization: Dark tourism and Ontario's penal history museums. *Punishment & Society, 13*(4), 451–472. doi:10.1177/1462474511414784

Walker, R. M. (2008). An empirical evaluation of innovation types and organizational and environmental characteristics: Towards a configuration framework. *Journal of Public Administration: Research and Theory, 18*(4), 591–615. doi:10.1093/jopart/mum026

Walmsley, D. J. (2003). Rural tourism: A case of lifestyle-led opportunities. *The Australian Geographer, 34*(1), 61–72. doi:10.1080/00049180320000066155

Walter, T. (1991). Modern Death: Taboo or not Taboo? *Sociology, 25*(2), 293–310. doi:10.1177/0038038591025002009

Walter, T., Littlewood, J., & Pickering, M. (1995). Death in the News: The Public Investigation of Private Emotion. *Sociology, 29*(4), 579–596. doi:10.1177/0038038595029004002

Wanda George, E. (2010). Intangible cultural heritage, ownership, copyrights, and tourism. *International Journal of Culture, Tourism and Hospitality Research, 4*(4), 376–388. doi:10.1108/17506181011081541

Wang, B., Yang, Z., Han, F., & Shi, H. (2017). Car tourism in Xinjiang: The mediation effect of perceived value and tourist satisfaction on the relationship between destination image and loyalty. *Sustainability (Basel), 9*(1), 22. Advance online publication. doi:10.3390u9010022

Wang, C.-J., Tsai, H.-T., & Tsai, M.-T. (2014). Linking transformational leadership and employee creativity in the hospitality industry: The influences of creative role identity, creative self-efficacy, and job complexity. *Tourism Management, 40*, 79–89. doi:10.1016/j.tourman.2013.05.008

Wang, G., Oh, I., Courtright, S., & Colbert, A. (2011). Transformational Leadership and Performance Across Criteria and Levels: A Meta-Analytic Review of 25 Years of Research. *Group & Organization Management, 36*(2), 223–270. doi:10.1177/1059601111401017

Wang, J., Deng, S., Chiu, D. K. W., & Chan, C. T. (2022). Social network customer relationship management for orchestras: A case study on Hong Kong Philharmonic Orchestra. In N. B. Ammari (Ed.), *Social customer relationship management (Social-CRM) in the era of Web 4.0* (pp. 250–268). IGI Global. doi:10.4018/978-1-7998-9553-4.ch012

Wang, N. (1999). Rethinking authenticity in tourism experience. *Annals of Tourism Research, 26*(2), 349–370. doi:10.1016/S0160-7383(98)00103-0

Wang, P., & Zhu, W. (2011). Mediating role of creative identity in the influence of transformational leadership on creativity: Is there a multilevel effect? *Journal of Leadership & Organizational Studies, 18*(191), 25–39. doi:10.1177/1548051810368549

Wang, W., Lam, E. T. H., Chiu, D. K. W., Lung, M. M., & Ho, K. K. W. (2021). Supporting Higher Education with Social Networks: Trust and Privacy vs. Perceived Effectiveness. *Online Information Review, 45*(1), 207–219. doi:10.1108/OIR-02-2020-0042

Wang, Y. (2021). From place to space: The development and construction of physical bookstores as urban public space: A case study of Hefei Physical Bookstores. *Journal of Changzhou Institute of Technology, 39*(05), 82–87.

Werewolf of Hanover Crime Tour . (2023). Booking.com. https://www.booking.com/attractions/de/pr6mcl2zmkq7-werewolf-of-hanover-crime-tour.hu.html?date=2023-02-24&start_time=20%3A00&ticket_type=OFuR59frVqNc

West, M., & Sacramento, C. (2012). Creativity and innovation: the role of team and organizational climate. In M. D. Mumford (Ed.), *Handbook of organizational creativity* (pp. 359–385). Academic Press. doi:10.1016/B978-0-12-374714-3.00015-X

West, P. (2004). *Conspicuous Compassion: Why Sometimes it really is Cruel to be Kind*. CIVITAS.

What's On. (2023). Fremantle Prison. https://fremantleprison.com.au/

White, L., & Frew, E. (2013). *Dark tourism and place identity: Managing and interpreting dark places*. Routledge. doi:10.4324/9780203134900

Whiting, J., & Hannam, K. (2014). Journeys of inspiration: Working artists' reflections on tourism. *Annals of Tourism Research*, *49*, 65–75. doi:10.1016/j.annals.2014.08.007

Wiener Kriminal Museum. (2023). Wiener Kriminal Museum. https://wien.kriminalmuseum.at/en/news/

Wight, A. C. (2006). Philosophical and methodological praxes in dark tourism: Controversy, contention and the evolving paradigm. *Journal of Vacation Marketing*, *12*(2), 119–129. doi:10.1177/1356766706062151

Wikimedia Commons. (2013). *Map of Maharashtra*. https://commons.wikimedia.org/wiki/Category:SVG_maps_of_Maharashtra#/media/File:Maharashtra_Divisions_Eng.svg

Williams, H. (2021). Communing with the Fictional Dead: Grave Tourism and the Sentimental Novel. *British Sociability in the European Enlightenment: Cultural Practices and Personal Encounters*, 41-62.

Winter, C. (2009). Tourism social memory and the Great War. *Annals of Tourism Research*, *36*(4), 607–626. Advance online publication. doi:10.1016/j.annals.2009.05.002

Winter, C. (2011). Battlefield visitor motivations: Explorations in the Great War town of Ieper, Belgium. *International Journal of Tourism Research*, *13*(2), 164–176. doi:10.1080/14616688.2011.575075

Winter, C. (2012). Commemoration of the Great War on the Somme: Exploring personal connections. *Journal of Tourism and Cultural Change*, *10*(3), 248–263. doi:10.1080/14766825.2012.694450

Wong, A. K.-k., & Chiu, D. K. W. (2023). Digital transformation of museum conservation practices: A value chain analysis of public museums in Hong Kong. In R. Pettinger, B. B. Gupta, A. Roja, & D. Cozmiuc (Eds.), *Handbook of Research on the Digital Transformation Digitalization Solutions for Social and Economic Needs* (pp. 226–242). IGI Global. doi:10.4018/978-1-6684-4102-2.ch010

Wong, S., & Pang, L. (2003). Motivators to creativity in the hotel industry: Perspectives of managers and supervisors. *Tourism Management*, *24*(5), 551–559. doi:10.1016/S0261-5177(03)00004-9

World Heritage Convention. (2010). *Silk Road Sites in India - UNESCO World Heritage Centre*. UNESCO. https://whc.unesco.org/en/tentativelists/5492/

World Tourism Organization. (2016). The transformative power of tourism: A paradigm shift towards a more responsible traveller (Affiliate Members Global Reports, Volume 14). Madrid: UNWTO.

World Travel and Tourism Council. (2020). To Recovery and Beyond: The future of travel andtourism in the wake of Covid-19. In *World Travel and Tourism Council*. https://wttc.org/Portals/0/Documents/Reports/2020/To Recovery and Beyond-The Future of Travel Tourism in the Wake of COVID-19.pdf?ver=2021-02-25-183120-543

World Travel Awards. (2017). *World's Leading Destination 2017*. Available at: https://www.worldtravelawards.com/award-worlds-leading-desti nation-2017

Wright, D. W. M. (2018). Terror Park: A Future Theme Park in 2100. *Futures*, *96*, 1–22. doi:10.1016/j.futures.2017.11.002

WTTC. (2019). *Travel & Tourism: Economic Impact 2019 World*. World Travel & Tourism Council.

Wu, F., Mahajan, V., & Balasujbramanian, S. (2003). An analysis of e-business adoption and its impact on business performance. *Journal of the Academy of Marketing Science*, *31*(4), 425–447. doi:10.1177/0092070303255379

Wulf Betencourt, E. (2018). *Responsabilidad Social Empresarial. Un desafío corporativo.* Editorial Universidad de la Serena.

Xiang, Z., & Gretzel, U. (2010). Role of social media in online travel information search. *Tourism Management*, *31*(2), 179–188. doi:10.1016/j.tourman.2009.02.016

Xie, P. F., Lee, M. Y., & Wong, J. W. C. (2020). Assessing community attitudes toward industrial heritage tourism development. *Journal of Tourism and Cultural Change*, *18*(3), 237–251. doi:10.1080/14766825.2019.1588899

Xue, B., Lam, A. H. C., & Chiu, D. K. W. (2023). Redesigning Library Information Literacy Education with the BOPPPS Model: A Case Study of the HKUST. In R. Taiwo, B. Idowu-Faith, & S. Ajiboye (Eds.), *Transformation of Higher Education Through Institutional Online Spaces*. IGI. Global.

Yang, J., Yang, R., Chen, M., Su, C., Zhi, Y., & Xi, J. (2021). Effects of rural revitalization on rural tourism. *Journal of Hospitality and Tourism Management*, *47*(June), 35–45. doi:10.1016/j.jhtm.2021.02.008

Yang, M. C., & Hsing, W. C. (2001). Kinmen: Governing the culture industry city in the changing global context. *Cities (London, England)*, *18*(2), 77–85. doi:10.1016/S0264-2751(00)00059-7

Yang, W., & Deng, S. (2018). A look at the countermeasures for supporting antique bookstores from the perspective of bookstores integration. *View on Publishing*, (12), 10–13.

Yankholmes, A., & Hoque, N. (2020). Socio-cultural impacts of dark tourism: A Critical Review of Literature. *Journal of Heritage Tourism*, *15*(6), 601–618.

Yekimov, S., Sobirov, B., Turdibekov, K., Aimova, M., & Goncharenko, M. (2022). Using the Digital Ecosystem in Tourism Clusters in Green Tourism. In *Ecosystems Without Borders: Opportunities and Challenges* (pp. 105–111). Springer International Publishing. doi:10.1007/978-3-031-05778-6_11

Yew, A., Chiu, D. K. W., Nakamura, Y., & Li, K. K. (2022). Quantitative Comparison of LIS Programs Accredited by ALA and CILIP. *Library Hi Tech*, *40*(6), 1721–1745. doi:10.1108/LHT-12-2021-0442

Yi, Y., & Chiu, D.K.W. (2023). Public information needs during the COVID-19 outbreak: A qualitative study in mainland China. *Library Hi Tech*, ahead-of-print. doi:10.1108/LHT-08-2022-0398

Yi, X., Lin, V. S., Jin, W., & Luo, Q. (2017). The authenticity of heritage sites, tourists' quest for existential authenticity, and destination loyalty. *Journal of Travel Research*, *56*(8), 1032–1048. doi:10.1177/0047287516675061

Yoo, K. H., & Gretzel, U. (2011). Influence of personality on travel-related consumer-generated media creation. *Computers in Human Behavior*, *27*(2), 609–621. doi:10.1016/j.chb.2010.05.002

Yoon, Y., & Uysal, M. (2005). An examination of the effects of motivation and satisfaction on destination loyalty: A structural model. *Tourism Management*, *26*(1), 45–56. doi:10.1016/j.tourman.2003.08.016

Yuan, F., & Woodman, R. (2010). Innovative Behavior in the Workplace: The Role of Performance and Image Outcome Expectations. *Academy of Management Journal*, *53*(2), 323–342. doi:10.5465/amj.2010.49388995

Yu, H. H. K., Chiu, D. K. W., & Chan, C. T. (2023). Resilience of symphony orchestras to challenges in the COVID-19 era: Analyzing the Hong Kong Philharmonic Orchestra with Porter's five force model. In W. Aloulou (Ed.), *Handbook of Research on Entrepreneurship and Organizational Resilience During Unprecedented Times* (pp. 586–601). IGI Global.

Yu, H. Y., Tsoi, Y. Y., Rhim, A. H. R., Chiu, D. K. W., & Lung, M. M. W. (2022). Changes in habits of electronic news usage on mobile devices in university students: A comparative survey. *Library Hi Tech*, *40*(5), 1322–1336. doi:10.1108/LHT-03-2021-0085

Yu, J., & Egger, R. (2022). Looking behind the scenes at dark tourism: A comparison between academic publications and user-generated-content using natural language processing. *Journal of Heritage Tourism*, *17*(5), 548–562. doi:10.1080/1743873X.2022.2097011

Yukl, G. (2013). *Leadership in organizations* (8th ed.). Pearson Education.

Yu, P. Y., Lam, E. T. H., & Chiu, D. K. W. (2023). Operation management of academic libraries in Hong Kong under COVID-19. *Library Hi Tech*, *41*(1), 108–129. doi:10.1108/LHT-10-2021-0342

Zhang, J. (2022, November 3). Suzhou released white paper on the development of cultural industries. *Xinhua Daily*. https://www.doi.org/10.28872/n.cnki.nxhrb.2022.006001

Zhang, J., Morrison, A. M., & Chen, Y. (2018). *How Country Image Affects Tourists' Destination Evaluations: A Moderated Mediation Approach*. doi:10.1177/1096348016640584

Zhang, C. X., Fong, L. H. N., Li, S., & Ly, T. P. (2019). National identity and cultural festivals in postcolonial destinations. *Tourism Management*, *73*, 94–104. doi:10.1016/j.tourman.2019.01.013

Zhang, J. J. (2010). Of Kaoliang, bullets and knives: Local entrepreneurs and the battlefield tourism enterprise in Kinmen (Quemoy), Taiwan. *Tourism Geographies*, *12*(3), 395–411. doi:10.1080/14616688.2010.494685

Zhang, J., Lam, A. H. C., & Chiu, D. K. W. (2023). Evaluating the Effectiveness of Learning Commons as Third Space with the 5E Usability Model: The Case of Hong Kong University of Science and Technology Library. In C. Kaye, J. H. Writer, & J. Batsaikhan (Eds.), *Third-Space Exploration in Education*. IGI Global.

Zhang, X. (2021). Scene integration, community activation, and "Experimental field"—Research on the value of physical bookstores as knowledge production space from the perspective of urban communication. *Dongyue Tribune*, *42*(04), 131–138.

Zhao, Y., & Agyeiwaah, E. (2023). Understanding tourists' transformative experience: A systematic literature review. *Journal of Hospitality and Tourism Management*, *54*, 188–199. doi:10.1016/j.jhtm.2022.12.013

Zhou, J., Lam, E. T. H., Au, C. H., Lo, P., & Chiu, D. K. W. (2022). Library café or elsewhere: Usage of study space by different majors under contemporary technological environment. *Library Hi Tech*, *40*(6), 1567–1581. doi:10.1108/LHT-03-2021-0103

Zhou, J., & Shalley, C. E. (2003). Research on employee creativity: a critical review and proposal for future research directions. In J. J. Martocchio & G. R. Ferris (Eds.), *Research in personnel and human resource management*. Elsevier. doi:10.1016/S0742-7301(03)22004-1

Zhou, L. (2014). Online rural destination images: Tourism and rurality. *Journal of Destination Marketing & Management*, *3*(4), 227–240. doi:10.1016/j.jdmm.2014.03.002

Zhu, W., Avolio, B. J., & Walumbwa, F. O. (2009). Moderating role of follower characteristics with transformational leadership and follower work engagement. *Group & Organization Management*, *34*(5), 590–619. doi:10.1177/1059601108331242

Zhu, Y., & Sui, W. (2020). Exploration and analysis of urban physical bookstore curating and communication based on the perspective of knowledge reproduction. *Editorial Friend*, (11), 46–51.

Zilihona, I. J., & Mamboya, S. F. (2018). Contribution of Spice Tourism to Youth Employment: A Case of Kijichi Spices Farms at Kijichi Shehia in Unguja-Zanzibar. *Rural Planning Journal*, *20*(2), 81–89. https://repository.irdp.ac.tz/bitstream/handle/123456789/344/RPJ%20Vol20_Issue2_7.pdf?sequence=1&isAllowed=y

Zodiac Killer. (2004). Presidio Heights. https://www.zodiackiller.com/PresidioHeights2004.html

Zukin, S. (1995). *The Cultures of the Cities*. Blackwell.

Zuo, Y., Lam, A. H. C., & Chiu, D. K. W. (2023). Digital protection of traditional villages for sustainable heritage tourism: A case study on Qiqiao Ancient Village, China. In A. Masouras, C. Papademetriou, D. Belias, & S. Anastasiadou (Eds.), *Sustainable Growth Strategies for Entrepreneurial Venture Tourism and Regional Development*. IGI Global. doi:10.4018/978-1-6684-6055-9.ch009

About the Contributors

Maria Antónia Rodrigues is a senior lecturer at the business school of Polytechnic Institute of Porto. She is a director of Bachelor in Marketing, member of the school's Scientific Council, member of the scientific committee of the International Journal of Marketing, Communication and New Media, a researcher at CEOS.PP, and SIIS Porto. She has published several papers. Her main research interests are services, consumer behavior, and business relationships. Maria Antónia Rodrigues has also professional experience in business and services.

Maria Amélia Carvalho is a PhD in Business and Management Studies with specialization in Marketing and Strategy from the Faculty of Economics (FEP), University of Porto, in 2017. In 2012, she completed the Master in Economics and City Management and graduated in Economics, in 2010 Both academic degrees were obtained by the Faculty of Economics (FEP), University of Porto. Invited Adjunct Professor at ISCAP since 2018. Her main research areas are related to brands and consumer behavior. She has specific interests in city branding, customer engagement, customer experience, city marketing and services marketing.

* * *

Habib Alipour (PhD) is a professor of tourism policy and planning at the Eastern Mediterranean University, Faculty of Tourism. His research and scholarly interests focused on sus- tainable development, tourism planning, institutional analysis and smart growth. His recent research addressed Edu-tourism as a sustainable option in island states.

Bráulio Alturas is Associate Professor of Department of Information Science and Technology of Iscte-Instituto Universitário de Lisboa (University Institute of Lisbon), Lisbon, and researcher of the Information Systems Group of ISTAR_Iscte (Information Sciences, Technologies and Architecture Research Center). He holds a PhD in Management with specialization in Marketing, a MSc in Management

Information Systems, and a BSc in Business Organization and Management, all from ISCTE-IUL. His scientific research interests and publications are in acceptance and use of technology, digital marketing, social media, e-commerce, information management, information systems, direct selling and consumer behavior. ORCID: 0000-0003-0142-3737.

Francesco Bifulco is a Full Professor in Management at University of Naples Federico II. His main areas of interest are focused on cultural heritage (branding enhancement, phygital journey, sustainable business models, innovation ecosystem). He published papers and books about these themes on top journal and publisher. He chaired and participated sessions in international conferences. He led, as Scientific coordinator (University of Naples Federico II) in projects PON Research & Competitiveness Program (High-tech districts and related networks).

Francesco Carignani is a PhD student in Management at the Department of Economics, Management and Institutions at the University of Naples Federico II. He earned a Master's degree in Cultural Heritage Management. His main areas of interest are cultural management, social impact of culture, social innovation, audience development.

Oindrila Chakraborty is a gold medalist in her MBA programme from University of Calcutta. She has 14 years of teaching experience in management education. She has completed her PhD from University of Calcutta and has several publications to furnish in National and International Journals of high repute.

Dickson K. W. Chiu received the B.Sc. (Hons.) degree in Computer Studies from the University of Hong Kong in 1987. He received the M.Sc. (1994) and Ph.D. (2000) degrees in Computer Science from the Hong Kong University of Science and Technology (HKUST). He started his own computer consultant company while studying part-time. He has also taught at several universities in Hong Kong. His teaching and research interest is in Library & Information Management, Service Computing, and E-learning with a cross-disciplinary approach involving library and information management, e-learning, e-business, service sciences, and databases. The results have been widely published in around 300 international publications (most of them have been indexed by SCI/-E, SSCI, and EI, such as top journals MIS Quarterly, Computer & Education, Government Information Quarterly, Decision Support Systems, Information Sciences, Knowledge-Based Systems, Expert Systems with Application, Information Systems Frontiers, IEEE Transactions, including many taught master and undergraduate project results and around 20 edited books. He received a best paper award at the 37th Hawaii International Conference on System

Sciences in 2004. He is an Editor (-in-chief) of Library Hi Tech, a prestigious journal indexed by SSCI. He is the Editor-in-chief Emeritus of the International Journal on Systems and Service-Oriented0 Engineering (founding) and International Journal of Organizational and Collective Intelligence, and serves on the editorial boards of several international journals. He co-founded several international workshops and co-edited several journal special issues. He also served as a program committee member for around 300 international conferences and workshops. Dr. Chiu is a Senior Member of both the ACM and the IEEE and a life member of the Hong Kong Computer Society. According to Google Scholar, he has over 6,000 citations, h-index 40, i-10 index 132, ranked worldwide 1st in "LIS," "m-learning," and "e-services." He received nearly 900 citations in 2022.

Denise Capela dos Santos is a Pharmacist, with a MSc in Management and a PhD in Economics, has a high capacity for leadership, communication, proactivity, and emotional resilience, likes teamwork, and is result oriented. Is an Associate Professor in the Department of Economic and Business Sciences at Universidade Autónoma de Lisboa, since 2010, and a Coordinating Professor at Escola Superior de Enfermagem S. Francisco das Misericórdias, since 2021. Is a member of RICH - Research and Innovation Center for Health. Its main areas of research interest are health economics, hospital administration, and cancer management.

Ebru Düşmezkalender is working as Associate Professor in the Department of Tourism Management, within the Faculty of Tourism at Eskişehir Osmangazi University in Turkey. Her research areas are tourism, tourism management, alternative tourism, and organizational

Xuechen Gao received the Honours Bachelor of Commerce in Management Information Systems and Analytics from the University of Ottawa in Canada. She worked as an intern at Ipsos for four months in 2023 and at Capgemini for three months in 2021. Her research interest is in consumer behavior/psychology, brand image, and digital/media marketing, and she expects to work in the consulting industry after graduating with an MSc (Library and Information Management) at HKU.

Jorge Gomes is a researcher at ADVANCE, ISEG, School of Economics & Management of the Universidade de Lisboa. He holds a PhD in Management from ISEG and a Masters in Management Sciences from ISCTE-IUL, He also have a post-graduation in Project Management from INDEG/ISCTE, and a degree in Geographic Engineering from the Faculty of Sciences of the Universidade de Lisboa. During the past 30 years, he has worked as an engineer, project manager, quality auditor and consultant. Teaches Management at ULHT, Lisboa. His research interests include Benefits Management, Project Management, Project Success, Maturity Models, IS/IT Investments, IS/IT in Healthcare and IS/IT Management.

Marjan Kamyabi is a tourism expert with over 10 years of experience in the industry. She holds a PhD in Tourism Management from Eastern Mediterranean University, where her research focused on accessible tourism and the impact of disability on travel experiences. She is currently faculty member in Tourism Management at Cyprus International University, where she teaches courses on Niche tourism, tourism policy and planning, tourism marketing and sustainable tourism. Her passion for accessible tourism stems from her belief that everyone should have the opportunity to travel and experience different cultures, regardless of their physical abilities. She is dedicated to promoting tourism that is inclusive and accessible, and her research and advocacy have helped to create positive changes in the tourism industry.

Maximiliano E. Korstanje is editor in chief of International Journal of Safety and Security in Tourism (UP Argentina) and Editor in Chief Emeritus of International Journal of Cyber Warfare and Terrorism (IGI-Global US). Korstanje is Senior Researchers in the Department of Economics at University of Palermo, Argentina. In 2015 he was awarded as Visiting Research Fellow at School of Sociology and Social Policy, University of Leeds, UK and the University of La Habana Cuba. In 2017 is elected as Foreign Faculty Member of AMIT, Mexican Academy in the study of Tourism, which is the most prominent institutions dedicated to tourism research in Mexico. He had a vast experience in editorial projects working as advisory member of Elsevier, Routledge, Springer, IGI global and Cambridge Scholar publishing. Korstanje had visited and given seminars in many important universities worldwide. He has also recently been selected to take part of the 2018 Albert Nelson Marquis Lifetime Achievement Award. a great distinction given by Marquis Who´s Who in the world.

Apple Hiu Ching Lam obtained her degree of Bachelor of Business Administration (Honours) in International Business from City University of Hong Kong (2016) and degree of Master of Science in Library and Information Management with distinction from the University of Hong Kong (2020). She is a doctoral student in Education at the University of Hong Kong. Her current research interests are social media in library, user education, and the 5E Instructional Model.

Cláudia Leitão holds a Master's degree in Business Management from Universidade Autónoma, Lisbon (2020), a Bachelor's degree in Hotel Management from Estoril Higher Institute for Tourism and Hotel Studies (2013) and a Pedagogical Aptitude training course (2019). She has a special interest in areas such as Innovation Management and Digital Marketing. She has been an hotel professional, working in several international hotel brands in Portugal and Spain.

Bruno Melo Maia holds a Ph.D. in Mathematics by the University of Warwick, UK (2008). He is an auxiliary professor in Universidade Autónoma de Lisboa and member of the board of CICEE - Research Center in Economics and Business Sciences.

Sérgio Moro is Associate Professor with Habilitation at Instituto Universitário de Lisboa (ISCTE-IUL), and Deputy Director and Coordinator of the Information Systems Group of ISTAR (Information Sciences, Technologies and Architecture Research Center). He is an interdisciplinary data scientist that envisions the development of innovative predictive systems through data science approaches in distinct domains such as marketing, management, tourism, and education. His publication portfolio includes more than 40 Quartile 1 Scopus indexed journal articles (e.g., European Journal of Marketing, Tourism Management, Journal of Business Research, Telematics & Informatics, International Journal of Information Management). His work has received more than 2,000 Scopus citations. He is the Director of the Master in Data Science at ISCTE. ORCID: 0000-0002-4861-6686.

Yunus Özhasar completed his undergraduate education at Akdeniz University and his master's degree at Adnan Menderes University. He is currently working as a research assistant at Eskişehir Osmangazi University, Faculty of Tourism, where he is studying for a doctorate. He continues to work on issues such as crisis management, sustainability and social media marketing.

Árpád Papp-Váry, PhD habil, is the Head of the Marketing and Tourism Programme at the Doctoral School of the Faculty of Economics of the University of Sopron, Hungary. In addition to that, he is a senior research associate at the Economic Geography and Urban Marketing Centre of John von Neumann University, Kecskemét, Hungary, and he leads the Marketing MSc programme at the biggest Hungarian business university, Budapest Business School. Árpád is the author of several well-known marketing and branding books, the most recent of which is "Country Branding – Creating a Competitive Identity and Image". A large part of his publications is also available on his website in pdf format: Besides university education, he regularly holds training sessions and provides branding consultancy, and his company, Márkadoktor Kft., is an accredited consultant of the Hungarian Multi Program. He is a jury member of more than ten marketing, advertising and PR competitions. He has been Vice President of the Hungarian Marketing Association (MMSZ) for three terms.

Hamed Rezapouraghdam, Ph.D., is an Assistant Professor in the Faculty of Tourism at Eastern Mediterranean University (Gazimagusa, TRNC, via Mersin 10, 99628, TURKEY). He obtained his Ph.D. degree in Tourism Management from Eastern Mediterranean University. His research interests are in the areas of sustainable tourism, destination management, environmental psychology, and corporate sustainability.

Paulo Rita, PhD in Marketing (Cardiff University, UK) and Post-Doc in E-Marketing (University of Nevada Las Vegas, USA), is Full Professor of Marketing at NOVA Information Management School (NOVA IMS), Universidade NOVA de Lisboa, Portugal. Professor Rita is Director of the Master in Data-driven Marketing and Postgraduate programs in Marketing Intelligence, Marketing Research & CRM, Digital Marketing & Analytics, Data Science for Marketing at NOVA IMS. His scientific research interests and publications are in digital marketing, social media, marketing analytics, consumer behavior and tourism marketing. He has published more than one hundred scientific articles in indexed journals such as in Journal of Business Research, European Journal of Marketing, Journal of Retailing and Consumer Services, Decision Support Systems, International Journal of Information Management, Tourism Management, Journal of Hospitality Marketing and Management. His work has received more than 9,200 Google Scholar citations. ORCID: 0000-0001-6050-9958.

Julieta Salazar-Echeagaray is a full-time teaching career professor of the Faculty of Sciences Economic Administrative in Mazatlán of the Autonomous University of Sinaloa. Degree in Accounting, Degree in Management. Master's degree in corporate finance. Doctorate in administration.

Teresa Salazar-Echeagaray is a research professor at Faculty of Administrative Economic Sciences of Mazatlan.

Marta Parente dos Santos holds a BSc in Economics from ISCTE-IUL and has just finished a double degree in MSc International Management from ISCTE Business School, Lisbon, & Nottingham Trent University. Her scientific research interests encompass Tourism, International Markets and Marketing. She is currently working as a Marketing Specialist at Greenvolt Comunidades.

Titanilla Virág Tevely is a PhD student at the University of Sopron, Alexandre Lámfalussy Faculty of Economics, Hungary. She studied Tourism and Hospitality BSc, Marketing and Commerce BSc, and Tourism Management MSc at the Budapest Metropolitan University. She received two first prizes at the Annual Scientific Students' Associations Conference with her works on "The Perceptions of Consumers About Dark Tourism and Attractions Connected to Serial Killers" and "Traveling in the Shadow of the Pandemic – How Travel Habits and Consumer Decisions of the Hungarian Population Changed During the Pandemic". As part of the Lámfalussy Research Center, she is involved in teaching the Basics of Marketing and PR, and conducting internal defenses and conferences. Her dissertation is about Marketing Communication for Sustainable Heritage Tourism in the context of dark tourism.

Sheng-Hshiung Tsaur is a professor, Department of Marketing and Tourism Management, National Chiayi University, Chiayi City, Taiwan. His research interests include construct development and theory building in the disciplines of tourism, hospitality and leisure.

Hong-Ru Wu is in the Department of Marketing and Tourism Management, National Chiayi University, Taiwan. His areas of research include tourism management.

Chenyang Xu got the Bachelor's degree in Information Management and Information System from Shanghai University and the Master's degree in Library and Information Management from the University of Hong Kong. Her research interest is in social media analytics and Big Data research.

Chang-Hua Yen is a professor, Department of Leisure and Recreation Management, National Taichung University of Science and Technology, Taiwan. His areas of research include tourism management, travel marketing, and hospitality management.

Selçuk Yücesoy has completed his undergraduate studies at Anadolu University and his master's degree at Erciyes University. He is currently a doctoral student in the Department of Tourism Management at Eskişehir Osmangazi University, Institute of Social Sciences. Selçuk Yücesoy's academic interests encompass sustainable tourism, organizational behavior, innovation, and dark tourism, and he has authored numerous publications in these areas, both at national and international levels.

Harshada Satghare is working as an Assistant Professor in the department of travel and tourism at Vishwakarma University, Pune (India). Having almost a decade's experience in the field of Travel & Tourism, she has specialization in e-tourism, tourism marketing, internet marketing and sustainable tourism. A university topper in Master of Tourism Administration, she has completed PhD in the area of internet marketing in tourism under the guidance of Dr Madhuri Sawant. She was awarded with the reputed UGC NET JRF and SRF fellowships. Prime areas of the research are Socio-economic impacts of tourism, tourism marketing, destination websites, social media marketing, smart destinations etc. Both the quantitative analysis techniques and qualitative analysis techniques were applied.

Index

P

R

S

T

U

W

Ensure Quality Research is Introduced to the Academic Community

Become an Evaluator for IGI Global Authored Book Projects

The overall success of an authored book project is dependent on quality and timely manuscript evaluations.

Applications and Inquiries may be sent to:
development@igi-global.com

Applicants must have a doctorate (or equivalent degree) as well as publishing, research, and reviewing experience. Authored Book Evaluators are appointed for one-year terms and are expected to complete at least three evaluations per term. Upon successful completion of this term, evaluators can be considered for an additional term.

If you have a colleague that may be interested in this opportunity, we encourage you to share this information with them.

www.ingramcontent.com/pod-product-compliance
Lightning Source LLC
LaVergne TN
LVHW080629011026
849160LV00014B/1162